RADIOECOLOGY

RADIOECOLOGY

Edited by V. M. KLECHKOVSKII, G. G. POLIKARPOV,
and R. M. ALEKSAKHIN

Translated from Russian by N. Kaner and H. Mills

Translation edited by D. Greenberg

A HALSTED PRESS BOOK

JOHN WILEY & SONS

New York · Toronto

ISRAEL PROGRAM FOR SCIENTIFIC TRANSLATIONS

Jerusalem · London

Sole distributors for the Western Hemisphere and Japan

HALSTED PRESS, a division of
JOHN WILEY & SONS, INC., NEW YORK

Library of Congress Cataloging in Publication Data
Main entry under title:

Radioecology.

  Translation of *Radioékologiia*.
  "A Halsted Press book."
  1. Radioecology. I. Klechkovskiĭ, Vsevolod
Mavrikievich, 1900– ed. II. Polikarpov, Gennadiĭ
Grigor'evich, ed. III. Aleksakhin, Rudol'f Mikhaĭlo-
vich, ed.
QH543.5.R3313   574.5'222   73-4697
ISBN 0-470-49035-7

Distributors for the U.K., Europe, Africa and
the Middle East

JOHN WILEY & SONS, LTD., CHICHESTER

Distributed in the rest of the world by

KETER PUBLISHING HOUSE JERUSALEM LTD.

ISBN 0 7065 1291 X
IPST cat. no. 22051

This book is a translation from Russian of
RADIOEKOLOGIYA (SOVREMENNYE PROBLEMY
RADIOBIOLOGII, Tom II)
Atomizdat
Moscow 1971

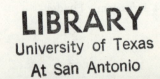

Contents

PART II. RADIOECOLOGY OF AQUATIC BIOCENOSES

INTRODUCTION

Radioecology is a comparatively young branch of science, owing its
origin and rapid development to the steadily increasing use of nuclear
energy in various spheres of human economic activity. The early radioecological works of 1920 —1930 were related to studies of the biological
effect of ionizing radiation on living organisms in regions with high
concentrations of natural radionuclides (uranium, radium, and thorium).
However, radioecological research has proceeded rapidly over the last
15 —20 years, when it became clear that ionizing radiation was becoming an environmental ecological factor whose effects were largely obscure, calling for
extensive investigation. The necessity of such work became especially urgent
in the wake of nuclear tests, which were accompanied by global dispersion of
artificial radioactive substances in the biosphere.

Artificial radionuclides in the atmosphere enter the biogeochemical
migration cycles of chemical elements on the earth's surface. Consequently, besides irradiation from the natural radiation background
(cosmic radiation and radiation from natural radionuclides, i. e.,
isotopes of uranium, radium, thorium, and their decay products as well
as ^{40}K and other natural radioisotopes), all living organisms have become
affected by artificial radionuclides (external irradiation as well as irradiation from radionuclides incorporated in tissues of plants, animals,
and people). Comprehensive studies of the transformation of artificial
radionuclides in the biosphere and evaluation of the possible consequences
of further irradiation of plants, animals and people have become an urgent
task of radioecology.

Radioecological investigations during this period concentrated on
solving the two fundamental problems of radioecology: migration of
radionuclides in natural biogeocenoses and the effect of ionizing
radiation on communities of microorganisms, plants, and animals.
Numerous experimental investigations conducted in various countries
have yielded a vast amount of information on patterns associated with
the migration of radionuclides along biological and food chains under
a variety of biogeochemical environmental conditions (such as different
climatic and soil zones) and the effects of ionizing radiation on plants
and animals in their natural habitats. Besides their considerable
theoretical significance, the results of radioecological studies also possess
practical importance: assessment of the biological and ecological consequences of introducing artificial radionuclides into the external environment.

Studies of the radiobiological effect at the level of populations, communities and cenoses have developed somewhat less rapidly. Nevertheless,
such experiments have yielded data concerning the effect of ionizing radiation on forest communities, crop stands, and natural meadows and pastures.

Extensive radioecological information is available in periodicals and journals, and has been generalized in reports of the U. N. Scientific Committee on the Effects of Atomic Radiation as well as in such works as Radioactivity and Human Diet, edited by R. S. Russell [New York, Pergamon Press. 1966]. These publications do not at all satisfy the need for monographs reviewing the current status of radioecological problems, especially as work done by Soviet scientists has been ignored in published material outside the USSR. This book, which is one volume in the series "Current problems of radiobiology," is an attempt to fill this gap.

Modern radioecology is divided with respect to the specific properties of the environment (in which the migration of radionuclides and ionizing radiation take place) into the radioecology of land biogeocenoses and the radioecology of hydrobiocenoses (marine and freshwater radioecology). This book follows such an approach.

Chapter 1 describes the behavior of natural radionuclides in soils, which is one of the best-developed topics of radioecology. The traditional subject matter of this branch is related to studies dealing with the migration of natural radioisotopes of heavy elements (uranium, thorium, radium, and their daughter products) in the soil-plant cover with the purpose of devising biogeochemical prospecting techniques for these elements as raw material for the nuclear industry. Much interest has recently been displayed in these investigations in connection with the biological role of the natural radiation background, especially in regions with high concentrations of natural radionuclides. Of great importance are the natural radiation background and the effects of small radiation doses on biological objects, including evaluation of the roles played by additional irradiation from artificial radiation sources. These aspects are currently very topical and further research in this direction is mandatory.

The atmospheric source of radionuclide fallout on the soil-plant cover represents the starting stage of migration in many radioactive contamination situations. The behavior of artificial radionuclides in the atmosphere, their fallout and distribution on the earth's surface are described in Chapter 2.

After being deposited from the atmosphere on the land surface, radionuclides contaminate the soil-plant cover. Their migration and participation in biogeochemical cycles of matter in the soil-plant cover depend upon the specific geographical and landscape conditions. These aspects of the problem are dealt with in Chapter 3, chiefly by the example of the behavior of ^{90}Sr, a long-lived fission product.

Soil is an important link in the complex chain of biogeocenotic systems which incorporate components of radioactive fallout on land. The high sorptivity of soils with respect to trace amounts of the majority of radioactive fission products generally results in the accumulation of long-lived artificial radionuclides in the upper soil horizons. This circumstance possesses dual significance. On the one hand, sorption of artificial radionuclides in soils considerably reduces their migration and facilitates the accumulation of long-lived fallout components in the upper horizons; this creates a long-lived source from which the radionuclides are absorbed by plants and proceed along biological and food chains into the organisms of animals and humans. On the other hand, strong sorption of certain long-lived nuclides (^{137}Cs) in soil is liable to reduce their absorption by plants from the soil via their roots (this attenuation of the migration of radionuclides is variable and depends on the soil and plant properties). These and other aspects of the behavior of radionuclides in soils are examined in Chapters 4 and 5.

Migration of radionuclides in various types of land biogeocenoses is of considerable interest. An instance is provided by their peculiar behavior in forest stands, described in Chapter 6. Forest communities are mostly multistoried, their various components interacting among themselves. The complexity of the migration paths of radionuclides in forest biogeocenoses is conditional on the presence of a considerable number of trophic chains. Some transport features of radionuclides are introduced by the perennial nature of arboreal vegetation involving a cumulative type of radionuclide accumulation by the presence of forest floor delaying their penetration into the soil, and by their incorporation into the cycle stages including their absorption by plants through the root systems. The transport of radionuclides in other types of land biogeocenoses also displays certain specific features.

The tasks of radioecology are closely interrelated with the general problem of ecology, radiobiology, biogeochemistry, agrochemistry, etc. Contacts also exist between radioecology on the one hand, and radiation hygiene and radiation public health on the other. As regards radioactive contamination of foodstuffs in the human ration, both radioecology and radiation public health are concerned with various aspects of the migration of radionuclides along biological and food chains. At the same time, radioecology cannot but be concerned with analysis of radiation conditions in the human environment. These topics are examined in Chapter 7.

As already pointed out, studies into the effects of ionizing radiation as regards cenoses deserve close attention, and this is one of the fundamental, though far from resolved, radioecological problems. There is no doubt that ionizing radiation is capable of producing primary and secondary effects, related to the direct influence of radiation on living organisms constituting a cenosis and to the resultant changes in the functioning of biogeocenoses as self-regulating systems. These changes are most distinct in cases of radiation damage sustained by biogeocenoses. Relatively small irradiation doses may therefore produce slight changes in the most radiosensitive cenosis components (reduction in growth and development rates, appearance of slight morphological variations, etc.). Larger irradiation doses may totally eliminate such components from the cenoses and destroy the balance of coupling and feedback among the cenosis constituents.

Studies of long-term effects of radiation exposure on cenoses, especially under prolonged irradiation, also consider radiation-genetic alterations occurring in the irradiated populations. The effect of radiation on plants and animals that are highly heterogeneous with respect to their radiosensitivity alters the rate of the natural mutative process and natural resistance to radiation on the population level. Chapter 8 provides information concerning the effect of ionizing radiation on populations and cenoses with respect to radiation-genetics.

The magnitude and nature of a radiation effect may largely depend upon the type of natural biogeocenoses. Arboreal vegetation, especially some of its species, possesses extremely high radiosensitivity (this was discovered in the early 1960s). Initial signs of radiation damage to forests were observed for irradiation doses which do not as a rule produce any significant changes in herbaceous vegetative cenoses. Subsequently, radiobiologists studied the causes of high radiosensitivity of arboreal species, while radioecologists showed that forest stands are among the types of natural biogeocenoses that are least resistant to radiation damage. The processes involved in radiation damage to forest phytocenoses are also being studied at present. Chapter 9 describes some of these radioecological studies as well as the aims of further research concerned with the effect of ionizing radiation on forest biogeocenoses.

Progress in any branch of science is closely related to the introduction and extensive use of quantitative methods. As applied to radioecology this means that its progress will largely depend upon that of ecological dosimetry, the fundamental task of which consists in evaluating irradiation doses created in the environment. The tasks of ecological dosimetry become especially complicated when the source of ionizing radiation comprises radionuclides migrating through biocenoses and the overall effect is produced by a combination of external irradiation with the influence of incorporated radiators. The direct experimental determination of irradiation doses of different components of biogeocenoses due to radionuclides emerging into the environment plays a role in calculations of irradiation doses. Another important role is played by the accumulation of information concerning the dynamics of radionuclide migration through the cenosis components. In this respect, ecological dosimetry is closely related to regularities governing the cycle of radionuclides, and their incorporation and distribution in organisms. Dosimetric aspects of radioecology are treated in Chapter 10.

Migration of artificial radionuclides in biological and food chains with the participation of wild and farm animals, and some aspects of the effect of ionizing radiation on animals are dealt with in Chapters 11 and 12. The practical importance of investigating laws governing the buildup of artificial radionuclides in farm animals lies mainly in the fact that the products of animal husbandry (milk, meat, etc.) act as sources of radionuclides in the human ration.

The migration rates of radionuclides in the environment are controlled by numerous biogeochemical and biogeophysical factors (meteorological conditions, soil cover, vegetation, hydrological conditions, etc.). In certain cases the combined effects of diverse environmental ecological factors bring about the rapid incorporation of radionuclides into the food and biological chains, with a resultant, relatively large buildup of radionuclides in plants and animals, and higher radiation doses. Regions of the Far North provide examples of areas with a higher incorporation rate of the principal long-lived fission products (^{90}Sr and especially ^{137}Cs) and certain natural radionuclides (^{210}Po) into the plants-animals-man chain. The specific features of these regions are due, in particular, to the existence of the lichen-deer-man radionuclide migration chain. Lichen, being a perennial formation, acts as an unusual accumulator of long-lived radionuclides from atmospheric fallout. After accumulating in the lichens, such radionuclides migrate into deer and then find their way into human food. An important fact is that the radionuclides falling from unit area are concentrated in a relatively small mass of lichen. The special radioecological conditions in the Far North and the related specific features of the radiation background in these regions have prompted large-scale radioecological studies in the Far North by the USSR, USA, Canada, and the Scandinavian countries. Various radioecological aspects of Far North landscapes are described in Chapter 13.

One field of modern radiobiology is the radioecology of aquatic life. The treatment of marine and freshwater radioecology as an independent branch is closely related to the specific features of radionuclide migration in the aquatic environment in comparison to the cycle of radioactive substances in continental biogeocenoses. The aquatic environment with its radionuclides acting as irradiation source for hydrobionts possesses certain peculiar features. Some countries use rivers, seas and oceans for the disposal of radioactive wastes from their nuclear industries. The hydrosphere has likewise received most of the global radioactive products of nuclear explosions.

The radioecology of aquatic biocenoses aims at studying features pertaining to the migration of radionuclides in aquatic environment, their incorporation in hydrobionts, and the effect of ionizing radiation on aquatic life. From a practical viewpoint, marine and freshwater radioecology must provide a blueprint for protecting the hydrosphere against radioactive pollution. The radioecology of marine and freshwater plants and animals is dealt with, respectively, in Chapters 14 and 15. Various aspects of radionuclide migration in seas and oceans, including the part played by the form of radionuclides in water during their migration in the aquatic environment, are described in Chapter 16. The buildup of artificial radionuclides in fish is examined in Chapter 17.

At present, prospects for the peaceful uses of atomic energy are most encouraging. In the coming decades, nuclear power must assume a prominent place among sources of power. Wide prospects are opened up by more extensive employment of atomic energy in heavy earthmoving operations (construction of canals and dams, mining of solid and liquid mineral resources). Nuclear power is being increasingly used in the conquest of space (use of isotopes in space). Plans are under way for agricultural and industrial complexes powered by nuclear energy.

Expansion of the peaceful uses of atomic power must allow for safeguards to guard against appreciable penetration of the environment by radioactive substances. Acceptance of this premise does not free us from the need of conducting painstaking and comprehensive studies of all possible ecological consequences of such environmental pollution. Such studies also play an important role in the drafting of suitable restrictions.

A certain "polarization" presently characterizes the development of advanced biological research. On the one hand, impressive achievements have been attained by molecular biology, genetics, biochemistry, and biophysics. There is reason to believe that in the near future the transmission mechanisms of genetic information, and the regulation of metabolisms and morphogenesis will be better understood. On the other hand, it must be acknowledged that besides deep studies into biological phenomena on a microscopic scale, investigations undertaken on the level of natural populations and communities of living organisms in biogeocenoses and the terrestrial biosphere in its entirety are likewise of extreme interest and practical importance. The exceptional urgency of studies into these forms of the "macromanifestation" of life on Earth is due to man's ever-expanding influence on his environment. Among the sciences active in this field of study, an important place belongs to radioecology, which deals with the ecological consequences of practical applications of atomic power, one of the greatest achievements of the twentieth century.

PART I

RADIOECOLOGY OF LAND BIOGEOCENOSES

Section 1. Radionuclide Migration
in Land Biogeocenoses

Chapter 1

BEHAVIOR OF NATURAL RADIONUCLIDES IN SOILS

Introduction

The radioactivity of soils in the USSR was first studied in 1936 by
Vernadskii /17, 19/ in connection with the comprehensive investigation of
soils as a separate entity and a special layer of the biosphere.

Current research into the natural radioactivity of soils is proceeding
along the following lines:

1. Determination of the natural radiation background of the habitats
of living organisms in the course of studying pollution of the earth's
surface by products of nuclear explosions and wastes from the nuclear
industry /65/.

2. Radiometric, metallometric, radiohydrochemical and geochemical
surveys in prospecting for deposits of nuclear raw materials /35, 49, 54,
57, 59, 60, 68/.

3. Studies of the geochemistry of radionuclides in the soil-plant cover
/5, 7 — 16, 23, 25, 26, 28, 31, 36, 37, 39, 41, 46, 47, 50, 51, 61, 62, 76/.

4. The use of natural radionuclides as soil-forming process indicators
/42, 43/.

5. Investigation into the influence of natural radioactivity on the
development of living organisms (radioactive biogeochemical aspects)
/24, 27, 32, 45/.

This chapter deals mainly with the geochemical aspect of the natural
radioactivity of the soil cover of the biosphere.

The biological role of natural radioactivity requires further clarification.
There are two opposite viewpoints concerning the role of the natural background of ionizing radiation as an ecological factor in the vital activities
of organisms. One maintains that the significance of the natural radiation
background for living organisms has on the whole been overestimated
/21, 22/. The other claims that the natural radiation background should
be regarded as a major ecological factor capable of exercising a significant
influence over populations of living organisms.

The main approach in the study of the geochemistry of natural radionuclides in soils is via radiochemical and physicochemical analysis /16/.
Radiometric spectroscopic techniques can sometimes be used to determine
individual radionuclide concentrations. Radiometric measurements of soil
radioactivity using the total intensity of alpha, beta and gamma radiations
cannot be used to obtain quantitative information. The results of measurements stated in different units are incommensurable /1, 2, 4, 6, 20, 27, 29, 30,
34, 44, 52, 53, 56, 58, 63, 67, 73 — 75/. Data on individual radionuclide

TABLE 1.1. Concentration of natural radioactive elements in soils and underlying rocks in various countries

Country, region	Nature of soil or rock (number of samples)	Concentration, wt %				Method of determination	Reference
		U, 10^{-4}	Ra, 10^{-10}	Th, 10^{-4}	K		
Czechoslovakia	Various soils (35)	—	0.1 – 3.8 (1.5)	—	—	Unknown	/72/
British Isles: Ireland	Soil (2)	—	1.3 – 2.9 (2.0)	—	—	Unknown	/55/
England	Soil (2)	—	0.08	—	—	Unknown	/70/
New Zealand, Nayau Island in the Pacific Ocean	Soil on volcanic ash (1)	60	—	—	—	Chemical	/64/
USA: Maryland, Washington County	Soil	0.9 – 4.7	—	6.6 – 13	0.7 – 5.7	Chemical	/68/
	Underlying rocks	0.2 – 3.2	—	0.53 – 17	0.19 – 7.0	Chemical	/60/
	Soil	2.58	—	5.23	1.53	Gamma-spectroscopy	/60/
Yellow Cat, Grand County, Utah	Sandy soils in deluvium (background areas)	0.5	—	—	—	Chemical, fluorometric	/54/
	Sand	4	—	—	—	Chemical, fluorometric	
	Argillites	4 – 30	—	—	—	Chemical, fluorometric	
	Soils on areas with ore aureoles	12 – 47	—	—	—	Chemical, fluorometric	
France: Esterel	Soils on background areas	1.2 – 6.2	—	—	—	Chemical, fluorometric	/57/
	Soils on areas with ore aureoles	7.3 – 37.6	—	—	—	Chemical, fluorometric	
Gabon: Mounana	Tropical soils on background areas	1	—	—	—	Chemical, fluorometric	/59/
	Tropical soils on areas with ore aureoles	50 – 100	—	—	—	Chemical, fluorometric	

TABLE 1.1. Contd.

Country, region	Nature of soil or rock (number of samples)	Concentration, wt %				Method of determination	Reference
		U, 10^{-4}	Ra, 10^{-10}	Th, 10^{-4}	K		
Poland	Various soils	0.5 — 1.6	–	–	0.4 — 3.5	Beta pulse for K; U determined by the beta-pulse method and from the ratio Th:U = 10:1, obtained alpha-radiographically by Zmyslowska	/66/
Mt. Chelm	Podzolic soil on sand:						
	A_1-horizon	1.54	–	–	–	Alpha-radiography	
	B-horizon	2.28	–	–	–	Alpha-radiography	/69/
	C-horizon	1.9	–	–	–	Alpha-radiography	
	Podzolic soil on loess:						
	A_1-horizon	5.88	–	–	–	Alpha-radiography	
	B-horizon	9.02	–	–	–	Alpha-radiography	
	C-horizon	12.5	–	–	–	Alpha-radiography	
Spain	Soil	–	1.5	–	–	Unknown	/71/

TABLE 1.2. Concentration of natural radioactive elements in soils and underlying rocks in the USSR

Republic, region, location	Nature of soil or rock	Concentration, wt %					Method of determination	Reference
		U, 10^{-4}	Ra, 10^{-10}	Th, 10^{-4}	K	Th:U		
	Soils on plains							
Moscow Region	Podzolic, A-horizon	–	0.9	5.0	–	–	Radiochemical for Ra and Th, chemical for U	/23/
	Podzolic, A-horizon	2.6	1.0	12.0	0.3	4.6	Flame photometry	/23/
	Sod-podzolic, sandy-loamy	–	–	–	1.07	–	Flame photometry	/4/
	Sod-podzolic, light-loamy	–	–	–	1.66	–	Flame photometry	/4/
	Sod-podzolic, medium-loamy	–	–	–	–	–	Flame photometry	/4/
Leningrad Region	Podzolic on varved clays, A-horizon	3.0	0.8	10	–	3.3	Flame photometry	/23/
	Soils on water divides							
Belorussian SSR	Podzolic on eolian sands	0.2	–	–	–	–	Fluorometry	/25/
	Sod-podzolic on morainic loam	1.2	–	–	–	–	Fluorometry	/46/
	Soils in depressions							
	Sod-podzolic on aquatic-glacial loams	3.5–5	–	–	–	–	Fluorometry	/25/
	Sod-podzolic, sandy	1.8	–	–	–	–	Fluorometry	/47/
	Sod-podzolic, gley	2.4	–	–	–	–	Fluorometry	/47/
	Peat-gley	2.7	–	–	–	–	Fluorometry	/47/
Ukrainian SSR, Zhitomir Region	Sod-podzolic	1.47*	–	12.3	–	8.4	Fluorometric for U, radiochemical for Ra and Th	/27/
Tatar ASSR, Kazan	Gray forest soil	–	0.5–0.86	6–10	–	–	Ditto	/23/
	subsoil	–	0.93	14.5	–	–	Ditto	/23/
RSFSR, Tula Region	Gray forest soil	2.3–3.2	1.0–1.1	9–12	–	3.8	Ditto	/23/
	subsoil	2.7	1.1	12	–	4.5	Ditto	/23/
Ukrainian SSR, Vinnitsa Region	Podzolized gray forest soil	0.77*	0.5	9.3	–	12	Ditto	/27/
RSFSR: Kursk Reservation	Deep fertile chernozem	4.0	–	–	1.0	–	Fluorometry for U, flame photometry for K	/4/
Rostov Region	Strongly leached Ciscaucasian chernozem	–	–	–	1.08–1.18	–	Ditto	/4/
Voronezh Region	Ordinary loamy chernozem	4.0	0.8	10	–	2.5	Chemical for U, radiochemical for Ra and Th	/23/
Various regions in Ukraine	Various chernozems	0.77–2.52*	0.8–2.1	7.3–16.4	–	3.4–6.5	Chemical, radiochemical	/27/

TABLE 1.2. Contd.

Republic, region, location	Nature of soil or rock	Concentration, wt %					Method of determination	Reference
		U, 10^{-4}	Ra, 10^{-10}	Th, 10^{-4}	K	Th:U		
	Soils in mountain regions							
Khibiny Mts. Crimea	Soils of alpine tundra	3.0	1.1	8	—	2.7	Chemical, radiochemical	/23/
Mt.Chamny-Burun-Uraga	Brown mountain forest soil on diorite	—	0.32	2.2	—	—	Radiochemical	/48/
	rock							
Mt. Kastel'	Brown mountain forest soil on diorite	—	0.11	0.6	—	—	Radiochemical	/48/
	rock	—	0.92	3.2	—		Radiochemical	
Nikitskii Botanical Gardens	Red earth (krasnozem) on limestone:	—	0.21	4.0	—		Radiochemical	/48/
	0 – 11 cm	—	1.9	6.5	—		Radiochemical	/48/
	20 – 30 cm	—	1.34	12.1	—		Radiochemical	
	35 – 50 cm	—	0.83	9.1	—		Radiochemical	
	Limestone	—	0.29	0.15	—		Radiochemical	
Mt. Kara-Dar	Leached chernozem on igneous rocks:							
	soil 0 – 11 cm	—	0.41	5.6	—			/48/
	20 – 26 cm	—	0.35	4.7	—			
	Limestone	—	0.52	3.3	—			
RSFSR, North Caucasus	Light-chestnut	3.0	1.0	10.0	—	3.3	Chemical for U and Th, radiochemical for Ra	/23/

TABLE 1.2. Contd.

Republic, region, location	Nature of soil or rock	Concentration, wt %					Method of determination	Reference
		$U, 10^{-4}$	$Ra, 10^{-10}$	$Th, 10^{-4}$	K	Th:U		
Georgian SSR, Batumi	Red earth	4.0	1.1	10.0	–	2.5	Chemical for U and Th, radiochemical for Ra	/23/
Asenauli	Red earth	–	–	–	1.12	–	Flame photometry	/4/
Armenian SSR	Brown soils	–	1.4	–	–	–	Radiochemical	/3/
	Chernozems	–	1.7 (0.8 – 3.6)	–	–	–	Radiochemical	/3/
Dilizhan	Forest soil	–	0.8	–	–	–	Radiochemical	/3/
	Mountain chestnut soils	–	0.9 – 2.1	–	–	–	Radiochemical	/3/
Siberia	Various soils	0.9 – 5.0	0.8 – 1.7	2.3 – 14.0	1.6 – 2.4	–	Fluorometric for U, radio-chemical for Ra and Th, flame photometry for K	/31/
Transbaikalia	Various soils	–	1.0 (0.6 – 1.7)	11.4 (8.8 – 17.5)	–	–	Radiochemical	/39/
Amur River basin	Brown forest soil 0 – 7 cm	–	–	–	2.04	–	Flame photometry	/4/
	83 – 93 cm	–	–	–	1.66	–	Flame photometry	/4/
Golodnaya Step (Hunger-steppe) in Central Asia Mira-Rabat	Serozem	1.4 – 13.0	–	–	1.75	–	Flame photometry	/4/
Different districts	Desert sandy soil	23	–	–	–	–	Fluorometry	/35/
	Soils on areas with ore aureoles	–	–	32	–	1.4	Fluorometric for U, chemical for Th	/35/
Global average	Various soils	1.0	0.8	6.0	1.36	–	Various methods	/23/

* Grodzinskii's data /27/ on U concentrations in soils are too low, probably as a result of losses in analysis.

concentrations in soils can be used to calculate the contribution made by their radiation to the total irradiation dose from the edaphic source.

When natural radionuclides are present in soil in their Clarke concentrations, the contribution of alpha radiation represents approximately 65% of the total radiation dose due to soil, the contributions from radioisotopes of the uranium and thorium series being approximately equal. Beta radiation comprises 28% of the total dose, and the contribution of ^{40}K is twice that of the radioisotopes of the uranium and thorium series combined. Finally, gamma radiation contributes approximately 7% to the total dose, the bulk of it (approximately 5%) being due to ^{40}K.

The natural radioactivity of soils has been studied inadequately outside the USSR, and many genetic soil types in various countries have not been characterized with respect to their concentration of major radionuclides (Table 1.1).

Data on the concentration of natural radionuclides and their distribution among the soil genetic horizons in the USSR territory illustrate that many genetic soil types have not been studied and that the distribution of natural radionuclides over the soil profiles have not yet been established (Table 1.2). Consequently, the distribution of natural radionuclides in the soils of different regions of European USSR was investigated.

Concentration and distribution of natural radionuclides
in major soils of some soil-climatic zones of the USSR

The distribution of natural radionuclides in genetic soil types was examined by studying soils in the forest, forest-steppe, meadow steppe, steppe and dry steppe zones, as well as some mountain soils in the North Caucasus and Crimea in European USSR. In Asian USSR (Turkmen SSR), only sierozems of the semidesert zone were investigated. A more detailed study was undertaken in Estonian SSR, and the first radiogeological soil mapping was conducted.

Altogether 675 soil and parent rock samples were analyzed for total contents of Ra, Th and K /16/. Ra and Th were determined radiochemically from the radionuclides of ^{226}Ra and ^{224}Ra (ThX) using 20 g samples of calcined soil, which were decomposed by fusion with alkalis. Ra isotopes were separated from the dissolved fused samples by coprecipitation with $BaSO_4$. After a second fusion and dissolving, the salts were subjected to determination by emanation. The analytical sensitivity was 10^{-13} and 10^{-7} g/g for Ra and Th, respectively, with corresponding errors of $7-10$ and $20-25\%$. K was determined by flame photometry after Poluektov /40/; the method's sensitivity was 10^{-4} g/g, the analytical error being $2-3\%$.

Determination of U concentration in various soils is essentially an independent task. Separate analyses for U in soils were performed only for evaluation of the shift of radioactive equilibrium with Ra. No universal methods for routine tests of soils for U are available, while methods developed for rocks and minerals do not yield reliable data on U concentration in soils /5, 26, 36/.

TABLE 1.3. Average and extreme concentrations of natural radioactive elements in major soils and parent rocks in different climatic zones of European USSR and in Central Asian serozems (extreme values are in parentheses), wt %

Soil type and variety	Soil, rock	Number of samples	In air-dry soil			Th: RaU°	In calcined soil		
			Th, 10^{-4}	Ra, 10^{-10}	K		Th, 10^{-4}	Ra, 10^{-10}	K
Soils in the southern taiga zone									
Podzolic soils									
Typical podzolic sandy	Soil	25	2.0 (0.2—4.0)	0.2 (0.1—0.4)	0.7 (0.4—1.1)	3.2 (0.5—10)	2.0 (2.0—5.0)	0.3 (0.1—0.9)	0.7 (0.42—1.2)
	Rock	4	3.0 (2.0—4.0)	0.25 (0.2—0.3)	0.9 (0.57—1.16)	3.5 (2.2—6.7)	3.0 (2.0—4.0)	0.25 (0.2—0.3)	0.9 (0.57—1.16)
Sod-strongly podzolic, on aquatic-glacial sand	Soil	4	6.3 (4.3—8.3)	0.8 (0.5—1.7)	1.27 (1.08—1.52)	2.9 (1.6—3.8)	6.7 (5.0—9.0)	0.9 (0.6—1.8)	1.34 (1.26—1.52)
	Rock	1	4.7	0.5	1.5	3.1	5.0	0.5	1.5
Sod-strongly podzolic loamy, on morainic and mantle loams	Soil	21	8.5 (2.6—15.0)	0.95 (0.6—3.1)	2.2 (1.39—3.30)	3.0 (1.0—4.3)	8.6 (3.0—16.5)	1.0 (0.65—3.4)	2.36 (1.55—3.49)
	Rock	7	8.7 (5.0—13.0)	0.7 (0.5—1.0)	2.2 (1.74—2.63)	4.0 (2.8—5.0)	8.7 (5.0—13.0)	0.7 (0.5—1.0)	2.2 (1.74—2.63)
Sod-medium podzolic, ranging from sandy-loamy to light-loamy, on morainic deposits and mantle loams	Soil	37	6.4 (2.0—11.0)	0.8 (0.3—1.4)	2.1 (1.38—3.0)	3.1 (1.5—4.0)	6.4 (2.0—11.0)	0.8 (0.32—1.46)	2.1 (1.38—3.05)
	Rock	7	7.0 (5.0—10.0)	0.85 (0.5—1.3)	2.0 (1.38—2.52)	3.0 (2.2—4.0)	7.0 (5.0—10.0)	0.86 (0.55—1.3)	2.0 (1.38—2.55)
Sod-medium podzolic heavy-loamy, on varved clays	Soil	5	7.4 (3.0—13.0)	0.9 (0.3—1.4)	2.11 (1.02—2.80)	2.8 (2.5—3.3)	8.0 (6.0—14.0)	1.0 (0.65—1.6)	2.42 (1.87—2.97)
	Rock	1	9.0	1.3	2.94	2.3	9.0	1.3	3.0
Sod-lightly podzolic heavy-loamy, on varved clays	Soil	2	7.7 (5.8—9.7)	1.1 (0.7—1.6)	2.55 (2.53—2.56)	2.5 (2.1—2.9)	8.1 (6.0—10.3)	1.3 (0.7—1.65)	2.69 (2.68—2.70)
	Rock	1	7.0	1.3	2.36	1.8	7.5	1.4	2.54
Sod-slightly podzolic, on fluvio-glacial sands	Soil	4	8.8 (4.0—1.40)	1.0 (0.7—1.3)	1.8 (1.46—2.49)	1.9 (1.9—3.7)	9.4 (5.0—14.5)	1.0 (0.8—1.35)	1.9 (1.55—2.58)
	Rock	2	7.0 (5.0—9.0)	0.7 (0.4—0.9)	1.4 (1.36—1.43)	3.7 (3.3—4.2)	7.7 (5.5—10.0)	0.7 (0.45—1.0)	1.56 (1.53—1.59)
Bog-podzolic soils									
Sod-podzolic-gleyey and gley sandy-loam and loamy, underlain by clays	Soil	22	4.6 (2.0—10.6)	0.6 (0.3—1.2)	2.27 (1.20—2.76)	2.7 (1.3—5.0)	4.9 (2.0—11.5)	0.6 (0.3—1.25)	2.4 (1.26—2.82)
	Rock	10	11.7 (5.0—20.5)	1.0 (0.6—1.95)	2.33 (1.75—2.75)	3.8 (1.9—7.6)	12.2 (5.0—20.5)	1.1 (0.6—2.1)	2.35 (1.78—2.75)

TABLE 1.3. Contd.

Soil type and variety	Soil, rock	Number of samples	In air-dry soil Th, 10⁻⁴	In air-dry soil Ra, 10⁻¹⁰	In air-dry soil K	Th : RaU°	In calcined soil Th, 10⁻⁴	In calcined soil Ra, 10⁻¹⁰	In calcined soil K
Sod-podzolic-gleyey and gley sandy-loam and loamy, underlain clays	Soil	10	8.0 (1.0—10.0)	0.9 (0.1—1.4)	—	2.9 (2.1—3.7)	10.0 (9—15)	1.3 (1.0—1.5)	—
	Rock	3	7.3 (6.0—9.0)	1.0 (0.9—1.1)	—	2.4 (2.2—2.7)	8 (6.0—9.0)	1.1 (0.9—1.2)	—
Peaty-podzolic-gley on sandy loams and sands	Soil	11	4.5 (2.0—9.0)	0.45 (0.2—1.0)	—	3.3 (2.2—4.4)	5.5 (2.0—14.0)	0.6 (0.2—1.6)	—
	Rock	5	5.1 (2.0—8.5)	0.4 (0.2—0.7)	—	4.2 (3.3—5.3)	5.1 (2.0—8.5)	0.4 (0.2—0.7)	—
Calcareous soils									
Humus-calcareous shallow soils on areas with mineralization aureoles	Soil	7	9.6 (5.0—14.0)	4.1 (1.8—5.8)	1.95 (1.8—2.2)	0.8 (0.5—1.1)	11.8 (7.5—18.0)	5.1 (2.7—6.6)	2.54 (2.1—2.84)
	Rock	6	2.0 (<2.0—5.0)	0.9 (0.7—1.5)	0.52 (0.39—0.75)	0.7 (0.4—2.4)	2.0 (<2—6.5)	0.7 (0.7—1.6)	0.53 (0.39—0.77)
Humus-calcareous shallow loamy and sandy-loamy, on Ordovivian limestones (North Estonian plateau)	Soil	6	7.0 (6.4—7.3)	1.2 (0.8—1.7)	—	2.1 (1.4—2.9)	8.8 (7.5—12.0)	1.6 (0.9—2.6)	—
	Rock	4	1.0 (<1.0—2.0)	0.2 (0.1—0.3)	—	2.2	2.0 (<2.0—2.3)	0.3 (0.1—0.5)	—
Humus-calcareous loamy shallow soils on Devonian limestones (southern Estonia)	Soil	3	2.1 (1.7—2.5)	0.5 (0.23—0.67)	—	1.6 (1.1—2.5)	2.5 (2.0—3.0)	0.6 (0.3—0.8)	—
	Rock	1	1.0	0.15	—	2.2	1.5	0.2	—
Sod-calcareous typical loamy shallow soils on calcareous moraine and calcareous fluvioglacial deposits	Soil	9	6.3 (1.47—11.0)	1.0 (0.7—1.2)	1.3	2.0 (0.41—3.3)	7.1 (1.7—12.0)	1.2 (0.85—1.45)	1.4
	Rock	11	2.2 (1.0—5.0)	0.4 (0.1—0.7)	0.55 (0.39—0.70)	2.1 (0.73—3.3)	3.3 (<2—5.0)	0.45 (0.15—0.8)	0.6 (0.39—0.73)
Sod-calcareous typical loamy medium deep soils on calcareous moraine and fluvioglacial deposits	Soil	11	5.0 (3.0—7.0)	1.0 (0.7—1.4)	1.7 (1.36—2.0)	1.7 (0.9—2.9)	5.7 (4.—7.5)	1.1 (0.8—1.45)	2.1 (1.94—2.18)
	Rock	7	2.4 (2.0—3.5)	0.47 (0.3—0.7)	1.4 (1.30—1.46)	1.8 (1.0—2.2)	3.0 (2.0—4.0)	0.57 (0.3—0.85)	1.4 (1.3—1.5)
Sod-calcareous leached sandy-loamy, on calcareous moraine and fluvio-glacial deposits	Soil	7	6.8 (5.0—11.0)	0.8 (0.6—1.0)	—	2.9 (2.2—3.7)	7.3 (5.0—12.0)	0.9 (0.6—1.05)	—
	Rock	5	6.0 (3.0—13.0)	0.7 (0.4—0.9)	—	3.0 (1.4—4.8)	6.0 (3.0—13.5)	0.7 (0.4—0.9)	—

TABLE 1.3. Contd.

Soil type and variety	Soil, rock	Number of samples	In air-dry soil Th, 10^{-4}	In air-dry soil Ra, 10^{-10}	In air-dry soil K	Th : Ra$_U$*	In calcined soil Th, 10^{-4}	In calcined soil Ra, 10^{-10}	In calcined soil K
Sod-calcareous leached loamy, on calcareous moraine and fluvio-glacial deposits	Soil	7	7.0 (3.0—12.0)	1.1 (0.9—1.4)	1.7 (1.34—1.93)	2.2 (1.1—3.3)	7.8 (3.5—13.0)	1.1 (1.0—1.45)	2.0 (1.38—2.04)
	Rock	2	5.0 (3.0—7.0)	0.8	0.4	2.1 (1.3—2.9)	5.5 (4.0—7.0)	0.9 (0.8—1.0)	0.4
Sod-calcareous podzolized loamy, on calcareous moraine	Soil	29	9.6 (4.0—17.0)	1.1 (0.6—1.7)	2.2 (1.90—2.5)	3.2 (1.0—5.2)	10.0 (4.0—18.0)	1.1 (0.7—1.8)	2.3 (2.01—2.59)
	Rock	8	6.4 (5.0—10.0)	0.7 (0.5—1.1)	2.0	3.0 (2.1—4.4)	7.0 (6.0—10.5)	0.8 (0.6—1.15)	2.0
Hydromorphic soils									
Sod-gley clayey, on varved clays	Soil	8	12.0 (1.0—16.0)	1.4 (0.2—2.5)	—	2.8 (1.7—3.9)	14.5 (7.8—18.0)	1.8 (1.2—2.6)	—
	Rock	4	11.5 (9.0—13.0)	1.4 (1.2—1.5)	—	2.8 (2.5—3.1)	12.6 (9.0—14.4)	1.5 (1.2—1.7)	—
Sod-gley clayey, on calcareous clays	Soil	10	7.7 (1.9—12.0)	1.1 (0.2—1.6)	3.1 (2.85—3.28)	2.3 (1.3—3.3)	8.8 (2.5—13.5)	1.3 (0.25—1.8)	3.3 (3.13—3.44)
	Rock	3	6.8 (4.0—9.2)	1.1 (0.7—1.5)	2.9	2.0 (1.9—2.1)	7.7 (5.0—10.0)	1.3 (0.8—1.6)	3.0
Humus-sod-gley, on kuckersite moraine underlain by "rikhk"† moraine	Soil	3	6.3 (4.0—9.0)	1.5 (1.2—2.0)	—	1.4 (1.1—1.5)	12.8 (7.7—18.0)	3.1 (2.3—4.2)	—
	Rock	2	5.0 (4.0—6.0)	0.95 (0.9—1.0)	—	1.8 (1.3—2.2)	8.0 (7.2—8.7)	1.5 (1.3—1.8)	—
Humus-sod-gley saturated loamy, on calcareous moraine	Soil	3	4.0 (3.0—4.8)	0.9 (0.75—1.2)	1.4 (1.0—1.99)	1.4 (1.3—1.7)	6.0 (3.0—8.0)	1.4 (0.8—1.75)	2.1 (2.03—2.15)
Humus-sod-gley unsaturated loamy, on calcareous moraine	Soil	4	8.0 (5.0—10.0)	1.1 (0.8—1.2)	2.5 (2.12—2.72)	2.5 (1.5—3.0)	9.5 (7.0—10.0)	1.2 (0.8—1.8)	2.73 (2.56—2.82)
	Rock	1	9.0	0.8	2.65	3.8	9.0	0.9	2.8
Peaty-humus-gley sandy-loamy, underlain by varved clays	Soil	4	3.7 (<1.0—6.0)	0.5 (0.1—0.9)	1.6 (0.65—3.05)	2.7 (2.0—3.3)	4.9 (1.0—7.5)	0.7 (0.15—1.25)	2.1 (1.38—3.18)
	Rock (varved clay)	1	7.4	1.3	2.9	1.0	8.0	1.4	3.2

† [Rikhk, or richk, consists of clay, lumps or broken plates of limestone and northern boulders.]

TABLE 1.3. Contd.

Soil type and variety	Soil, rock	Number of samples	In air-dry soil			Th : Ra$_U$*	In calcined soil		
			Th, 10^{-4}	Ra, 10^{-10}	K		Th, 10^{-4}	Ra, 10^{-10}	K
Bog soils									
Lowland bog soils of high (> 15 %) ash content	Peat	6	2.2 (0.6—5.7)	0.41 (0.17—1.0)	0.10 (0.09—0.15)	1.9 (1.0—3.3)	7.3 (3.3—12.0)	1.3 (1.0—2.1)	0.58 (0.5—0.81)
Lowland bog soils of low (< 15 %) ash content	Peat	9	0.25 (0.13—0.45)	0.13 (0.05—0.23)	0.08 (0.05—0.13)	0.8 (0.33—1.8)	2.5 (1.0—4.2)	1.2 (0.78—2.43)	0.58 (0.44—1.3)
Upland bog soils	Peat	6	0.06 (<0.04—0.15)	0.014 (0.005—0.022)	0.056	1.9 (<1—4.5)	3.6 (3.5—<7.0)	0.80 (0.47—1.34)	3.2
Soils of river floodplains and inundated lake shores									
Sod-alluvial sandy-loamy soils of near-channel floodplain	Soil	6	5.0 (3.0—6.8)	0.8 (0.5—1.13)	—	2.1 (1.3—3.2)	5.3 (3.6—7.1)	0.85 (0.16—1.16)	—
	Rock	1	5.0	0.9	—	1.9	6.2	1.1	—
Sod-alluvial sandy-loamy granular soils of central floodplain	Soil	6	11.5 (2.1—18.0)	1.4 (0.28—2.5)	2.0 (1.72—2.34)	2.8 (2.0—3.6)	12.4 (2.5—19.0)	1.5 (0.34—2.55)	2.5 (2.47—2.55)
	Rock	2	14.5 (14.0—15.0)	1.6 (1.5—1.8)	2.64 (2.30—2.98)	2.9 (3.6—3.3)	15.7 (15.0—16.5)	1.8 (1.65—1.95)	2.75 (2.58—3.08)
Sod-alluvial gley soil of bogged floodplain	Soil	3	7.0 (3.0—10.0)	1.0 (0.55—1.4)	2.0 (1.9—2.08)	2.2 (1.9—2.4)	8.7 (3.6—12.5)	1.3 (0.65—1.8)	2.2 (2.02—2.43)
Sod-meadow sandy-loamy soils of inundated lake shores	Soil	8	6.1 (5.0—7.0)	0.9 (0.7—1.2)	—	2.2 (1.4—2.9)	6.6 (5.0—9.3)	1.1 (0.8—1.4)	—
	Rock	4	6.0 (5.0—7.0)	0.9 (0.5—1.2)	—	2.2 (1.7—3.3)	6.0 (5.0—7.0)	0.9 (0.8—1.2)	—
Sod-gley light-loamy soils of inundated lake shores	Soil	3	7.8 (6.4—9.2)	1.1 (0.85—1.4)	—	2.4 (2.2—2.5)	8.7 (6.8—11.0)	1.2 (0.9—1.7)	—
Silty bog soil	Soil	1	9.6	1.2	—	2.6	10.5	1.35	—
Soils of the forest-steppe and steppe zones									
Loamy gray forest soils	Soil	29	8.0 (4.0—12.0)	0.9 (0.5—1.4)	2.13	3.0 (1.7—4.4)	10.0 (4.5—13.0)	1.1 (0.6—1.5)	2.56
	Rock	4	9.0 (6.0—12.0)	1.0 (0.9—1.6)	—	2.7 (2.1—3.3)	9.0 (6.0—12.0)	1.1 (0.9—1.6)	—
Heavy-loamy deep chernozems	Soil	26	8.5 (6.0—14.0)	0.9 (0.5—1.4)	1.6 (1.0—2.3)	3.1 (2.2—4.7)	9.5 (7.0—16.0)	1.0 (0.6—1.6)	1.9 (1.20—3.19)
	Rock	8	10.0 (8.0—12.0)	1.0 (0.8—1.2)	1.7 (1.6—1.8)	3.3 (2.4—5.0)	10.0 (8.0—13.0)	1.0 (0.8—1.3)	1.8 (1.72—1.87)

TABLE 1.3. Contd.

Soil type and variety	Soil, rock	Number of samples	In air-dry soil: Th, 10^{-4}	In air-dry soil: Ra, 10^{-10}	In air-dry soil: K	Th:Ra_U*	In calcined soil: Th, 10^{-4}	In calcined soil: Ra, 10^{-10}	In calcined soil: K
Ordinary heavy-loamy chernozems	Soil	23	8.0 (5.0—10.0)	0.9 (0.6—1.2)	1.7 (1.6—1.81)	3.0 (1.7—4.2)	9.0 (6.0—13.0)	1.0 (0.8—1.3)	1.9 (1.71—2.10)
	Rock	9	8.0 (7.0—10.0)	0.96 (0.8—1.2)	1.4 (1.25—1.47)	2.8 (1.9—3.3)	8.5 (7.0—11.0)	1.0 (0,8—1.3)	1 5 (1.25—1.62)
Loamy southern chernozems	Soil	3	10.0 (10.0—11.0)	1.0 (0.7—1.3)	—	3.3 (2.6—4.8)	11.0	1.1	—
Heavy-loamy azov chernozems	Soil	16	8.0 (6.0—11.0)	1.0 (0.7—1.3)	—	2.7 (1.7—3.8)	9.0 (6.0—11.0)	1.1 (0.7—1.5)	—
	Rock	5	7.0 (6.0—9.0)	0.8 (0.8—0.9)	—	2.7 (2.2—3.3)	8.0 (7.0—9.0)	0.9 (0.8—1.0)	—
Soils of the dry steppe zone									
Light-chestnut slightly solonetzous, light-loamy (Kalmyk ASSR)	Soil	4	6.5 (5.0—8.0)	0.7 (0.6—0.8)	1.40 (1.01—1.80)	3.1 (2.5—3.8)	7.3 (6.0—8.0)	0.8 (0.7—0.9)	1.57 (1.21—1.80)
	Rock	2	6.0 (5.0—7.0)	0.7	1.42 (1.36—1.48)	2.9 (2.4—3.3)	6.0 (5.0—7.0)	0.7	1.42 (1.36—1.48)
Medium-columnar loamy solonetz of the chestnut soil zone (Kalmyk ASSR)	Soil	4	7.5 (6.0—10.0)	0.8 (0.5—1.1)	1.32 (1.24—1.40)	3.2 (2.6—4.6)	8.0 (6.0—11.0)	0.8 (0.5—1.2)	1.44 (1.42—1.50)
	Rock	1	6.0	0.6	1.28	3.3	6.0	0.6	1.28
Mountain soils									
Loamy gray mountain-forest soils (Caucasus)	Soil	5	7.0 (6.0—9.0)	0.76 (0.7—0.8)	—	3.2 (2.5—3.8)	—	—	—
	Rock	2	6.5 (5.0—8.0)	0.65 (0.6—0.7)	—	3.3 (2.8—3.8)	—	—	—
Brown mountain-forest heavy-loamy (Caucasus)	Soil	5	7.0 (6.0—8.0)	0.7 (0.6—0.9)	—	3.2 (2.6—4.4)	—	—	—
	Rock	1	7.0	1.0	—	2.3	—	—	—

TABLE 1.3. Contd.

Soil type and variety	Soil, rock	Number of samples	In air-dry soil				In calcined soil		
			Th, 10^{-4}	Ra, 10^{-10}	K	Th:RaU*	Th, 10^{-4}	Ra, 10^{-10}	K
Brown mountain-forest heavy-loamy (Crimea)	Soil	5	9.6 (9.0—10.0)	0.8 (0.8—0.9)	1.9	3.8 (3.3—4.2)	10.5 (9.0—11.0)	0.9 (0.9—1.0)	2.1
	Rock	1	8.0	0.8	—	3.3	8.0	0.8	—
Cinnamon mountain loamy, underlain by limestone (Crimea)	Soil	4	7.0 (6.0—8.0)	0.7 (0.5—0.9)	1.4	3.3 (2.9—4.0)	8.2 (6.0—11.0)	0.85 (0.5—1.2)	1.6
	Rock	2	5.5 (5.0—6.0)	0.6	—	3.0 (2.8—3.3)	5.5 (5.0—6.0)	0.60	—
Cinnamon mountain heavy-loamy soil underlain by marl (Caucasus)	Soil	2	8.0 (7.0—9.0)	0.75 (0.7—0.8)	—	3.6 (2.9—4.3)	—	—	—
	Rock	2	4.0	0.55 (0.5—0.6)	—	2.5 (2.2—2.7)	—	—	—

Soils of the semidesert zone
(Asian USSR)

Soil type and variety	Soil, rock	Number of samples	In air-dry soil				In calcined soil		
			Th, 10^{-4}	Ra, 10^{-10}	K	Th:RaU*	Th, 10^{-4}	Ra, 10^{-10}	K
Loamy solonchakous serozems (Kazakhstan)	Soil	4	9.0 (9.0—10.0)	1.1 (1.1—1.2)	—	2.7 (2.5—3.0)	—	—	—
Sandy-loamy and light-loamy solonchakous serozems (Turkmenia)	Soil, A-horizon	49	4.3 (3.1—6.0)	0.7 (0.4—1.1)	1.7 (1.31—2.04)	2.0 (1.21—3.42)	—	—	—

* RaU denotes Ra expressed in equilibrium U units.

The error of analytical results obtained by direct determination of U by existing methods is, in the majority of cases, below the fluctuation limits of radioactive equilibrium between U and Ra. Therefore, the data presented in this chapter represent Ra concentration determinations by the emanation method, repeated for all the investigated samples. For general conclusions, Ra concentrations will be converted to equilibrium U concentrations.

Analysis of the distributions of natural radionuclides in 50 genetic soil types and subtypes revealed several important features. The concentration of natural radionuclides in soils depends on their concentration in soil-forming rocks and the degree of alteration of the parent rock in the course of soil formation (Table 1.3). The average overall content of radionuclides over the entire soil profile of mineral soils formed on a thick mantle of unconsolidated sedimentary rocks is close to their concentration in the parent rocks.

The opposite characteristics are found in soils differing markedly from underlying rocks in their physicochemical and structural properties. The concentrations of Ra, Th and K in bog soils of lowland and upland peat bogs are lower by factors of tens and hundreds when calculated per air-dry matter, and almost the same as underlying unconsolidated sedimentary rocks when calculated per calcined matter.

The concentration of natural radionuclides is several times higher in soils formed on eluvium of calcareous rocks than in calcareous rocks, i. e., the weathering of calcareous rocks results in a relative concentration of radioactive elements in soils in the course of soil formation.

The concentrations of natural radioactive elements (calculated per air-dry matter) in the investigated soils varied fairly widely: Ra, $n \cdot 10^{-13} - n \cdot 10^{-9}\%$; Th, $n \cdot 10^{-6} - n \cdot 10^{-3}\%$; K, $n \cdot 10^{-2} - n\%$.

These variations are much smaller (from 1 to 1.5 orders of magnitude) within the following three, arbitrarily delineated large groups of soil:

1) Upland and lowland peat bogs have the following concentrations: Ra, $n \cdot 10^{-13} - n \cdot 10^{-11}\%$; Th, $n \cdot 10^{-6} - n \cdot 10^{-5}\%$; K, $n \cdot 10^{-2} - n \cdot 10^{-1}\%$. The concentrations of Ra and Th and the Th:Ra$_U$ ratio in peat bogs are directly related to their ash contents.

2) In mineral soils on background areas: Ra, $n \cdot 10^{-11} - n \cdot 10^{-10}\%$; Th, $n \cdot 10^{-4} - n \cdot 10^{-3}\%$; K, $n \cdot 10^{-1} - n\%$. Average concentrations of the same elements in soils of light and heavy textures vary by factors of 7, 6 and 5 for Ra, Th and K, respectively. In soils of heavy texture only (loamy and clayey), the factors are 3 for Ra and Th, and 2 for K.

3) In soils located on areas with ore aureoles, the concentration of Ra varies from $n \cdot 10^{-10}$ to $n \cdot 10^{-9}\%$, that of Th varies from $n \cdot 10^{-3}\%$ and higher, while the concentration of K is of the order of $n\%$.

Spatial features in the distribution of natural radioactive elements in the soil cover are of a regional nature. Their concentrations in soils depend on the relief. Variations of the combined Ra + Th concentration in soils of different genetic types but similar texture, that are constituents of different elementary landscapes of the same large region, are scarcely noticeable.

Differences between soils of different landscapes are best revealed by comparing the Th:Ra ratio, which serves as an index of the trend of weathering processes and the nature of soil formation, testifying to a higher mobility of Ra compared with Th in soil cover. The Th:Ra (or Th:U) ratio is therefore proposed as a geochemical parameter.

Mineral soils of flushing regimes (in eluvial and transeluvial landscapes) of all genetic types in different climate zones display a relatively larger loss of Ra and accumulation of Th; in their case, Th:Ra$_U \geqslant 3$.

A characteristic feature of soils in accumulation (hydromorphic and semihydromorphic) areas is the relative accumulation of Ra over Th, in which case Th:Ra$_U$ < 3. The mean value of Th:Ra$_U$ over the entire soil profile is close to its value in rocks, but it varies from one genetic horizon to another, increasing in upper (A_0, A_1, A_2) horizons, especially in acid soils, and decreases in the illuvial B horizon, pointing to preferential eluviation in the illuvial horizon. The vertical distribution of Ra and Th over soil horizons depends on the soil-forming process.

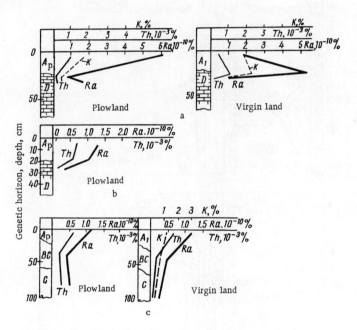

FIGURE 1.1. Distribution of Ra, Th and K over the profile of calcareous soils in Estonian SSR:

a, b — humus-calcareous loamy soils underlain by limestone; c — sod-calcareous typical shallow loamy soils on "rikhk" moraine.

Calcareous soils on residual weathering crust possess maximum concentration of natural radionuclides in the upper humus horizon, decreasing down toward the rock (Figure 1.1). Podzolization, solonetzization and gleyzation entail leaching of Ra and Th from the upper A_1 and A_2 horizons and their accumulation in the illuvial B and gley G, B_g horizons, where their concentration increases 1.5 — 3 and 1.2 — 2 times in comparison to their respective concentrations in the humus A_1 and rock C horizons (Figure 1.2).

The distribution of natural radioelements over the profiles of gray forest, chernozem, chestnut and various mountain soils is only weakly differentiated (Figure 1.3).

As a rule, the distribution of Ra and Th over the soil profile is in agreement with the distribution of silty and clayey particles and also with the
distribution of sesquioxides.

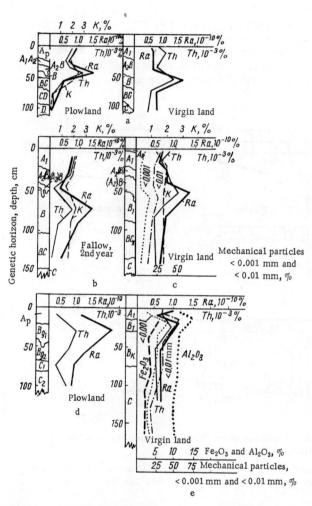

FIGURE 1.2. Distribution of Ra, Th and K, mechanical particles and
sesquioxides over profiles of different soils:

a — sod-calcareous podzolized loamy soils on calcareous moraine,
Estonian SSR; b — sod-medium podzolic soils, ranging from sandy-
loamy to loamy, on binary deposits, Estonian SSR; c — sod-strongly
podzolic light-loamy soil on red-brown moraine, Estonian SSR;
d — sod-humus-gley loamy medium-cultivated soil on kuckersite
moraine, Estonian SSR; e — light-loamy medium-columnar solonetz
on loess-like loam in the zone of chestnut soils, Kalmyk ASSR.

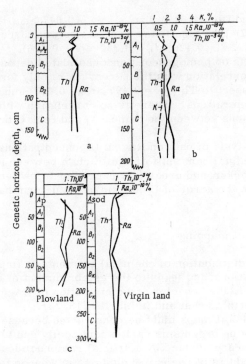

FIGURE 1.3. Distribution of Ra, Th and K over soil profiles in forest-steppe and steppe zones:

a — heavy-loamy dark-gray forest soil, Tula Region;
b — ordinary heavy-loamy chernozem on a loess-like
loam, Voronezh Region; c — extra-deep mycelial-cal-
careous heavy-loamy azov chernozems, Krasnodar
Territory.

Tillage of soils does not introduce any appreciable changes in the distribution of natural radioactive elements over the soil profile, provided its
depth exceeds 1 m (see Figures 1.1c, 1.2 and 1.3). Breaking of very
shallow soils (10 — 30 cm) underlain by consolidated bedrock alters the
original distribution pattern. For instance, maximum Ra and Th concentration occurs in the lower part of the humus horizon at contact with limestone
bedrock in the case of virgin shallow humus-calcareous loamy soils, but it
shifts to the upper part of the humus horizon in similar soils (see Figure 1a);
regular plowing of such soil over a prolonged period causes Ra and Th to
become distributed uniformly over the entire soil layer (see Figure 1.1b).
   Statistical processing of small samples (16 — 18 soil profiles) points to
a correlation, in the majority of cases, between Ra and Th concentrations
in soil horizons and underlying rocks. The correlation between Ra and Th
concentration in soils and rocks depends on the type of soil-forming processes controlling the degree of migration of these elements.
   Calcareous soils display a statistically significant positive correlation
with the underlying calcareous rocks only with respect to Th concentration

in the humus A_1 and illuvial B horizons ($r_{A,C} = 0.647 > r_{0.01} = 0.623$; $r_{BC} = 0.547 > r_{0.05} = 0.497$)*; no significant correlation was discovered with respect to Ra ($r_{A,C} = 0.147 \ll r_{0.1} = 0.426$; $r_{BC} = 0.280 \ll r_{0.1} = 0.426$).

Sodpodzolic soils on homogeneous unconsolidated deposits display a significant positive correlation with the parent rocks only for the illuvial B horizon with respect to Th ($r_{BC} = 0.497 > r_{0.05} = 0.468$) and Ra ($r_{BC} = 0.572 > r_{0.02} = 0.543$) concentrations. There is no correlation with respect to Ra and Th concentrations between the humus A_1 and rock C horizons in sodpodzolic soils.

Painstaking analysis of data on Ra and U concentrations in various soils reveals only an insignificant shift of equilibrium between Ra and U in mineral soils in background areas (not exceeding a few tens of percent), mostly lying within the error of U concentration determination.

Radiogeological soil mapping

Mapping of the distribution of chemical elements is important in biogeochemical investigations, and leads to the establishment of important geochemical patterns in the behavior and distribution of trace and radioactive elements in the earth's crust and its shells /18, 23/.

Radiogeological soil maps must be constructed because the radioactivity in the habitat of living organisms varies continually due to pollution of the earth's atmosphere and surface by artificial radionuclides (on a global scale), such as products of nuclear explosions in weapon testing. Monitoring of environmental radiation levels calls for accurate information on the natural background due to radiation from natural radionuclides.

Construction of radiogeological soil maps is necessary for the following purposes; 1) establishment of the general behavior of natural radionuclides in the planetary soil cover and determination of the role played by the soil-forming process in the geochemistry of natural radionuclides; 2) striking the geochemical balance in the cycle of natural radionuclides on Earth; 3) establishment of criteria in biogeochemical and aureole prospecting methods for radioactive ores; 4) determination of accurate data on natural radiation background, necessary for monitoring any variation in the environmental radiation level due to human activity; 5) examination of the influence of natural radiation on the development of biogeocenoses.

The construction of a map is controlled by requirements to be met by its data. Maps for estimating the effect of radionuclides on plants and animals must differ from analogous agrochemical maps of the concentrations of trace elements (nonradioactive). Besides featuring the deficiency or excess of radionuclides in their mobile forms (water-soluble, absorbed and acid-soluble), they must also feature their total content, since the reactions of organisms to ionizing radiation do not depend solely on the quantity of incorporated radionuclides controlling the internal irradiation dose, but also on their total content in the environment, controlling the external irradiation doses. Therefore the total content of natural radioactive elements must be

* $r_{A,C}$ and r_{BC} are coefficients of correlation between the concentrations in the A_1C and BC horizons.

determined when investigating soil radioactivity and constructing maps of
their distribution in the soil cover.

Dosimetric evaluation of radiation effects on plants and animals is sometimes performed by means of radiometric maps marked with isopleths of
the combined intensity of beta or gamma radiation. However, the usefulness
of these radiometric maps is very limited owing to the variety of properties
and the genesis of such radiation sources. The above considerations testify
that radiogeological soil mapping must be based on a biogeochemical
approach.

Maps of the natural radionuclide distribution in soils do not yet exist,
and their construction has not yet been discussed. Some general techniques
used in the mapping of trace elements in soils can also be applied to the
natural radioactive elements.

In view of the comparatively uniform distribution of trace elements and
natural radioactive elements in various sedimentary rocks and soils over
very large areas, the method of key areas [clue method] was adopted for
field observations. Execution of a large-scale survey over a predeter-
mined grid is limited by the cumbersomeness and inefficiency of analyzing
samples for several trace and radioactive elements.

Features revealed by investigations are: 1) dependence of radionuclide
concentration in soils on their concentration in soil-forming rocks; 2) dependence of radionuclide distribution over the soil profile on the type of
soil-forming process and its conformity to the distribution of clay fractions
and sesquioxides; and 3) influence of geomorphological factors on the migration and distribution of radionuclides in soils. These features suggest that
techniques for mapping the distribution of radionuclides in soils should be
based on the same principles as soil mapping. A territory planned for soil
mapping is divided into landscape-geographical units characterized by the
same combination and interaction of the different natural factors (climate,
relief, vegetation, soil types, lithological and stratigraphic properties of
parent rocks, etc.).

Our first attempt at radiogeological soil mapping was carried out in
Estonian SSR /37, 38/. During the field period our observations were performed by the method of key areas in the following sequence:

1. Investigation of all the geomorphological regions (eluvial, accumul-
ative and other characteristic types of landscape).

2. Studies of the principal soil types in the geomorphological regions.

3. Studies of the major textural varieties (sandy, loamy, clayey) within
every soil species.

4. Additional investigation of soils of the same species and very similar
textures on parent rocks of different genesis.

Samples were taken in the field from every genetic horizon of the soil
profile, following the standard method adopted in soil science. Radiometric
techniques were used in the field only as auxiliary when choosing suitable
locations for soil profiles.

The distribution of natural radioactive elements in Estonia was studied
with the aid of 87 soil profiles, from which 386 samples were taken, according to genetic horizons, for the determination of Ra, Th and K. The method
of key areas was used to investigate nearly all the varieties of plowlands and
virgin automorphic and hydromorphic soils in the major types of landscapes, such as limestone plateaus, limestone plains, morainic hills, drum-

lins, sandur plains, marginal formations and eskers, plains on varved clays, lacustrine-glacial plains, lakeshore plains, river floodplains, and bogs.

Maps of the distribution of natural radionuclides in the soil cover were based on our data concerning Ra and Th concentrations in soils (calculated for air-dry matter) from key areas by extrapolation based on other natural maps of Estonian SSR, e. g., the soil map on the scale 1 : 1,500,000 compiled by Lillema /33/, the geologico-lithological map of Quaternary deposits, the geologico-morphological map and geological map of bedrock.

A taxonomic scale was developed for maps of the Ra and Th distributions in Estonian soils, based on the following criteria: 1) estimates of average Ra and Th concentrations and the Th:Ra ratio in the soil cover in Estonia; 2) variation of the concentration of these elements in soil varieties; 3) occurrence frequencies of their concentrations and accuracy of analytical methods.

The mean concentration of radionuclides (or weighed mean K) for the soil cover of the entire Estonian SSR was calculated by the formula

$$K = \frac{\sum_{i=1}^{n} c_i S_i}{100},$$

where c_i is the mean concentration of radionuclides in the genetic horizon (or soil layer) of every i-th soil type occurring on area S_i (%), the entire Estonian territory being taken as 100%.

The taxonomic scale for Ra and Th concentrations contains seven intervals (Table 1.4), while the scale of the Th:Ra$_U$ ratio contains four intervals (Table 1.5). The gradations are not all equal, owing to the wide range of Ra and Th concentrations in all soil types (as much as three orders of magnitude) and their small variation in the most common mineral soils.

TABLE 1.4. Taxonomic intervals of Ra and Th concentrations (adopted for compiling schematic maps of the distribution of natural radioactive elements in the soil cover of Estonian SSR)

Concentration, taxonomic intervals	Content in air-dry soil, %		Concentration, taxonomic intervals	Content in air-dry soil, %	
	Ra, 10^{-10}	Th, 10^{-4}		Ra, 10^{-10}	Th, 10^{-4}
High	> 3.0	—	Slightly low	0.4−0.7	4−7
Rather high	1.5−3.0	15−30	Rather low	0.1−0.4	2−4
Slightly high	1.0−1.5	10−15	Low	<0.1	<2
Normal, average	0.7−1.0	7−10			

TABLE 1.5. Taxonomic gradations of Th:Ra$_U$ ratios in soils (adopted for compiling schematic maps of the distribution of natural radioactive elements in the soil cover of Estonian SSR)

Relative concentration, gradation	Th:Ra$_U$ ratio	Relative concentration, gradation	Th:Ra$_U$ ratio
Excessive, Ra	≤1	Normal, Ra and Th	2.0−3.0
Predominant, Ra	1.0−2.0	Predominant, Th	> 3.0

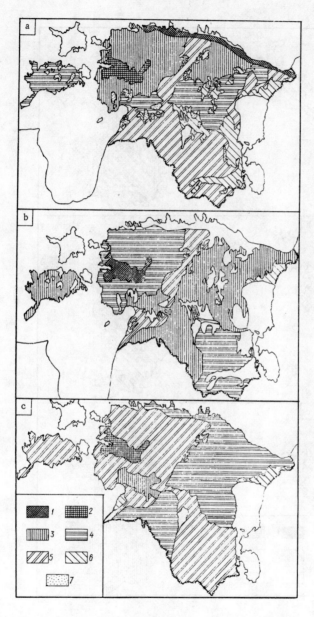

FIGURE 1.4. Schematic map of the distribution of Ra concentrations in the humus (a), illuvial (b) horizons and the parent rock (c)
of soils in the Estonian SSR.

Ra concentrations (10^{-10}%): > 3 (1); 1.5−3 (2); 1−1.5 (3);
0.7−1.0 (4); 0.4−0.7 (5); 0.1−0.4 (6); and < 0.1 (7).

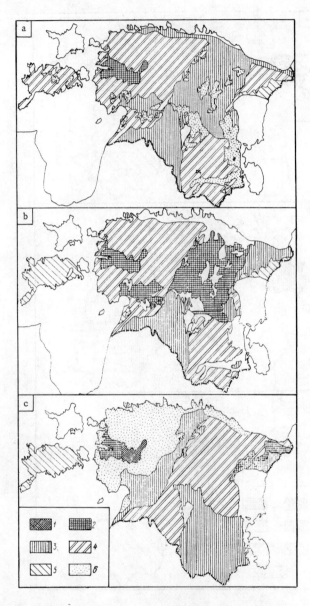

FIGURE 1.5. Schematic map of the distribution of Th concentrations in the humus (a), illuvial (b) horizons and the parent rock (c)
of soils in the Estonian SSR.

Th concentrations ($10^{-4}\%$): 15—30 (1); 10—15 (2); 7—10 (3);
4—7 (4); 2—4 (5); and < 2 (6).

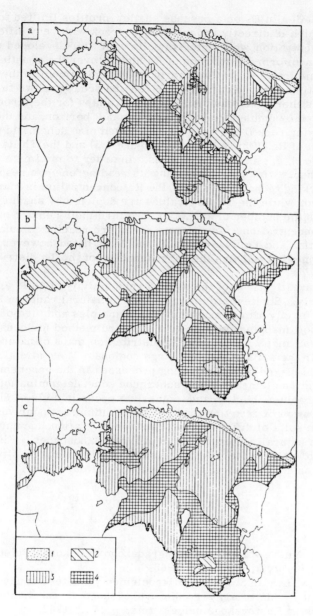

FIGURE 1.6. Schematic map of the distribution of Th:Ra$_U$ ratios
in the humus (a), illuvial (b) horizons of soils and the parent
underlying rocks (c) in the Estonian SSR.

Th:Ra$_U$ ratios: < 1 (1); 1.0—2.0 (2); 2.0—3.0 (3), and 3.0—5.0 (4).

Some soil varieties in Estonia possess shallow profiles limited to a humus horizon which is directly underlain by bedrock with very different radionuclide concentration, while mineral soils with well-developed profile display a very nonuniform distribution of Ra and Th over the genetic horizons, with a minimum in the A_0A_1 horizon and a maximum in the B horizon (several times that in the A_0A_1 horizon and occupying two taxonomic intervals). It was therefore feasible to construct maps for the three major genetic horizons: the humus A horizon, the illuvial B horizon, and the parent rock C or underlying rock D horizon. We compiled nine schematic maps of the distribution of Ra (Figure 1.4), Th (Figure 1.5) and the Th:Ra ratio (Figure 1.6) in the A, B * and $C(D)$ horizons. In order to minimize the number of map sheets without complicating its reading, one can resort to another data-marking technique, namely, the Ra concentration is marked on the map in color while the Th:Ra$_U$ values are depicted by shading. This technique reduces the number of sheets to one-third their number otherwise. The Th concentrations for different areas are readily calculated from the Ra concentration and Th:Ra$_U$ ratio. It should be noted, however, that this technique somewhat impairs the representation of the Th distribution in soil cover.

Present-day investigations of the natural radioactivity of soils are conducted inadequately. Studies into the migration of natural radionuclides in the soil cover mainly aim at examining the principles and method of radiogeological soil mapping, developing a standard method for constructing radiogeological soil maps, and at plotting distribution maps of natural radioactive elements in the soils of various large regions. More attention should be paid to the role played by soil-forming processes in the geochemistry of natural radioactive elements in the supergene zone, determination of the forms in which uranium, thorium and potassium occur in soils of different genesis, the migration of decay products of the uranium, actinouranium and thorium series in soils of different types, and the migration, accumulation and dispersion of natural radionuclides in different landscapes. Finally, it is necessary to evaluate the role of soils in the migration and circulation of radionuclides on a global scale in the biosphere of our planet.

References

1. Alekseev, Yu. V. — Zapiski Leningradskogo Sel'skokhozyaist-vennogo Instituta, 98(3):44. 1965.
2. Ananyan, V. L. — In Sbornik: "Mikroelementy i estestvennaya radio-aktivnost'." Materialy 3-go Mezhvuzovskogo soveshchaniya. Izdatel'stvo Rostovskogo universiteta, p. 175. 1962.
3. Ananyan, V. L. and A. Sh. Avetisyan. — Agrokhimiya, No. 6:154. 1964.
4. Andreeva, E. A. — Pochvovedenie, No. 5:21. 1960.
5. Baeva, A. I. — Izvestiya Akademii Nauk AzSSR. Seriya Biologiches-kaya, Vol. 5:89. 1964.

* In schematic maps of the distribution of radio-elements in the illuvial horizon, white spots correspond to shallow soils devoid of this horizon (B).

6.   B a e v a, A. I. — Trudy Instituta Pochvovedeniya i Agrokhimii Akademii
        Nauk AzSSR, Vol. 22:149. 1964.
7.   B a r a n o v, V. I. and S. G. T s e i t l i n. — DAN SSSR, Vol. 30:328. 1941.
8.   B a r a n o v, V. I. et al. — Geokhimiya, No. 6:530. 1959.
9.   B a r a n o v, V. I. — In Sbornik: "Mikroelementy i estestvennaya radio-
        aktivnost' pochv." Materialy 3-go Mezhvuzovskogo soveshchaniya.
        Izdatel'stvo Rostovskogo universiteta, p. 162. 1962.
10. B a r a n o v, V. I. et al. — Ibid., p. 165.
11. B a r a n o v, V. I. et al. — Pochvovedenie, No. 8:11. 1963.
12. B a r a n o v, V. I. — Izvestiya Akademii Nauk SSSR. Seriya Biologiches-
        kaya, No. 1:159. 1964.
13. B a r a n o v, V. I. et al. — In Sbornik: "Problemy geokhimii," p. 550.
        Moskva, "Nauka." 1965.
14. B a r a n o v, V. I. et al. — In Sbornik: "Radioaktivnost' pochv i metody ee
        opredeleniya," p. 195. Moskva, "Nauka." 1966.
15. B a r a n o v, V. I. et al. — Sbornik Nauchnykh Trudov Estonskoi Sel'sko-
        khozyaistvennoi Akademii, Vol. 49:182. 1966.
16. B a r a n o v, V. I. and N. G. M o r o z o v a. — In Sbornik: "Fiziko-
        khimicheskie metody issledovaniya pochv," p. 5. Moskva,
        "Nauka." 1966.
17. V e r n a d s k i i, V. I. — Pochvovedenie, No. 1:15. 1936.
18. V e r n a d s k i i, V. I. Izbrannye sochineniya (Selected Works), Vol. 1.
        Moskva, Izdatel'stvo Akademii Nauk SSSR. 1957
19. V e r n a d s k i i, V. I. Ibid., Vol. 5. 1960.
20. V e s h k o, E. I. — In Sbornik: "Radioaktivnost' pochv i metody oprede-
        leniya," p. 212. Moskva, "Nauka." 1966.
21. V i n o g r a d o v, A. P. — DAN SSSR, Vol. 110:375. 1956.
22. V i n o g r a d o v, A. P. and V. V. K o v a l' s k i i. — DAN SSSR,
        Vol. 113:315. 1957.
23. V i n o g r a d o v, A. P. Geokhimiya redkikh i rasseyannykh elementov v
        pochvakh (Geochemistry of Rare and Dispersed Elements in Soils).
        Moskva, Izdatel'stvo Akademii Nauk SSSR. 1957.
24. V l a s y u k, P. A. and D. M. G r o d z i n s k i i. — DAN SSSR, Vol. 106:562.
        1956.
25. V o i t e k h o v s k a y a, Ya. V. et al. — In Sbornik: "Materialy I Nauchnoi
        konferentsii molodykh geologov Belorussii," p. 98. Minsk. 1965.
26. G o l u b k o v a, M. G. — In Zbirnyk: "Fiziolohychno-biokhimichni osnovy
        pidvyshchennya produktyvnosti roslyn," p. 363. Kiev, Dershsil'-
        hospvydav. 1963. [In Ukrainian.]
27. G r o d z i n s k i i, D. M. Estestvennaya radioaktivnost' rastenii i pochv
        (Natural Radioactivity of Plants and Soils). Kiev, "Naukova
        dumka." 1965.
28. G u r s k i i, G. V. — In Sbornik: "Materialy I Nauchnoi konferentsii
        molodykh geologov Belorussii," p. 95. Minsk. 1965.
29. G u s' k o v a, V. I. and A. N. B r a g i n a. — Gigiena i Sanitariya,
        No. 7:103. 1962.
30. K a v t e l a d z e, B. M. — Soobshcheniya Akademii Nauk GruzSSR,
        Vol. 28:583. 1962.
31. K o v a l e v s k i i, A. L. — In Sbornik: "Trudy Konferentsii pochvovedov
        Sibiri i Dal'nego Vostoka," p. 53. Novosibirsk. 1964.

32. Lagunov, L. L. — Vestnik Dal'nevostochnogo Filiala Akademii Nauk
    SSSR, Vol. 9:105. 1934.
33. Lillema, A. I. — In Sbornik: "Ezhegodnik Estonskogo geograficheskogo
    obshchestva," p. 66. Tallin. 1960.
34. Maldav, Z. A. et al. — Agrokhimiya, No. 2:123. 1965.
35. Malyuga, D. P. Biogeokhimicheskii metod poiskov rudnykh mesto-
    rozhdenii (Biogeochemical Method of Prospecting for Ore Deposits).
    Moskva, Izdatel'stvo Akademii Nauk SSSR. 1963.
36. Movlyanov, G. A. and K. Kh. Mirzoeva. — In Sbornik: "Mikro-
    elementy i estestvennaya radioaktivnost' pochv." Materialy 3-go
    Mezhvuzovskogo soveshchaniya, p. 61. Izdatel'stvo Rostovskogo
    universiteta. 1962.
37. Morozova, N. G. — Sbornik Nauchnykh Trudov Estonskoi
    Sel'skokhozyaistvennoi Akademii, Vol. 49:165. 1966.
38. Morozova, N. G. Estestvennaya radioaktivnost' pochv Evropeiskoi
    chasti SSSR i opyt sostavleniya kart rasprostraneniya radioaktiv-
    nykh elementov (na primere pochv Estonskoi SSR) (Natural Radio-
    activity of Soils in European USSR and Experience in the Compila-
    tion of Distribution Maps of Radioactive Elements (for Soils of
    Estonian SSR)). Thesis. Moskva. 1967.
39. Nikoforova, E. M. — Vestnik MGU. Seriya Geograficheskaya,
    No. 3:99. 1968.
40. Poluektov, N. S. Metody analiza po fotometrii plameni (Flame-
    photometric Analysis). Moskva, Goskhimizdat. 1959.
41. Rubtsov, D. M. — Pochvovedenie, No. 3:55. 1966.
42. Rusanova, G. V. — Pochvovedenie, No. 9:85. 1962.
43. Rusanova, G. V. — Pochvovedenie, No. 3:63. 1964.
44. Sokolov, A. P. — Zhurnal Russkogo Fiziko-Khimicheskogo
    Obshchestva, Seriya Fizicheskaya, Vol. 37:101. 1905.
45. Tagizade, A. Kh. and S. G. Guseinov. — Izvestiya Akademii Nauk
    AzSSR, Seriya Biologicheskaya, Vol. 6:91. 1964.
46. Tikhonov, S. A. — DAN BSSR, Vol. 7:190. 1963.
47. Tikhonov, S. A. — Ibid., p. 405.
48. Tseitlin, S. G. — Trudy Biogeokhimicheskoi Laboratorii Akademii
    Nauk SSSR, Vol. 7:127. 1944.
49. Alexander, L. T. et al. Sympos. Radioisotopes in Biosphere, 1959.
    Minneapolis, Minn. 1960.
50. Baranov, V. I. and S. G. Zeitlin. Compt. rend. Acad. Sci.,
    Vol. 30:330. 1941.
51. Baranov, V. I. et al. — In: 3rd United Nations International Conference
    on the Peaceful Uses of Atomic Energy. A/Conf. — 28/P/385,
    USSR. 1964.
52. Bohn, J. L. J. — Franklin Inst., Vol. 210:461. 1930.
53. Burton, J. D. et al. — Soil Sci. Soc. Amer. Proc., 28:500. 1964.
54. Cannon, H. L. — U. S. Geological Survey Bull., 1176:127. 1964.
55. Fletcher, A. L. — Philos. Mag., Vol. 23. 1952.
56. Gibbs, H. S. and G. J. McCallum. — N. Z. J. Sci. Tech., Vol. 3B:354.
    1955.
57. Goldstein, M. — Bull. Soc. franç. Minéralogie, Crist., Vol. 80:318.
    1957.
58. Gorham, E. — Canad. J. Bot., Vol. 41:1309. 1963.

59. Grimbert, A. — Chron. Mines. Rech. Minière, ann. 32 (326):3. 1964.
60. Gustafson, P. F. Natural Radiation Environment. University of
      Chicago Press. 1964.
61. Hansen, R. O. et al. Radioisotopes in Soils: Physical-Chemical
      Composition. Sympos. Radioisotopes in Biosphere, 1959.
      Minneapolis, Minn. 1960.
62. Hoogteiling, P. J. and G. J. Sizou. — Physica, Vol. 14:357. 1948.
63. Marsden, E. and C. Watson-Munro. — N. Z. J. Sci. Tech.,
      Vol. :26 B:99. 1944.
64. Marsden, E. — Nature, Vol. 183:924. 1959.
65. Marsden, E. — Nature, Vol. 187:192. 1960.
66. Maskal, S. and K. Czerwinska. — Roczniki gleboznawcze,
      Vol. 12:331. 1962.
67. McCallum, G. J. — N. Z. J. Sci. Tech., Vol. 37B:172. 1955.
68. Moxham, R. M. — Geophysics, Vol. 28:262. 1963.
69. Niwiński, Z. and S. Zmysiowska. — Sylwan, Vol. 108:59. 1964.
70. Satterly, J. — Proc. Cam. Phil. Soc., Vol. 16:514. 1912.
71. Serrano, S. and B. Goyanes. — Rev. geofis. Madrid, Vol. 5. 1946.
72. Stoklasa, J. Biologie des Radiums und Uraniums, Vol. 1. 1932.
73. Sizou, G. J. and P. J. Hoogteiling. — Physica, Vol. 13:517. 1947.
74. Talibudeen, O. — Soils and Fertilizers, Vol. 27:347. 1964.
75. Talfair, D. et al. — Science, Vol. 131:727. 1960.
76. Zmyslowska, S. and S. Wilgain. — Nucleonika, Vol. 6:813. 1961.

Chapter 2

ARRIVAL AND DISTRIBUTION OF RADIOACTIVE PRODUCTS
OF NUCLEAR EXPLOSIONS ON THE EARTH'S SURFACE

Introduction

Extensive data have been accumulated on the intensity of radioactive
fallout and the concentration of individual radionuclides on the earth's
surface /34, 47, 52/, but this information has not been adequately studied.
Therefore this chapter treats the arrival and distribution of radioactive
products of global fallout on the earth's surface. The majority of investigations into environmental radioactivity deal with the behavior of ^{90}Sr.
Any examination of existing observation data and analyses pertaining to
the arrival of radionuclides from the atmosphere must make allowance for
the different methods used to gather radioactive fallout in different countries.
Comparison of even the most common techniques (based on ion-exchange
resins, high-walled tanks and other collectors, followed by sample evaporation) revealed differences averaging about 20% /47, 50/. The lower values
obtained with ion-exchange resins arise because radionuclides may be
present in fallout in the form of not only cations and anions, but also neutral
compounds /42, 95/.

Global radioactive fallout

Fallout in the northern and southern hemispheres. Figure 2.1, illu-
strating the arrival intensity of ^{90}Sr in 1956—1966, reveals that the fallout
of this radionuclide from the atmosphere is extremely nonuniform.
Although fallout intensities differ for different countries in the northern
hemisphere, on the whole the fallout of ^{90}Sr for the entire hemisphere is
definitely heavier than in the southern hemisphere. The effect of experimental explosions is more pronounced in the northern hemisphere, as
indicated by considerable fluctuations in fallout levels during years of
nuclear tests and moratoriums. The effect in the southern hemisphere is
less distinct. Unlike the northern hemisphere, maximum fallout in the
southern hemisphere occurred in 1964.
Curves of cumulative ^{90}Sr fallout in both hemispheres differ markedly,
notwithstanding their common upward trend. For instance, in 1957—1966
the cumulative fallout in the southern hemisphere increased gradually and
smoothly, while the curve for the northern hemisphere is more irregular
(Figure 2.2).

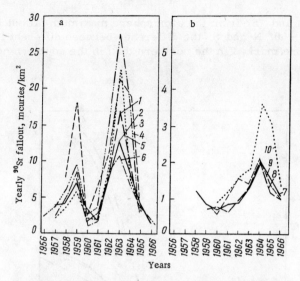

FIGURE 2.1. Intensity of yearly ^{90}Sr fallout in some countries of the northern (a) and southern (b) hemispheres:

1—Canada /98, 120, 135, 136, 140/; 2 —USA /135, 136/; 3—Norway /97—101, 104—106, 135, 136, 140/; 4—United Kingdom /97—101, 104—106, 140/; 5—USSR /33, 35, 72—75/; 6 —Japan /135, 136, 140, 148, 149, 158/; 7 — Australia /123, 135, 136/; 8 — Brazil /135, 136, 140/; 9 — Argentine /91, 135, 136/; 10 — New Zealand /88, 118, 119, 123, 135, 136/.

The arrival of fission products is nonuniform not only on a global scale, but also within individual countries and even at single observation sites. Figure 2.3 and Table 2.1 show that the fallout intensity of ^{90}Sr and ^{137}Cs differs significantly in different cities and depends on location and time of observation. However, notwithstanding these differences in the fallout levels of fission products, they display the same characteristic behavior as for the entire country; maximum fallout occurred in 1959 and 1963 in the northern hemisphere and in 1964 in the southern hemisphere, cumulative fallout curves displaying similar shapes (Figure 2.4). In examining data on cumulative fallout for different regions it must be borne in mind that environmental radioactivity observations commenced at different times. Therefore it is more correct to compare the shapes of curves reflecting fallout rates of radionuclides on the earth's surface, rather than their absolute values.

Latitudinal effect and seasonality in the fallout of nuclear explosion products. The dependence of fallout intensity on the location of observation sites reflects the latitudinal effect in the fallout of radioactive products of nuclear explosions /1, 30, 34, 35, 51, 52, 73 — 75, 102, 103, 126, 132, 175, 176, 180/. Figure 2.5 and Table 2.2 show that for prolonged observations of the

arrival of ^{137}Cs and ^{90}Sr from the atmosphere maximum fallout levels
occurred at 40 −50° N and S, the difference between different latitudinal
zones being more marked in the northern than in the southern hemisphere.

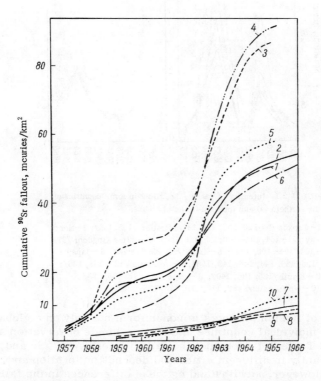

FIGURE 2.2. Cumulative ^{90}Sr fallout in some countries of the northern and southern hemispheres.

Notation is the same as in Figure 2.1.

The ratio of ^{90}Sr fallout intensities in the same latitudinal zones in the
northern and southern hemispheres is not constant, there being a downward
trend in all the zones during a moratorium (Table 2.3).

The above differences in the quantity of ^{90}Sr fallout in the northern and
southern hemispheres are largely dependent on the geographical latitude of
experimental nuclear explosions /30, 34/. The latter were conducted in
the zones 30 −40° N and 20 −30° S in near-equatorial latitudes and the cir-
cumpolar zone of the northern hemisphere /52 /. When explosions occurred

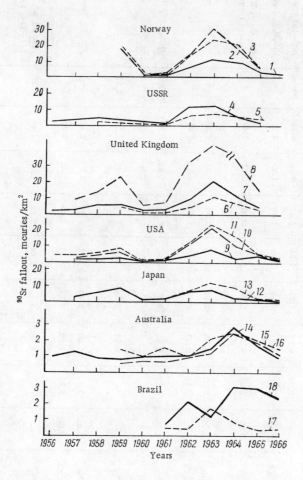

FIGURE 2.3. Strontium-90 fallout intensities in different cities in the northern and southern hemispheres:

Norway /97−101, 104−106, 135, 136, 140/; USSR /13, 14, 26−28/; United Kingdom /97, 99−101, 104−106, 170/; USA /135, 136/; Japan /135, 136, 148, 149, 158−161/; Australia /97, 99−101, 135, 136, 140/; Brazil /135, 136/; 1 − Oslo; 2 − Tromsö; 3 − Bodö; 4 − Leningrad; 5 − Moscow; 6 − Milford Haven; 7 − Abingdon; 8 − Snowden; 9 − Los Angeles; 10 − New York; 11 − Birmingham; 12 − Nagasaki; 13 − Misawa; 14 − Melbourne; 15 − Adelaide; 16 − Sydney; 17 − San Jose dos Campos; 18 − Rio de Janeiro.

TABLE 2.1. Ratios of ^{90}Sr and ^{137}Cs concentrations in fallout during 1958–1966

Country	Observation site	Radio-nuclide	1958	1959	1960	1961	1962	1963	1964	1965	1966	Reference
Norway	Tromsö / Bodö	^{90}Sr	–	0.93	0.93	1.30	1.04	1.25	0.92	0.85	–	/97–101, 104–106, 135, 136, 140/
		^{137}Cs	–	0.81	0.93	1.44	1.22	1.49	1.23	0.81	–	
United Kingdom	Milford Haven / Snowden	^{90}Sr	0.37	0.25	0.33	0.36	0.29	0.30	0.31	0.40	–	/97, 99 –101, 104–106, 170/
	Milford Haven / Abingdon	^{90}Sr	2.04	1.37	1.35	2.47	2.11	2.08	1.61	1.56	–	
		^{137}Cs	–	1.18	0.87	2.06	1.98	1.67	1.73	1.67	–	
USA	New York / Westwood	^{90}Sr	1.03	0.95	0.84	0.82	0.90	0.94	1.44	0.97	1.00	/135, 136/
	New York / Birmingham	^{90}Sr	–	–	0.97	0.85	1.04	1.07	1.45	1.11	0.92	
USSR	Leningrad / Moscow	^{90}Sr	2.02	1.73	1.00	0.40	1.71	1.55	1.06	0.52	–	/13, 14, 26–28/
		^{137}Cs	2.68	2.86	1.27	0.74	1.86	1.65	1.46	0.81	–	
Italy	Florence / Milan	^{90}Sr	–	–	1.22	1.22	1.46	1.52	1.09	1.44	0.82	/90, 93, 94, 111–114, 135,136/
	Florence / Ispra	^{90}Sr	–	–	–	0.64	0.62	0.47	0.71	0.86	–	

TABLE 2.1. (continued)

Country	Observation site	Radio-nuclide	1958	1959	1960	1961	1962	1963	1964	1965	1966	Reference
Japan	Nagasaki / Misawa	^{90}Sr	–	–	–	1.10	1.11	0.61	0.37	0.73	1.23	/135, 136, 149, 158–161/
	Nagasaki / Tokyo	^{90}Sr	0.99	1.09	–	0.85	0.94	–	–	–	–	
Australia	Melbourne / Sydney	^{90}Sr	–	0.56	0.99	0.64	1.08	0.70	1.13	0.84	0.70	/97, 99–101, 135, 136, 140/
	Melbourne / Adelaide	^{90}Sr	–	1.34	1.29	1.27	1.11	1.20	1.19	1.24	0.84	
Kenya	Kikuyu / Nairobi	^{90}Sr	–	–	1.00	1.36	0.80	0.99	–	0.39	0.53	/135, 136/

N o t e : ^{90}Sr or ^{137}Cs ratios are given for the indicated pairs of cities.

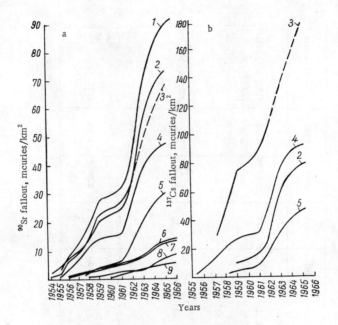

FIGURE 2.4. Cumulative ^{90}Sr (a) and ^{137}Cs (b) fallout in some cities in the northern and southern hemispheres:

1 – New York /135, 136/; 2 – Milford Haven / 97–101, 104–106, 140, 170/; 3 – Tokyo /148, 149, 158, 160, 161/; 4 – Leningrad /13, 14/; 5 – Moscow /26–28/; 6 – Melbourne /135, 136/; 7 – Wellington /135, 136/; 8 – Buenos Aires /135, 136/; 9 – Salisbury /135, 136/.

in temperate and high latitudes, most of the radionuclides fell within the same hemisphere, but in the case of explosions in near-equatorial latitudes about one-third and more of the fission products reached the other hemisphere.

TABLE 2.2. Cesium-137 fallout intensity in the USSR during 1962–1965, mcuries/km^2/65/

Degrees North	1962			1963			1964			1965
	Jan.–June	July–Dec.	Jan.–Dec.	Jan.–June	July–Dec.	Jan.–Dec.	Jan.–June	July–Dec.	Jan.–Dec.	Jan.–June
70 – 60	9.4	7.0	20.1	9.0	10.9	19.9	13.0	8.1	21.1	2.7
60 – 50	12.0	9.4	21.4	7.4	11.2	18.6	8.6	7.3	15.9	3.0
50 – 40	12.0	6.9	18.9	20.9	19.2	40.1	23.5	10.5	33.5	5.6

TABLE 2.3. Ratio of ^{90}Sr fallout in the same latitudinal zones of the northern and southern hemispheres, calculated from data cited in /102, 103, 132, 135, 136, 140, 175/

Latitudinal zone, deg.	Observation year								
	1958	1959	1960	1961	1962	1963	1964	1965	1966
50 − 40	6.7	5.9	1.7	2.8	6.1	10.7	3.7	2.0	1.4
40 − 30	3.5	10.0	3.3	1.9	8.2	7.0	4.3	2.2	1.5
30 − 20	2.8	8.4	1.6	1.4	5.2	9.2	3.2	1.5	1.1
21 − 10	2.6	6.0	2.0	1.8	3.2	3.8	2.8	3.7	2.2
10 − 0	0.7	1.5	1.4	1.3	1.3	2.2	1.2	1.8	2.3
Mean	3.3	6.4	2.0	1.8	4.8	6.6	3.0	2.2	1.7

According to long-term observations, maximum fallout intensity is characteristic of the spring-summer period /14, 26 −28, 30, 34, 35, 41, 69, 71, 73, 88, 96, 99, 110, 141, 144, 154, 165, 167, 175, 176, 180/. Therefore, the heaviest fallout is observed during the 2nd and 3rd quarters of the year in the northern hemisphere, and in the 4th and 1st quarters in the southern hemisphere (Figure 2.6).

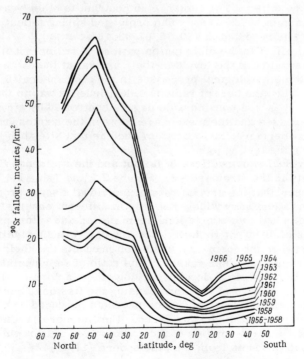

FIGURE 2.5. Latitudinal effect in cumulative ^{90}Sr fallout from the atmosphere /51, 102, 103, 132, 135, 136, 140, 175/

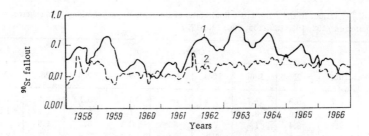

FIGURE 2.6. Strontium-90 (Mcuries/month) in the northern (1) and southern
(2) hemispheres /177/

Figure 2.7 depicts data on the fallout intensity of atmospheric precipitation, ^{90}Sr, ^{137}Cs and other fission products, as well as ^{54}Mn formed by irradiation of the structural materials of the bombs by neutrons. The spring
maximum is distinct, the pattern being even more distinct in the case of
long-lived radionuclides. However, the peak shapes and maximum fallout
times differ for the same radionuclides at different observations sites, as
well as for different nuclides at the same observation site. Even for a
single nuclide at one observation site, the maximum shifts over the months
and the peak shape differ. For instance, in contrast to other years, the
peak was less sharp in 1962 — 1965. The prolonged spring fallout maximum
in 1962 and 1963 was reported in /30, 35, 67, 83, 164/.

The latitudinal effect in the distribution pattern of falling-out fission
products and seasonality in the levels of their arrival at the earth's surface
are mainly due to meteorological processes in the atmosphere /6, 21 — 23,
29, 30, 34, 115/. When the bulk of radionuclides finds its way to the stratosphere, their fallout is to a considerable degree controlled by stratospheric
transport of air masses and their exchange between the stratosphere and
troposphere. Furthermore, the location of nuclear test sites has a significant effect during the testing period /30, 34/.

Ratios of individual radionuclides in fallout and the residence time of
explosion products in the stratosphere. Figure 2.7 and Table 2.4, and data
of Hardy /129/ show that the arrival rate of individual fission products
and their ratio in fallout vary widely not only in different observation sites,
but also at an individual site over a long-term period and within a single
year. The ratios of different radionuclides are controlled by their half-
lives, the time between the formation of fission products and their arrival
at the earth's surface, and other causes. The ratio of radionuclides at the
moment of formation therefore depends, among other factors, on the fission-
ing material, and the energy of neutrons causing the fission of nuclei
(Table 2.5). Frequent "mixing" of fission products produced by different
nuclear explosions must also be considered. Furthermore, fractioning of
radionuclides occurs during motion of the radioactive "mushroom" and the
fallout of explosion products from the atmosphere /11, 17, 146/.

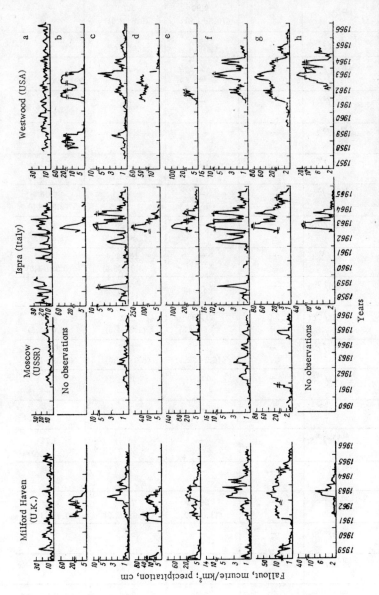

FIGURE 2.7. Fallout intensity of individual radionuclides at some observation sites / 13, 14, 26–28, 90, 93, 94, 97–101, 104–106, 111–114, 135, 136, 140, 170/:

a — precipitation; b — ^{89}Sr; c — ^{90}Sr; d — ^{95}Zr; e — ^{106}Ru; f — ^{137}Cs; g — ^{144}Ce; h — ^{54}Mn.

TABLE 2.4. Ratios of some radionuclides in fallout during 1957—1965

Observation site	Radionuclide	1957	1958	1959	1960
Ispra (Italy)	$^{137}Cs/^{90}Sr$	—	0.90 (0.5—1.6)	1.10 (0.5—2.0)	—
	$^{144}Ce/^{90}Sr$	—	—	—	—
	$^{106}Ru/^{90}Sr$	—	—	—	—
	$^{95}Zr/^{90}Sr$	—	—	—	—
	$^{89}Sr/^{90}Sr$	—	—	—	—
Milford Haven (U.K.)	$^{137}Cs/^{90}Sr$	2.06	1.69	1.61 (1.3—2.1)	1.56 (1.1—1.9)
	$^{144}Ce/^{90}Sr$	—	—	—	5.06 (2.3—8.0)
	$^{106}Ru/^{90}Sr$	—	—	6.63 (2.8—12.3)	3.02 (1.6—5.9)
	$^{95}Zr/^{90}Sr$	—	—	10.86 (0.9—28.4)	0.59 (0.09—1.0)
	$^{54}Mn/^{90}Sr$	—	—	—	—
Leningrad (USSR)	$^{137}Cs/^{90}Sr$	2.0	1.53 (0.5—3.3)	2.42 (1.9—4.8)	2.13 (0.6—3.5)
	$^{144}Ce/^{90}Sr$	—	12.7 (6.2—32.4)	27.48 (8.0—50.0)	6.9 (0.88—16.7)
Moscow (USSR)	$^{137}Cs/^{90}Sr$	—	1.2	1.5	1.7 (0.2—5.1)
Westwood (U.S.A.)	$^{137}Cs/^{90}Sr$	—	—	—	—
	$^{114}Ce/^{90}Sr$	—	—	—	3.72 (2.8—8.0)
	$^{95}Zr/^{90}Sr$	—	—	—	—
	$^{89}Sr/^{90}Sr$	—	28.40 (11.5—40.5)	7.79 (0.9—28.9)	0.45** (0.03—1.7)
	$^{55}Fe/^{90}Sr$	—	—	—	—
	$^{54}Mn/^{90}Sr$	—	—	—	—
Tokyo (Japan)	$^{137}Cs/^{90}Sr$	4.1 (2.8—6.3)	2.17 (1.2—5.2)	2.91 (1.5—5.2)	2.59

* After a series of nuclear tests.
** For the first half of the year.
† For the second half of the year
N o t e : Minimum and maximum values during the year are in parentheses.

Observation year					Reference
1961	1962	1963	1964	1965	
–	1.77	1.61	1.52	1.51	/90, 93, 94, 113,
–	(1.5–2.0)	(1.3–1.7)	(1.4–1.9)	(1.4–1.6)	114/
–	–	18.51	11.51	4.53	
		(12.2–29.1)	(8.1–15.9)	(2.6–7.6)	
–	–	12.88	3.70	–	
		(5.4–22.7)	(2.4–6.2)		
–	–	34.52	1.32	–	
		(5.4–10.60)	(0.1–3.2)		
–	–	6.32	–	–	
		(0.3–21.9)			
1.57	1.56	1.57	1.59	1.72	/97–101, 104–106/
(1.4–1.9)	(1.4–1.9)	(1.3–1.8)	(1.3–1.9)	(1.3–2.1)	
8.66	20.09	14.58	9.22	4.42	
(2.7–2.8)	(15–26)	(11–25)	(4.3–15.4)	(3.1–7.8)	
9.17	7.42	6.89	11.83	3.04	
(7.1–11.2)	(5.3–12.4)	(5.2–10.8)	(2.1–61.1)	(1.2–5.2)	
49.9*	13.96	7.93	0.23	0.21	
(25.9–79.4)	(2.7–34.5)	(1.1–34.4)	(0.09–0.9)	(0–0.7)	
0.80*	1.28	1.08	2.30	0.61	
(0.2–1.2)	(0.5–2.9)	(1.2–2.5)	(0.5–1.0)	(0.3–1.3)	
2.12	1.91	–	–	–	/13, 14/
(1.2–3.5)	(1.5–2.0)				
38.73	22.74	–	–	–	
2.7–142)	(7.2–45.2)				
1.1	1.7	1.9	1.5	1.2	/26–28/
(0.5–1.9)	(0.9–2.2)	(1.6–2.9)	(0.4–2.1)	(0.6–1.4)	
–	1.48	1.51	1.74	1.43**	/136/
	(1.2–1.8)	(1.2–2.1)	(0.2–2.6)	(1.3–1.7)	
6.42	22.5	20.9	13.04	3.57	
(2.0–37.0)	(18.9–33.6)	(13.9–26.8)	(0.8–21.5)	(2.1–5.9)	
–	32.26	14.59	5.82	0.33**	
	(12.1–73.1)	(3.1–46.9)	(0.2–1.5)	(0.06–0.74)	
57.67*	19.6	6.84	–	–	/136/
(28.5–84.0)	(5.4–45.9)	(0.3–27.6)			
–	4.6†	9.4	11.5	9.3	
	2.4–10.1)	(3.6–21.4)	(1.0–26.7)	(5.1–12.3)	
–	2.03**	7.14	5.94	8.3†	
	(1.2–3.2)	(2.7–12.8)	(1.3–13.6)	(4.8–10.1)	
3.55	2.71	–	–	–	/149, 158/
(1.2–6.0)	(1.7–4.7)				

TABLE 2.5. Ratio of individual radionuclides at the moment of formation by fission of nuclei of heavy elements caused by neutrons of different energies /16, 166/

Yield ratio	^{235}U	^{235}U	^{235}U	^{239}Pu	^{235}U	^{238}U
	n_{therm}	n_{fiss}			n_{14} MeV	
$^{144}Ce/^{90}Sr$	0.94	1.09	1.64	1.43	–	1.14
	1.05	1.00	1.40	–	0.73	0.75
$^{137}Cs/^{90}Sr$	1.07	1.41	2.08	2.75	–	1.96
	1.00	1.26	1.94	3.09	–	1.83
$^{106}Ru/^{90}Sr$	0.07	1.07	0.84	2.76	–	1.07
	0.06	–	0.85	–	0.35	0.67
$(^{95}Zr + {}^{95}Nb)/^{90}Sr$	1.07	–	1.78	–	1.11	1.47
$^{89}Sr/^{90}Sr$	0.76	–	0.90	–	1.00	0.67

N o t e : Calculated from the quantity of nuclei produced.

TABLE 2.6. Residence times of ^{90}Sr and ^{137}Cs in the stratosphere

Radionuclide	Observation period	Stratosphere residence time, months	Reference
^{90}Sr	Jan. — April, 1959	12	/76/
	March — June, 1959	17	
	May — Aug., 1959	67	
^{90}Sr	Nov. — Dec., 1958 and 1959	9.4	/142/
	Jan. — Feb., 1958 and 1960	8.0	
	March — April, 1959 and 1960	6.7	
	May — June, 1959 and 1960	8.0	
	July — Aug., 1959 and 1960	9.4	
	Sept. — Oct., 1959 and 1960	9.6	
^{137}Cs	1958 — 1960	14	/126/
	1959 — 1960	23	
^{90}Sr	Dec. 1958 — Oct. 1960	5	/176/
	May 1959 — April 1961	9	
	Jan. 1963 — Dec. 1964	20	
	Jan. 1964 — Dec. 1965	8	
	Jan. 1965 — Dec. 1966	10	
	June 1963 — May 1965	9	
	June 1964 — May 1966	9	

The ratio of radioactive fission products in fallout is influenced by their residence time in the stratosphere, which depends on the properties of radionuclides, the size of aerosol particles, the power of the explosion and its location, as well as the time elapsed since the explosion. For instance, it is seen from Table 2.6 that the residence time of ^{90}Sr and ^{137}Cs in the stratosphere varies widely in different observation periods, from 7 to 67 months for ^{90}Sr. According to Karol' and Malakhov /30/, the average residence time in different layers of the stratosphere amounts to 0.6 —4, 2.4, 1 and 0.6 —1.3 years, respectively, for ^{90}Sr, ^{137}Cs, ^{144}Ce, and ^{185}W. The average stratospheric residence time in the southern hemisphere is approximately twice that in the northern hemisphere, owing to slower exchange through the tropopause in the southern hemisphere.

The average residence time of explosion products in the stratosphere is expected to increase during a moratorium with increasing time after an explosion /30, 34, 76, 126/. This expectation is mainly based on decreasing gravitational deposition rates for aerosol particles due to their diminishing size. According to observations during two years after tests, the average stratosphere residence time of ^{90}Sr and ^{54}Mn for very small particles was taken as 10 months /121/. Furthermore, the residence time of fission products in the stratosphere has been reported to increase with increasing explosion power owing to the higher ejection height and small size of the produced aerosol particles (the latter diminish, on the average, proportionally to $W^{1/2}$, where W is the explosion power /34/).

The effect of aerosol particle size on the stratosphere residence time of the radionuclides is related to their distribution in particles of different sizes /17, 107, 116, 174/. Thus, coarse aerosols enriched in ^{95}Zr, ^{141}Ce, ^{185}W and ^{91}Y settle quicker than fine particles enriched in ^{90}Sr, ^{137}Cs, ^{144}Ce, and ^{106}Ru. Experimental data /34/ indicate that the stratosphere residence time of ^{91}Y, ^{95}Zr and ^{141}Ce amounts to about one-half that of ^{90}Sr and ^{137}Cs, while figures for ^{131}I, ^{140}Ba, ^{89}Sr, ^{144}Ce, ^{103}Ru and ^{106}Ru are intermediate between those for ^{95}Zr and ^{137}Cs, in agreement with data cited in /30/.

The average stratosphere residence time of explosion products (and hence their removal rate) is estimated at 4 —7 to 18 months, depending on the location of nuclear tests /138, 147/. Lower values correspond to explosions in polar latitudes, and higher values to tests in equatorial regions.

Fallout of radioactive fission products from the troposphere. The transport of radioactive products of nuclear explosions from the tropopause level to the layer washed by atmospheric precipitation, and farther down to the underlying surface, is very complicated compared to their migration in the stratosphere /30, 34/ because, in addition to vertical turbulent exchange and meridional motion of air masses, the fallout of radioactive aerosols is affected by the presence in the troposphere of water droplets, snowflakes, dust, and other particulate matter.

The average residence time of explosion products in the troposphere is estimated at 20 —40 days /29 -31, 34, 52, 76, 127, 145/. However, this time is shorter (only a few days) in the lower troposphere up to an altitude of 3 —4 km /29, 31, 37, 122/. Removal of radioactive aerosols from the troposphere comprises wet fallout in the presence of atmospheric precipitation and dry fallout on days without precipitation.

Washout of radionuclides from the atmosphere is affected by many factors, and the capture of aerosol particles by droplets in clouds and precipitation may proceed under the influence of various mechanisms /8, 9, 20, 23, 24, 36, 40, 52, 66, 77/.

Washout of radioactive aerosols depends on the intensity and type of precipitation, the form and altitude of clouds, the kind of atmospheric front, season of the year, geographical location of observation site, size of aerosol

particles, and so on /8, 9, 18, 24, 36, 39, 40, 52, 53, 66, 148/. Figure 2.8 shows
that snow possesses a higher washout capacity, on the average, than rain,
due to the former's specific properties (lower fall rate of snowflakes, their
larger surface area, presence of particles of coal, soil, etc.) /24, 40, 53, 66/.
The specific radioactivity of atmospheric precipitation decreases with
increased precipitation /8, 52, 124, 143/. The rate of this decrease and the
degree of washout differ for individual fission products. The ratio of con-
centrations of ^{103}Ru, ^{131}I, ^{140}Ba and ^{141}Ce to ^{95}Zr + ^{95}Nb in the initial part of
the rain was found to be 4.2, 2.4, 1.6 and 1.07, respectively, in comparison
to the average values for the entire duration of rain /39/, the difference
being due to the dependence of the degree of particle washout upon their size.

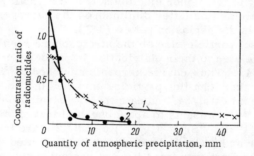

FIGURE 2.8. Concentration of artificial radionuclides
in air as a function of atmospheric precipitation /8/
(the ratio of mean radionuclide concentrations on wet
and dry days is marked on the y-axis):

1 — rain shower; 2 — snow shower.

   The effect of precipitation type and cloud form on the elimination of
radioactive aerosols from the atmosphere is illustrated in Figure 2.9. In
the case of light continuous snow and rain showers the increase in radionuclide fallout with increasing atmospheric precipitation ceases at some
limit value (5—7 mm/day), beyond which the fallout of radionuclides
increases only very slightly. In the case of snow showers, fallout intensity
depends linearly on the quantity of atmospheric precipitation /8/.
   The activity of precipitation is higher during the passage of cold fronts
/36/, because of tropopause discontinuities occurring over such fronts,
accompanied by the passage of explosion products from the stratosphere
to the troposphere, while the rapid exchange of air masses and especially
their intensive vertical transport are conducive to the migration of aerosols
in the troposphere. Moreover, the principal type of clouds in the presence
of cold fronts are cumulonimbus of extensive vertical development, so that
the upper part of a cloud occasionally reaches the troposphere boundary.

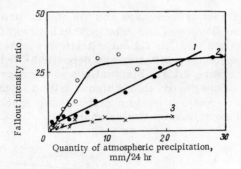

FIGURE 2.9. Ratio of radioactive fallout intensities on wet and dry days /8/:

1 — snow shower; 2 — rain shower; 3 — continuous snow.

It has been demonstrated that the intensity of ^{90}Sr fallout with atmospheric precipitation varies seasonally: spring > winter > autumn > summer /106, 151/; the relationship between fallout intensity and quantity of atmospheric precipitation is near-linear in spring. However, the shape of curves expressing this relationship differs for different seasons at the same site, and for the same season at different sites, on account of the seasonal nature of the arrival of this radionuclide from the stratosphere to the troposphere, and differences in the type of clouds and the kind of atmospheric precipitation. The linear relationship between fallout intensity and quantity of atmospheric precipitation is limited to the spring period probably due to the presence of favorable meteorological conditions for the fallout of ^{90}Sr at this time of year (intensified vertical exchange and precipitation from cumulonimbus which reaches the tropopause height) together with an increase in its tropospheric concentration.

The above data on the relationship between the fallout intensity of radioactive products of nuclear explosions from the atmosphere and meteorological conditions, including the quantity and type of atmospheric precipitation, point to the complex nature of this relationship, apparently responsible for the absence in some cases of any correlation between the intensity of radioactive fallout and the quantity of atmospheric precipitation. Thus, comparison of yearly ^{90}Sr fallout and atmospheric precipitation for the same period at different locations in the USA and Australia revealed different relationships between these quantities for these two countries situated in the northern and southern hemispheres (Figure 2.10). There was no reliable correlation in Australia during the entire observation period, and the correlation coefficients were negative for several years (1960, 1962, 1964, 1965). On the other hand, the relationship between atmospheric precipitation and ^{90}Sr fallout was distinct in the USA. The most significant correlation was observed in years with relatively low radioactive fallout levels (1960, 1961, 1965, 1966), when the correlation coefficients were 0.913, 0.695, 0.624 and 0.589, respectively. As the radioactive fallout intensity increases,

the experimental data scatter increases and the correlation coefficients
decrease, indicating weakening of the relationship between the quantity of
atmospheric precipitation and ^{90}Sr fallout.   A similar pattern was revealed
by comparing atmospheric precipitation and deposited ^{90}Sr at two sites in
the United Kingdom (Figure 2.11).   Fluctuations in the correlation coefficients and their low values testify that a higher fallout intensity is not
always related to heavier atmospheric precipitation (especially during, and
soon after periods of experimental explosions).

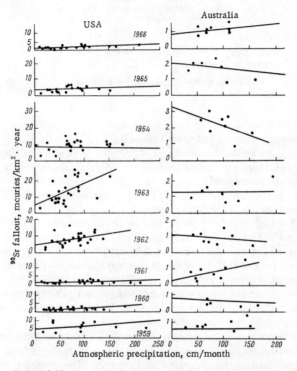

FIGURE 2.10. Strontium-90 fallout and atmospheric precipitation
at different sites in the USA and Australia during 1959–1966
/135, 136/

    It should be noted, however, that even in the presence of a linear relation-
ship between the quantity of atmospheric precipitation and ^{90}Sr and ^{137}Cs
fallout at different observation sites, any relationship need not necessarily
be proportional or even exist at all at every one of them in the course of
the year (Figure 2.12).   Absence of proportionality has been reported /25,
67, 92, 124, 165/, although, on the whole, the fallout intensity increases with
heavier atmospheric precipitation /25, 124/.
    Data on the relationship between the intensity of the influx of radioactive
products of global fallout and the quantity of atmospheric precipitation

indicate that the latter is not always the principal mechanism for the
elimination of radionuclides from the atmosphere.

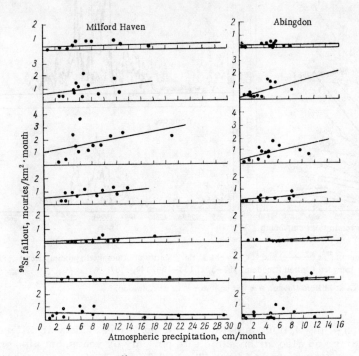

FIGURE 2.11. Monthly ^{90}Sr fallout and quantity of atmospheric precipitation
at two sites in the United Kingdom during 1959−1965 /97−101, 104−106,
140/

Fallout of radioactive products of nuclear explosions in the absence of
atmospheric precipitation occurs mainly due to gravity deposition on various
surface roughnesses during the motion of air, and consequently the level of
dry fallout is significantly affected by such factors as air dustiness, plant
cover, underlying surface, etc. Forest stands and herbaceous vegetation
possess a high capacity for interception of radioactive aerosols on their
surface /29, 34, 52, 169, 174/. Coniferous forests are more efficient in
scavenging the atmosphere than deciduous forests, owing to the larger surface of their needles in comparison to the latter's foliage /64, 152, 153, 171/.

The fraction of dry fallout was estimated in the majority of cases by
comparing the radionuclide concentration in fallout or soil with the quantity
of atmospheric precipitation /81, 130, 131, 139/. Moreover, the contents of
radioactive products of nuclear explosions in fallout were also measured
directly by collecting them in two receptacles that were automatically
closed alternately in dry and rainy weather /162, 163/.

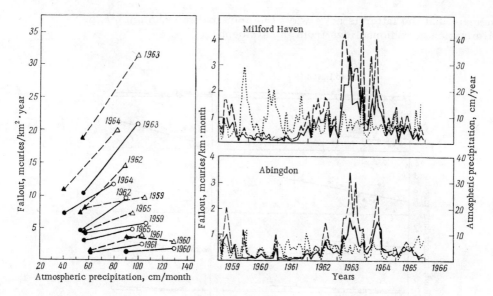

FIGURE 2.12. Fallout of ^{90}Sr (———) and ^{137}Cs (– –) and the quantity of atmospheric precipitation (· · ·) at Milford Haven and Abingdon (United Kingdom) /97—101, 104—106, 140, 170/:

○ — ^{90}Sr and △ — ^{137}Cs at Milford Haven; ● — ^{90}Sr and ▲ — ^{137}Cs at Abingdon.

   The fraction of dry fallout is not constant, but varies widely at different observation sites and also at the same site not only by years, but also within a single year, or even a single month (Table 2.7). In Japan, for instance, the fraction of fallout on dry days (without atmospheric precipitation) amounted to 66.8% of the total quantity of fallout, while days with light precipitation and rainy days, respectively, accounted for 20.3 and 12.9% /159/. According to Diaconescu /23/, the higher values of dry fallout in comparison to wet fallout are due to the turbulent motion of the atmosphere, being one of the major factors in fallout from the troposphere. The amount of deposited radionuclides may be insignificant in spite of atmospheric precipitation on days when turbulent motion is unimportant at the tropopause level or in clouds. On the other hand, there may be heavy dry fallout on days without atmospheric precipitation, provided the turbulent motion is significant.
   There have been reports of increased dry fallout in the moratorium /5, 39, 40, 52/. This has been explained by the variation in the size and spectrum of aerosol particles during a prolonged period of atmospheric scavenging. With the passage of time there is an increase in the stock of readily washed out aerosols (<0.1 and > 1μ), and the particles remaining in the atmosphere are those poorly washed out by atmospheric precipitation owing to their size and physicochemical properties (> 0.1 and <1μ) /10, 37 —40, 125, 157, 174/.

TABLE 2.7. Dry fallout of radioactive products of nuclear explosions

Observation site	Radionuclide	Observation period	Dry fall-out, %	Average, %	Reference
Norway	Mixture of fission products	1960	7 − 72	16	/156/
Denmark	^{90}Sr	Jan. − June 1963	56	29	/79/
		July − Dec. 1963	13	−	
German Federal Republic	Mixture of fission products	Nov. 1961	33	−	/163/
		May 1962	8	−	
German Federal Republic*	Ditto	Aug. − Dec. 1961	13 − 33	21	/162/
		Jan. − Dec. 1962	8 − 36	18	
		Jan. − Dec. 1963	11 − 30	19	
	^{95}Zr + ^{95}Nb	Jan. − Dec. 1962	19 − 36	26	
		Jan. − Aug. 1963	7 − 33	22	
	^{103}Ru	Jan. − Dec. 1962	7 − 48	21	
		Jan. − Aug. 1963	8 − 32	24	
	^{144}Ce	March − Aug. 1963	8 − 30	19	
USSR (Black Sea coast)	Mixture of fission products	Aug. − Dec. 1959	45 − 69	58	/5/
		Jan. − March 1960	6 − 16	11	
		Apr. − Oct. 1960	64 − 87	72	
Leningrad	Ditto	1956	−	15	/52/
		1957	−	8	
		1958	−	9	
		1959	−	6	
		1960	−	20.5	
		1961**	−	14.9	

* Fraction of activity in the dust with respect to the total activity deposited on the earth's surface, %.
** Data before the new series of tests in 1961.

According to Prawitz /157/, the relationship between wet and dry fallout varies in the following manner as a function of the size of aerosol particles:

Diameter of particles, μ	< 0.2	0.2	0.5	1	2	5
Wet fallout, %	88.3	88.6	88.2	87.5	95.5	99.8
Dry fallout, %	11.7	11.4	11.8	12.5	4.5	0.2

Consequently, old fission products are washed from the atmosphere slower than young products.

Distribution of radioactive fission products on the earth's surface

Contents of fission products in the soil cover in the northern and southern hemispheres. Data on ^{90}Sr concentrations in the soil cover show its distribution over the earth's surface to be nonuniform /3, 44, 47, 87, 132—134/; it varies fairly widely even within the boundaries of a single country (Table 2.8). The concentration of ^{90}Sr in 1963 was double that in 1959—1960 /44/. The ratio of ^{90}Sr contents in the soil cover of northern and southern hemispheres varies from one year to another; it was 3.7, 5.8, 3.8, 4.8 and 5.4, respectively, in 1957, 1958, 1959, 1960, and 1963.

TABLE 2.8. Concentration of ^{90}Sr in the surface soil horizon in 1963

Country	^{90}Sr concentration, mcuries/km^2	Reference	Country	^{90}Sr concentration, mcuries/km^2	Reference
USSR	19—54	/44/	Japan	49 — 51	/134/
USA	12—73	/134/	Denmark	26 —48	/79/
Norway	18—52	/134/	Australia	3 —12	/134/
Canada	20—63	/134/	Chile	0.5— 4.5	/134/
France	28—31	/134/	New Zealand	6 —17	/134/

TABLE 2.9. Latitudinal effect in the distribution of ^{90}Sr in the USSR soil cover, mcuries/km^2

Latitudinal zone, degrees	1959—1961*			Nov.—Dec. 1961
	0—5 cm /44/	0—15 cm** /44/	Mean†, 0—15 cm (calculated from data reported in /87, 133/)	0—5 cm /44/
70—60	11.2	12.8	11.4	11.7
60—50	13.8	17.2	17.7	19.2
50—40	16.2	20.3	23.6	21.0
40—30	—	—	24.8	—

TABLE 2.9. (continued)

Latitudinal zone, degrees	Apr.—May 1962	Aug.—Sept. 1963	Aug.—Sept. 1963	Summer 1963	Mean†, 0—15 cm (calculated from data reported in /134/)
	0—5 cm /44/	0—5 cm /44/	0—15 cm** /44/	0—15 cm /70/	
70—60	17.7	23.8	29.5	35	26.3
60—50	23.5	27.8	34.7	44	37.8
50—40	24.5	29.5	36.8	45	46.9
40—30	33.2	34.3	42.8	50	43.8

* Data prior to resumption of the series of nuclear tests.
** Extrapolated data assuming that the 0—5-cm layer retains an average of 80% ^{90}Sr (±15%, since this value is 66—87% depending on soil type).
† Mean global data.

The distribution of radioactive fission products over the earth's surface likewise displays the latitudinal effect, in spite of the above-mentioned fluctuations in radionuclide concentration in the soil cover (Tables 2.9 and 2.10, Figure 2.13). The highest content of fission products occurs in the zone 50 − 30° N and S, correlating with their contents in fallout.

TABLE 2.10. Latitudinal effect in the distribution of ^{95}Zr, ^{106}Ru, ^{125}Sb, ^{137}Cs and ^{144}Ce in the USSR soil cover, mcuries/km^2 /35, 65/

Latitudinal zone, degrees	^{95}Zr + ^{95}Nb	^{106}Ru + ^{106}Rh	^{125}Sb	^{137}Cs			^{144}Ce + ^{144}Pr
	1963	1963	1963	1960	1963	1965	1963
70 − 60	230	240	38	48.5	60	112	400
60 − 50	550	405	46	53.9	93	116	620
50 − 40	550	410	47	61.9	95	164	630
40 − 30	520	420	52	−	95	−	640

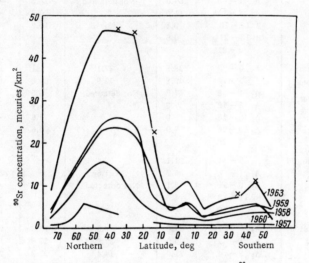

FIGURE 2.13. Latitudinal effect in the distribution of ^{90}Sr in soil cover /87, 133, 134/:

× − mean data for 1963 and beginning of 1964.

Distribution of radionuclides in soils. It must be borne in mind when considering the concentration of radioactive products of nuclear explosions on the earth's surface that they are incorporated in various physicochemical, biogeochemical and other processes taking place in the soil-plant cover, in the formation of removal and accumulation zones in the geochemical landscape, and that they are responsible for different penetration depths of radionuclides in soils depending on their physicochemical properties /4, 7, 32, 44, 47, 54, 57 − 63, 68, 89, 90, 117, 173, 178, 181/ (Table 2.11). The penetration of radionuclides is deeper in soils of lighter textures /7, 47, 128/. Thus, the respective concentrations of ^{90}Sr in a sandy and a loamy soil in the USA during 1963 were as follows: in the 0 — 5 cm layer, 56.4 and 84.5%; in the 5 — 15 cm layer, 38.4 and 13.7%; in the 15 — 30 cm layer, 5.2 and 1.8% of its total content in the 0 — 30 cm layer /128/; in 1965, as much as 7% of this radionuclide was discovered in the 30 — 45 cm layer of sandy soil /86/. Besides the variation of the relative distribution in surface horizons, there are also seasonal fluctuations of ^{90}Sr concentration in the soil profiles /32, 78, 82, 86/.

TABLE 2.11. Distribution of ^{90}Sr, ^{137}Cs and ^{144}Ce in the profiles of different soil types

Soil	Horizon, depth, cm	Distribution, %			Ratio	
		^{90}Sr	^{137}Cs	^{144}Ce	$\dfrac{^{137}Cs}{^{90}Sr}$	$\dfrac{^{144}Ce}{^{90}Sr}$
Loamy meadow soil on central floodplain alluvium	A_1 0 — 1	22.7	100.0	53.2	4.9	3.3
	A 1 — 6	54.0	Not detected	15.8	—	0.4
	A 6 — 11	11.3	"	20.8	—	2.6
	B 11 — 16	2.1	"	10.2	—	6.6
	B 16 — 21	9.9	"	—	—	—
	0 — 21				1.1	1.4
Strongly podzolic loamy soil on blanket loam	A_0 0 — 3	28.5	27.1	1.1	1.4	0.07
	A_1 3 — 4	42.2	72.9	69.8	1.8	3.0
	A_1 4 — 8	17.2	Not detected	5.5	—	0.6
	A_2 8 — 13	7.3	"	4.4	—	1.1
	A_2 13 — 18	1.9	"	9.6	—	8.8
	A_2 18 — 26	2.9	"	9.6	—	5.8
	0 — 26				1.2	1.8
Sandy podzolic soil on old alluvial sands	A_0 0 — 1	31.1	32.4	—	2.0	—
	A_2 1 — 6	56.2	67.6	—	2.3	—
	A_2 6 — 11	8.1	Not detected	—	—	—
	B_2 11 — 18	4.6	"	—	—	—
	0 — 18				1.9	—
Sod-podzolic gleyed sandy soil on alluvial sand	A_0 0 — 1	27.6	48.0	15.8	2.0	0.4
	A_1 1 — 2	39.0	37.6	22.8	1.2	0.4
	A_1 2 — 4	21.0	14.4	32.9	0.8	1.0
	A_1 4 — 12	8.5	Not detected	2.8	—	1.0
	A_2 12 — 18	3.9	"	15.7	—	2.7
	0 — 18				1.1	0.7
Peaty-gley sandy soil	T 0 — 1	11.7	33.4	16.0	3.5	6.6
	T 1 — 4	39.9	66.6	82.2	2.0	9.9
	T 4 — 9	24.2	Not detected	2.0	—	0.5
	T 9 — 19	24.2	"	—	—	—
	0 — 19				1.2	4.8

The distribution of fission products is more uniform in tilled soils than in virgin soils as a result of annual plowing (Figure 2.14). The distribution of ^{90}Sr in the plowing horizon is affected significantly by different tillage practices /45, 46/.

220511

Investigations conducted under natural and simulated conditions point to the lower migration capacity of ^{137}Cs and ^{144}Ce in comparison to ^{90}Sr /7, 12, 43, 55, 56, 70, 85, 155, 168, 178/. The lower migration of ^{137}Cs is borne out by the ^{137}Cs:^{90}Sr ratio in soil which was 1.3, 2.7, 3.5, 2.2 and 2.7, respectively, in the five years 1958—1962 /12/, and also by the lower values of its removal into rivers /150/. However, in certain cases ^{137}Cs may prove to be more mobile than ^{90}Sr.

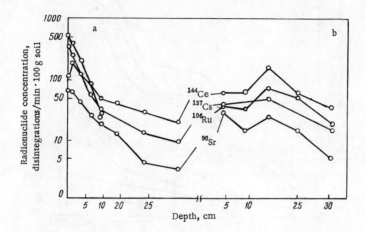

FIGURE 2.14. Distribution of ^{90}Sr, ^{106}Ru, ^{137}Cs and ^{144}Ce in virgin (a) and tilled (b) soils in the State of New Jersey, USA, in 1960 /178/

Accumulation rates and elimination of radioactive fission products from surface soil horizons. Accumulation of radioactive products of nuclear explosions falling out of the atmosphere onto the earth's surface proceeds rather nonuniformly, one of the major factors significantly affecting the accumulation rate of radionuclides in the soil cover being the intensity of experimental nuclear explosions. Figures 2.15 and 2.16 distinctly show the different ^{90}Sr accumulation rates during tests and moratoriums. In addition to the overall upward trend in ^{90}Sr concentration in soils during the moratorium, there is either a slight increase or a general decrease of its concentration in the surface horizons of the soil cover. Calculated ratios of ^{90}Sr concentrations in soils during 1960 with respect to 1959, and 1963 with respect to 1960, averaged 1.11 and 1.97, respectively, for ten countries in the northern and southern hemispheres /47/. This is also confirmed by data on ^{90}Sr concentration in the soil cover in the Urals /19/: 25, 32, 57, 68, 60 and 56 mcuries/km^2, respectively, in the six years 1961 — 1966.

Stabilization of ^{90}Sr concentration in the soil cover in the moratorium (1960—1961, prior to resumption of nuclear tests) was due to the near-equality of its elimination from the surface soil horizons and its influx from the atmosphere (Figures 2.15 and 2.16).

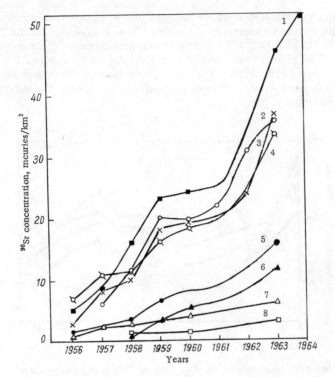

FIGURE 2.15. Accumulation of ^{90}Sr in soils of some countries in the
northern and southern hemispheres /3, 33, 44, 47, 87, 133, 134/:

1 — USA; 2 — Canada; 3 — USSR; 4 — Norway; 5 — USA (Alaska);
6 — New Zealand; 7 — Australia; 8 — Chile.

Comparison of data on the amount of fission products deposited in 1960 —
1963 and accumulated in soil indicates a difference between these two
quantities (Table 2.12).   The lower concentration of ^{90}Sr and ^{137}Cs in soils
in comparison to their cumulative fallout onto the soil cover was reported
in /12, 52, 80, 137, 139, 176/.   Differences between the amounts of deposited
fission products and their accumulation in the soil cover are not constant,
but vary quite widely (Table 2.13).   The observed difference between the
fallout fission products on the soil cover and their experimentally determined concentration in soils is due to the effect of a variety of factors pertaining to the horizontal and vertical migration of radionuclides.   One of
these factors is atmospheric precipitation, the relationship between the
quantity of the latter and the concentration of fission products in the soil
cover being even more complex than in the case of fallout.   On the one hand,
the more copious the atmospheric precipitation, the larger should be the
influx of radionuclides onto the soil surface.   However, it should be remembered that the specific radioactivity of atmospheric precipitation decreases

as its quantity increases. On the other hand, an increase in the quantity of atmospheric precipitation tends to increase the mobility of radionuclides and their migration in the soil-plant cover. The significance of the water factor in the processes of radioactive fission products in soil is treated in greater detail in /32, 48, 172/.

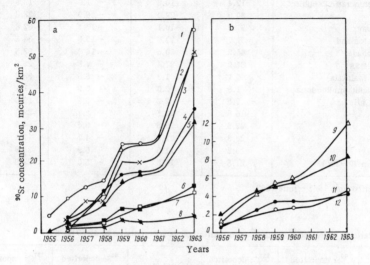

FIGURE 2.16. Accumulation rates of ^{90}Sr in soil in the vicinity of some cities in countries of the northern (a) and southern (b) hemispheres /3, 33, 44, 47, 87, 133, 134/:

1 — New York; 2 — Ottawa; 3 — Tokyo; 4 — Oslo; 5 — Paris; 6 — Panama (Canal Zone); 7 — Manila; 8 — Singapore; 9 — Brisbane; 10 — Adelaide; 11 — Salisbury; 12 — Huancaya (Peru).

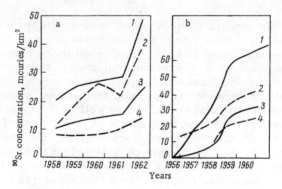

FIGURE 2.17. Cumulative fallout of ^{90}Sr and ^{137}Cs and their concentration in soil in the Leningrad Region (a) /12/ and Norway (b) /129/:

1 — amount of deposited ^{137}Cs; 2 — amount of ^{137}Cs in soil; 3 — amount of deposited ^{90}Sr; 4 — amount of ^{90}Sr in soil.

TABLE 2.12. Rate of influx and accumulation of ^{90}Sr in soils during 1960−1963 /44, 47/

Observation site	Influx of ^{90}Sr with fallout, mcuries/km^2	Increase of ^{90}Sr concentration in soil, mcuries/km^2	Difference	
			mcuries/km^2	%
German Democratic Republic*	12.4	10.8	− 1.6	12.9
USSR	30.9	22.0	− 8.9	28.8
Bodö, Norway	27.2	10.1	−17.1	62.8
Reykjavik, Iceland	31.7	20.9	−10.8	34.1
Ottawa, Canada	45.2	30.5	−14.7	32.5
New York (USA)	35.2	27.4	− 7.8	22.2
Singapore, Malaysia	8.1	1.3	− 6.8	83.9
Salisbury, Southern Rhodesia	1.8	2.0	+ 0.2	11.1
Santiago, Chile	2.5	2.2	− 0.3	12.0
70 − 60° N	16.8	13.1	− 3.7	22.0
60 − 50° N	22.9	19.5	− 3.4	14.8
50 −40° N	26.9	22.2	− 4.7	17.4
40 − 30° N	19.2	11.3	− 6.9	35.9
Mean within 70 −30° N	21.5	16.5	− 5.0	23.3

* In the period June 1961 − June 1963.

TABLE 2.13. Ratio of deposited ^{90}Sr and ^{137}Cs to their amounts accumulated in soil /12, 52/

Year of observation	$\dfrac{^{90}\text{Sr deposited}}{^{90}\text{Sr in soil}}$	$\dfrac{^{137}\text{Cs deposited}}{^{137}\text{Cs in soil}}$	Year of observation	$\dfrac{^{90}\text{Sr deposited}}{^{90}\text{Sr in soil}}$	$\dfrac{^{137}\text{Cs deposited}}{^{137}\text{Cs in soil}}$
1956	1.8	−	1960	1.9	1.0
1957	2.8	−	1961	1.5	1.3
1958	1.3	1.9	1962	1.8	1.3
1959	1.9	1.3	Mean	1.9	1.5

Comparison of ^{90}Sr concentration in the soil cover with the quantity of atmospheric precipitation shows these two quantities to be not always necessarily correlated (Figure 2.18); however, even in the presence of a reliable correlation the scatter of experimental points is very wide.

Redistribution of radionuclides within soil profiles depends not only on the total quantity of atmospheric precipitation, but also on the ratio of the relative contributions of dry and wet fallout as well as the forms of the individual nuclides in the latter, depending on the physical state of fallout, its chemical composition, the time elapsed after the nuclear explosion, and so on /42, 108, 162/. Table 2.14 reveals considerable differences in the distribution of individual radionuclides between the water-soluble and insoluble fractions of radioactive fallout. The following series of radioisotopes is presented in the order of their increasing relative contents in the soluble fraction: ^{144}Ce < ^{91}Y < ^{95}Zr < ^{137}Cs < ^{90}Sr. However, in spite of this series, the quantitative ratios between the fractions vary widely for every radionuclide in different samples. Besides the above-enumerated factors, the variation may likewise be related to fluctuations in the concentration of dust, since an increase in the latter is accompanied by an increase in the relative content of nuclides in the insoluble fractions /42, 108, 162/.

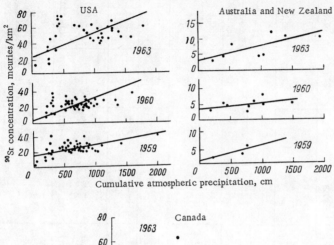

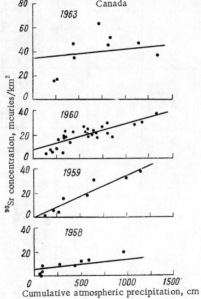

FIGURE 2.18. Strontium-90 concentration in soil cover and cumulative atmospheric precipitation (from January 1953 to the sampling time) in USA, Australia, New Zealand, and Canada /87, 133, 134/

TABLE 2.14. Distribution of radionuclides between soluble and insoluble fallout fractions, %

Fraction	Observation site	Observation time, year	Fallout type	^{89}Sr	^{90}Sr	^{91}Y	^{95}Zr	^{106}Ru	^{137}Cs	^{144}Ce	Reference
Soluble	New York	1958	Atmospheric precipitation	94.4	95.6	52.4	29.1	—	70.0	42.0	/179/
Insoluble				3.6	0.03	47.6	67.8	—	22.6	57.3	
Soluble	Moscow	1959—1961	Atmospheric precipitation Fallout	—	5.6	—	—	—	42.5	73.2	/2/
Insoluble				—	12.0	—	—	—	88.5	95.5	
Soluble	"	1963—1964	Snow	—	52.5 (35—70)	—	—	—	—	9.3 (4—13)	/42/
Insoluble				—	47.5 (30—65)	—	—	—	—	90.7 (87—96)	
Soluble	"	1963—1964	Fallout	—	92.2 (90—94)	—	—	—	—	21.8	
Insoluble				—	7.8 (6—10)	—	—	—	—	78.2	
Soluble	"	1963—1964	Rain	—	~100	—	—	—	—	—	
Insoluble				—	Not detected	—	—	—	—	—	
Soluble	"	1963—1964	Mean for all fallout types	—	81.6	—	—	—	—	15.5	
Insoluble				—	18.4	—	—	—	—	84.5	
Soluble	Leningrad	1961—1966	Atmospheric precipitation	—	91 (84—96)	16 (9—23)	23* (0—40)	70 (60—89)	72 (0—90)	12 (0—49)	/15/
Insoluble				—	9 (4—16)	84 (77—91)	77* (60—100)	30 (11—40)	28 (10—100)	88 (51—100)	
Soluble	Urals	1963—1964	Snow	80 (33—100)	71 (3—97)	—	—	—	— (36—50)	5.5 (0—15)	/19/
Insoluble				20 (0—77)	29 (3—40)	—	—	—	—	94.5 (85—100)	

Fraction	Region	Period	Type								Ref.
Soluble	Urals	1963–1964	Fallout	58 (20–89)	56.5 (20–92)	—	—	—	(36–50)	4.6 (0.2–16)	/19/
Insoluble				42 (11–80)	43.5 (8–80)	—	—	—	(50–64)	95.4 (84–100)	
Soluble	"	1963–1964	Mean for all fallout types	69	64	—	—	—	—	5.1	
Insoluble				31	36	—	—	—	—	94.9	
Soluble	Czecho-slovakia	1964	Fallout	—	73.7 (52–92)	—	—	—	44.9 (32–64)	—	/109/
Insoluble				—	26.3 (8–48)	—	—	—	55.1 (36–68)	—	
Soluble	Stuttgart	Aug.–Dec. 1961	Fallout and dust	—	76 (71–87)	—	—	—	—	—	/162/
Insoluble				—	24 (13–29)	—	—	—	—	—	
Soluble	"	1962	"	—	80 (54–93)	—	26* (8–36)	26** (8–35)	—	—	
Insoluble				—	20 (7–46)	—	—	—	—	—	
Soluble	"	Jan.–Oct. 1963	"	—	91.4 (81–95)	—	22 (13–27)	22** (7–33)	—	19 (8–27)	
Insoluble				—	8.6 (5–19)	—	—	—	—	—	

Note : Figures in parentheses are minimum and maximum values.

* ^{95}Zr + ^{95}Nb.

** ^{103}Ru + ^{102}Rh.

No less important a factor is represented by the forms in which the fission products are present in the soluble fraction of fallout, and which determine the strength of radionuclide retention by the soil in the first moments of their deposition on the earth's surface. It is possible that ^{90}Sr, ^{95}Zr, ^{106}Ru, ^{137}Cs and ^{144}Ce in fallout are not present exclusively as cations, but also as anions and neutral compounds /42, 95/. An indirect proof of the presence of fission products in noncation forms is provided by Baranov and Vilenskii /2/. Thus ^{90}Sr, ^{137}Cs and ^{144}Ce concentrations in fallout determined by concentrating them on KU-2 cation-exchange resin proved to be lower than the results obtained by evaporation of samples. The fission products in anionic form are only weakly combined with the soil and migrate fairly rapidly over the soil profile /49/.

Accumulation rates of individual radioactive products of nuclear explosions in the soil cover may be appreciably affected by a variation in the degree of their mobility in the insoluble fraction of fallout. The relative content of radionuclides in mobile form is liable to increase with increasing time elapsed after their formation and deposition on the earth's surface and with changes in the chemical composition and dispersity of the fallout particles /42/.

An important role in the redistribution of radionuclides in the soil cover is played by the physicochemical properties of soils /7, 32, 43, 155/. Elimination of ^{90}Sr from the top, $0 - 5$ cm soil layer is known to increase along the series sierozem < chernozem < sod-podzolic soil /44, 47/. The roles played by natural factors in the primary distribution of radionuclides deposited on the soil-plant cover and in their subsequent redistribution and migration, with allowance for landscape-geochemical local characteristics, are described in Chapter 3.

References

1. Baranov, V.I. and V.D. Vilenskii.— In Sbornik: "Issledovanie protsessov samoochishcheniya atmosfery ot radioaktivnykh izotopov," p. 141. Vil'nyus, "Mintis." 1968.
2. Baranov, V.I. and V.D. Vilenskii.— Radiokhimiya, Vol. 4:436. 1962.
3. Baranov, V.I. et al. — Atomnaya Energiya, Vol. 18:246. 1965.
4. Baranov, V.I. et al. — In Sbornik: "Problemy geokhimii," p. 556. Moskva, "Nauka." 1965.
5. Belyaev, L.I. et al. — In Sbornik: "Radioaktivnaya zagryaznennost' morei i okeanov," p. 58. Moskva, "Nauka." 1964.
6. Bolin, L. — In Sbornik: "Yadernaya geofizika," p. 206. Moskva, "Mir." 1964.
7. Bochkarev, V.M. et al. — Pochvovedenie, No. 9:56. 1964.
8. Burtsev, I.I. — In Sbornik: "Radioaktivnye izotopy v atmosfere i ikh ispol'zovanie v meteorologii," p. 217. Moskva, Atomizdat. 1965.
9. Burtsev, I.I. and S.G. Malakhov. — Fizika Atmosfery i Okeana, Vol. 4:328. 1968.

10. G a z i e v, Ya. I. — In Sbornik: "Radioaktivnye izotopy v atmosfere i ikh ispol'zovanie v meteorologii," p. 153. Moskva, Atomizdat. 1965.
11. G a z i e v, Ya. I. and L. E. N a z a r o v. — Ibid., p. 181.
12. G e d e o n o v, L. I. et al. — Ibid., p. 345.
13. G e d e o n o v, L. I. and M. I. Z h i l k i n a. — Ibid., p. 342.
14. G e d e o n o v, L. I. and M. I. Z h i l k i n a. Vypadenie produktov deleniya v okrestnostyakh Leningrada v 1957 — 1965 gg. (Fallout of Fission Products Around Leningrad in 1957 — 1965). Moskva, Atomizdat. 1967.
15. G e d e o n o v, L. I. et al. — In Sbornik: "Issledovanie protsessov samoochishcheniya atmosfery ot radioaktivnykh izotopov," p. 181. Vil'nyus, "Mintis." 1968.
16. G r e c h u s h k i n a, M. P. Tablitsy sostava produktov mgnovennogo deleniya U^{235}, U^{236}, Pu^{239} (Tables of the Composition of Instantaneous Fission Products of ^{235}U, ^{238}U and ^{239}Pu). Moskva, Atomizdat. 1964.
17. G r e c h u s h k i n a, M. P. and Yu. A. I z r a e l'. — In Sbornik: "Radioaktivnye izotopy v atmosfere i ikh ispol'zovanie v meteorologii," p. 164. Moskva, Atomizdat. 1965.
18. G r e i b i n, T. and G. L a n b e r. — In Sbornik: "Atmosfernye aerozoli i radioaktivnoe zagryaznenie vozdukha," p. 141. Leningrad, Gidrometeoizdat. 1964.
19. D i b o b e s, I. K. et al. Global'nye vypadeniya strontsiya-90 na territorii Urala v period 1961 — 1966 g. (Total Fallout of Strontium-90 Over the Urals During 1961 — 1966). Moskva, Atomizdat. 1967.
20. D m i t r i e v a, G. V. — In Sbornik: "Voprosy yadernoi meteorologii," p. 163. Moskva, Gosatomizdat. 1962.
21. D m i t r i e v a, G. V. In Sbornik: "Radioaktivnye izotopy v atmosfere i ikh ispol'zovanie v meteorologii," p. 283. Moskva, Atomizdat. 1965.
22. D m i t r i e v a, G. V. and V. N. K a s a t k i n a. — Ibid., p. 293.
23. D i a c o n i s c u, I. G. In "Radioaktivita atmosfery. Sborník Prací Hydrometeorologického Ústavu CSSR," Vol. 6:45. Praha. 1966.
24. Z i m i n, A. G. — In Sbornik: "Voprosy yadernoi meteorologii," p. 116. Moskva, Gosatomizdat. 1962.
25. Z o r i n, V. M. and E. I. K a b i s h c h e r. — Gigiena i Sanitariya, No. 11:114. 1965.
26. Z y k o v a, A. S. et al. Radioaktivnost' atmosfernogo vozdukha i nekotorykh produktov pitaniya v Moskve v 1965 i 1966 gg. (Radioactivity of Atmospheric Air and Some Foodstuffs in Moscow in 1965 and 1966). Moskva, Atomizdat. 1967.
27. Z y k o v a, A. S. et al. — Gigiena i Sanitariya, No. 9:5. 1963.
28. Z y k o v a, A. S. et al. Nekotorye dannye o zavisimosti mezhdu soderzhaniem strontsiya-90 i tseziya-137 v okruzhayushchei srede i organizme lyudei (Some Data on the Relationship Between Concentrations of Strontium-90 and Cesium-137 in the Environment and the Human Body). Moskva, Atomizdat. 1966.
29. K a r o l', I. L. and S. G. M a l a k h o v. Global'noe rasprostranenie v atmosfere i vypadenie na zemlyu radioaktivnykh produktov

yadernykh vzryvov (Obzor sovremennogo sostoyaniya problemy) (Global Distribution in the Atmosphere and Fallout on the Ground of Radioactive Products of Nuclear Explosions. Status Report on the Problem). Moskva, Izdatel'stvo Akademii Nauk, SSSR. 1960.

30. Karol', I.L. and S.G. Malakhov. — In Sbornik: "Radioaktivnye izotopy v atmosfere i ikh ispol'zovanie v meteorologii," p. 244. Moskva, Atomizdat. 1965.

31. Karol', I.L. and S.G. Malakhov. — In Sbornik: "Voprosy yadernoi meteorologii," p. 5. Moskva, Gosatomizdat. 1962.

32. Kvaratskhelia, N.T. and G.G. Glonti. — Pochvovedenie, No. 10:64. 1965.

33. Kurchatov, B.V. et al. — In Sbornik: "Sovetskie uchenye ob opasnosti ispytanii yadernogo oruzhiya," p. 66. Moskva, Atomizdat. 1959.

34. Lavrenchik, V.I. Global'noe vypadenie produktov yadernykh vzryvov (Global Fallout of Nuclear Explosion Products). Moskva, Atomizdat. 1965.

35. Malakhov, S.G. et al. — Atomnaya Energiya, Vol. 19:28. 1965.

36. Manolov, L. and M. Teneva. In "Radioaktivita atmosfery. Sborník Prací Hydrometeorologického Ústavu ČSSR," Vol. 6:52. Praha. 1966.

37. Makhon'ko, K.P. — "Izvestiya Akademii Nauk SSSR. Seriya geofizicheskaya," No. 4:596. 1964.

38. Makhon'ko, K.P. — Ibid., No. 11:1709. 1963.

39. Makhon'ko, K.P. — "Izvestiya Akademii Nauk SSR. Fizika atmosfery i okeana," No. 2:508. 1966.

40. Makhon'ko, K.P. — In Sbornik:"Radioaktivnye izotopy v atmosfere i ikh ispol'zovanie v meteorologii," p. 230. Moskva, Atomizdat. 1965.

41. Mukhin, I.E. and L.I. Nagovitsyna. Soderzhanie strontsiya-90 v global'nykh vypadeniyakh na territorii Ukrainskoi SSR v 1963 — 1966 gg. (Strontium-90 Content in Total Fallout Over Ukrainian SSR in 1963 — 1966). Moskva, Atomizdat. 1967.

42. Pavlotskaya, F.I. and L.N. Zatsepina. — Atomnaya Energiya, Vol. 20:333. 1966.

43. Pavlotskaya, F.I. et al. — In Sbornik: "Radioaktivnost' pochv i metody ee opredeleniya," p. 20. Moskva, "Nauka." 1966.

44. Pavlotskaya, F.I. and E.B. Tyuryukanova. — Atomnaya Energiya, Vol. 23:229. 1967.

45. Pavlotskaya, F.I. et al. — In Sbornik: "Radioaktivnost' pochv i metody ee opredeleniya," p. 174. Moskva, "Nauka." 1966.

46. Pavlotskaya, F.I. et al. — Gigiena i Sanitariya, No. 11:54. 1965.

47. Pavlotskaya, F.I. et al. Global'noe raspredelenie radioaktivnogo strontsiya po zemnoi poverkhnosti (Global Distribution of Radioactive Strontium Over the Earth's Surface). Moskva, "Nauka." 1970.

48. Polyakov, Yu.A. — In Sbornik: "Radioaktivnost' pochv i metody ee opredeleniya," p. 133. Moskva, "Nauka." 1966.

49. Polyakov, Yu.A. — Pochvovedenie, No. 8:57. 1956.

50. Pudovkina, I.B. Sravnenie rezul'tatov izmerenii atmosfernykh
     vypadenii strontsiya-90 v raznykh stranakh (Comparison of Mea-
     surements of Atmospheric Fallout of Strontium-90 in Different
     Countries). Moskva, Atomizdat. 1967.
51. Radioactive Fallout from Nuclear Explosions. — United Nations
     Document, A/AC. 82/R, 124. 1961.
52. Shvedov, V.P. and S.I. Shirokov (editors). Radioaktivnye
     zagryazneniya vneshnei sredy (Radioactive Environmental Pollu-
     tion). Moskva, Gosatomizdat. 1962.
53. Raifershaid, G. — In Sbornik: "Radioaktivnye chastitsy v atmosfere,"
     p. 194. Moskva, Gosatomizdat. 1963.
54. Rosyanov, S.P. et al. Raspredelenie Sr^{90} i Cs^{137} po profilyu pochv v
     prirodnykh usloviyakh v 1964 g. (Distribution of ^{90}Sr and ^{137}Cs
     Over Soil Profiles Under Natural Conditions in 1964). Moskva,
     Atomizdat. 1967.
55. Titlyanova, A.A. and N.A. Timofeeva. — "Pochvovedenie,"
     No. 3:86. 1959.
56. Timofeev-Resovskii, N.V. et al. — In Sbornik: "Radioaktivnost'
     pochv i metody ee opredeleniya," p. 46. Moskva, "Nauka," 1966.
57. Tyuryukanova, E.B. O metodike issledovaniya povedeniya radio-
     aktivnogo strontsiya v pochvakh razlichnykh geokhimicheskikh
     landshaftov (Investigation Techniques for the Behavior of Radio-
     active Strontium in Soils of Different Geochemical Landscapes).
     Moskva, Atomizdat. 1968.
58. Tyuryukanova, E.B. and F.I. Pavlotskaya. — Agrokhimiya,
     No. 9:50. 1967.
59. Tyuryukanova, E.B. et al. — In Sbornik: "Radioaktivnost' pochv i
     metody ee opredeleniya," p. 36. Moskva, "Nauka." 1966.
60. Tyuryukanova, E.B. et al. — Informatsionnyi byulleten' Radio-
     biologiya, No. 9:9. 1966.
61. Tyuryukanova, E.B. et al. Raspredelenie radioaktivnogo
     strontsiya v pochvakh razlichnykh zon (Distribution of Radioactive
     Strontium in Soils of Different Zones). Moskva, Atomizdat. 1967.
62. Tyuryukanova, E.B. et al. — Pochvovedenie, No. 8:88. 1964.
63. Tyuryukanova, E.B. et al. — Pochvovedenie, No. 10:66. 1964.
64. Facey, L. — In Sbornik: "Yadernaya geofizika," p. 299. Moskva,
     "Mir." 1964.
65. Fedorov, G.A. and N.E. Konstantinov. — In Sbornik: "Voprosy
     dozimetrii i zashchity ot izluchenii," No. 6:132. Moskva, Atomiz-
     dat. 1967.
66. Hintzpeter, M. — In Sbornik: "Radioaktivnye chastitsy v atmosfere,"
     p. 116. Moskva, Gosatomizdat. 1963.
67. Chalov, P.I. and M.A. Tsevelev. — Atomnaya Energiya,
     Vol. 19:470. 1965.
68. Chulkov, P.M. Soderzhanie strontsiya-90 v pochve i rastitel'nom
     pokrove Moskovskoi oblasti (Strontium-90 Concentration in Soil
     and Plant Cover of the Moscow Region). Moskva, Izdatel'stvo
     Akademii Nauk SSSR. 1960.
69. Chupka, Sh. et al. — Atomnaya Energiya, Vol. 18:496. 1965.

70. C h u r k i n, V. N. and V. F. B r e n d a k o v . — Informatsionnyi
       byulleten' Radiobiologiya, No. 9:11. 1966.
71. S h v e d o v, V. P. et al. — In Sbornik: "Radioaktivnye zagryazneniya
       vneshnei sredy," p. 163. Moskva, Gosatomizdat. 1962.
72. S h u b k o, V. M. Vypadenie strontsiya-90 na poverkhnost' territorii
       SSSR (Fallout of Strontium-90 on the Surface of the USSR). Moskva,
       Izdatelstvo Akademii Nauk SSSR. 1959.
73. S h u b k o, V. M. and A. M. E r e m i c h e v a . Vypadenie dolgozhivush-
       chikh produktov deleniya na territorii SSSR v 1961 g. (Fallout of
       Long-Lived Fission Products on the USSR in 1961). Moskva,
       Gosatomizdat. 1962.
74. S h u b k o, V. M. and A. M. E r e m i c h e v a . Vypadenie strontsiya-90
       na poverkhnost' territorii SSSR v chetvertom kvartale 1961 g. i
       pervoi polovine 1962 g. (Fallout of Strontium-90 on the Surface of
       the USSR in the Last Quarter of 1961 and the First Half of 1962).
       Moskva, Gosatomizdat. 1963.
75. S h u b k o, V. M. and B. V. K u r c h a t o v . Vypadenie dolgozhivushckikh
       produktov deleniya na territorii SSSR v 1959 — 1960 gg. (Fallout of
       Long-Lived Fission Products on the USSR in 1959 — 1960). Moskva,
       Gosatomizdat. 1961.
76. J u n g e, C. E. Air Chemistry and Radioactivity. New York, Academic
       Press. 1963.
77. J a c o b i, W. — In Sbornik: "Radioaktivnye chastitsy v atmosfere,"
       p. 129. Moskva, Gosatomizdat. 1963.
78. A a r k r o g, A. and J. L i p p e r t . Environmental Radioactivity in
       Denmark in 1960. — Danish Atomic Energy Research Establishment
       Risö. Risö Report, No. 23. 1961.
79. A a r k r o g, A. and J. L i p p e r t . Environmental Radioactivity in
       Denmark in 1963. — Ibid., Risö Report, No. 85. 1964.
80. A a r k r o g, A. and J. L i p p e r t . Environmental Radioactivity in the
       Faroes in 1964. — Ibid., Risö Report, No. 108. 1965.
81. A a r k r o g, A. and J. L i p p e r t . Environmental Radioactivity in
       Greenland in 1963. — Ibid., Risö Report, No. 87. 1964.
82. A a r k r o g, A. et al. Environmental Radioactivity in Denmark in
       1961. — United Nations Document A/AC. 82/G/L. 802. 1962.
83. A a r k r o g, A. et al. Environmental Radioactivity in Denmark in
       1962. — Danish Atomic Energy Research Establishment. Risö
       Report, No. 63. 1963.
84. A a r k r o g, A. et al. Environmental Radioactivity in Greenland in
       1962.—Ibid., Risö Report, No. 65. 1963.
85. A l e x a n d e r, L. T. — Health and Safety Laboratory. Fallout Program
       Quarterly Summary Report USAEC Report HASL-183, p. 16. 1961.
86. A l e x a n d e r, L. T. et al. Vertical Distribution of ^{90}Sr in Sandy Soils
       in May 1965. — Ibid., HASL-171, p. 370. 1966.
87. A l e x a n d e r, L. T. et al. Strontium-90 in the Earth's Surface.—
       Ibid., HASL-88, p. 195. 1960.
88. Annual Summary on Environmental Radioactivity in New Zealand. —
       Ibid., HASL-182, p. 61. 1967.
89. B a r a n o v, V. I. et al. Proceedings of the Third International Con-
       ference on the Peaceful Uses of Atomic Energy, Vol. 14:107. 1965.

90. B e n c o, A. et al. — Minerva nucleara, Vol. 4, No. 9. 1960.
91. B e n i n s o n, D. et al. Estudio de evolucion de materials radioactives
        en el medio terrestre. — United Nations Document, A/AC/.82/G/L.
        1033. 1965.
92. B e r g h, H. et al. — Svensk kem. tidskr., 71:695. 1959.
93. B o r t o l i, M. de, and A. M a l v i c i n i. — European Atomic Energy
        Commission, EUR-2965e. 1966.
94. B o r t o l i, M. et al. — Instrum. and Methods, Vol. 35:177. 1965.
95. B r e n a n, C. et al. — Nature, Vol. 211:68. 1966.
96. B r y a n t, F. J. et al. — Austral. J. Sci., 127:222. 1965.
97. C a m b r a y, R. S. et al. Radioactive Fallout in Air and Rain: Results
        to the Middle of 1965. — United Kingdom Atomic Energy Authority.
        Research Group Report, AERE-R- 4997. 1965.
98. C a m b r a y, R. S. et al. Radioactive Fallout in Air and Rain: Results
        to the Middle of 1966. — Ibid., AERE-R-5260. 1966.
99. C a m b r a y, R. S. et al. Radioactive Fallout in Air and Rain: Results
        to the Middle of 1962. — Ibid., AERE-R-4094. 1962.
100. C a m b r a y, R. S. et al. Radioactive Fallout in Air and Rain: Results
        to the Middle of 1963. — Ibid., AERE-R-4392. 1963.
101. C a m b r a y, R. S. et al. Radioactive Fallout in Air and Rain: Results
        to the Middle of 1964. — Ibid., AERE-R-4687. 1964.
102. C o l l i n s, W. R. Deposition on the Earth's Surface from 1958 through
        1963. — Health and Safety Laboratory. Fallout Program Quarterly
        Summary Report. USAEC Report, HASL-146, p. 241. 1964.
103. C o l l i n s, W. R. — Radiol. Health Data, Vol. 5:163. 1964.
104. C r o o k s, R. N. et al. Radioactive Fallout in Air and Rain: Results
        to the Middle of 1961. — United Kingdom Atomic Energy.
        Authority Research Group Report. AERE-R-3766. 1961.
105. C r o o k s, R. N. et al. The Deposition of Fission Products from
        Distant Nuclear Test Explosions:Results to the Middle of 1960. —
        Ibid., AERE-R-3349. 1960.
106. C r o o k s, R. N. et al. The Deposition of Fission Products from
        Distant Nuclear Test Explosions: Results to the Middle of 1959. —
        Ibid., AERE-R-3094. 1959.
107. C l a r k, R. S. et al. — J. Geophys. Res., Vol. 72-1793. 1967.
108. C s u p k a, St. — Kernenergie, Vol. 8:574. 1965.
109. C s u p k a, St. et al. — Jaderna Energia, Vol. 12:16. 1966.
110. S c u p k a, St. et al. — Nature, Vol. 213:1204. 1967.
111. Data on Environmental Radioactivity Collected in Italy (January —
        June 1961). — Comitato Nazionale Energia Nucleare. BIO/12/61,
        Rome. 1961.
112. Data on Environmental Radioactivity Collected in Italy (July —
        December 1961). — Ibid., BIO/06/62, Rome. 1962.
113. Data on Environmental Radioactivity Collected in Italy (January —
        June 1962). — Ibid., BIO/26/62, Rome. 1962.
114. Data on Environmental Radioactivity Collected in Italy (July —
        December 1961). — Ibid., BIO/03/63, Rome. 1963.
115. D y e r, A. J. — Nature, Vol. 189:905. 1961.
116. E d v a r s o n, K. et al. — Nature, Vol. 184:1771. 1959.

117. Ellis, F.B. and E.R. Mercer. Agricultural Research Council
    Radiobiological Laboratory. — Annual Report 1964 — 1965. ARCRL,
    No. 14, p. 72. 1965.
118. Fallout in New Zealand. — Health and Safety Laboratory. Fallout
    Program Quarterly Summary Report USAEC Report, HASL-161,
    p. 223. 1965.
119. Fallout in New Zealand. — Ibid., HASL-138, p. 209. 1963.
120. Fallout Measurement in Canada. October — December 1963. — Radiol.
    Health Data, Vol. 5:206. 1964.
121. Feely, H.W. et al. — Tellus, Vol. 18:316. 1966.
122. Fry, L. and K. Menon. — Science, Vol. 137:994. 1962.
123. Gibbs, W.J. et al. — Austral. J. Sci., Vol. 28:59. 1965.
124. Gibbs, W.J. and G.U. Wilson. — Austral. J. Sci., Vol. 28:44. 1965.
125. Greenfield, S. — J. Meteorol., Vol. 14:115. 1957.
126. Gustafson, P.F. et al. — J. Geophys. Res., Vol. 67:4611. 1962.
127. Hageman, F. et al. — Science, Vol. 130:542. 1959.
128. Hardy, E. Vertical Penetration of Strontium-90 in Three Soils,
    Sampled in March 1963. — Health and Safety Laboratory. Fallout
    Program Quarterly Summary Report. USAEC Report, HASL-138,
    p. 249. 1963.
129. Hardy, E. The Ratio of Cs^{137} to ^{90}Sr in Global Fallout. — Ibid.,
    HASL-182. 1967.
130. Hardy, E. and L.T. Alexander. — Science, Vol. 136:881. 1962.
131. Hardy, E. and L.T. Alexander. The Relationship Between Rain-
    fall and Strontium-90 Deposition in Clallam County, Washington. —
    United Nations Document A/AC. 82/G.L. 776. 1962.
132. Hardy, E. and W. Collins. — Radiol. Health Data, Vol. 4:9. 1963.
133. Hardy, E. et al. Strontium-90 on the Earth's Surface. II. Summary
    and Interpretation of the World-Wide Soil Sampling Program,
    1960—1961 Results. — United Nations Document, A/AC. 82/G/L. 822.
    1963.
134. Hardy, E.P. and J. Rivera. Sr-90 in World-Wide Soils Collected
    in 1963 — 1964. — Health and Safety Laboratory. Fallout Program
    Quarterly Summary Report. USAEC Report, HASL-149, p. 29.
    1964.
135. Health and Safety Laboratory. Ibid., HASL-171. 1966.
136. Health and Safety Laboratory. Ibid., HASL-183. 1967.
137. Holland, J.Z. United States Atomic Energy Commission,
    Technical Information Service. TID-5554. 1959.
138. Hvinden, T. Radioactive Fallout in Norway 1959. — Forsvarents
    Forkskningsinstitut Norwegian Defence Research Establishment.
    Intern. Rapport F-0394. 1959.
139. Hvinden, T. and A. Lillegraven. — Nature, Vol. 192:1144. 1961.
140. Kulp, J.L. and A.R. Schuler. Strontium-90 in Man and His En-
    vironment. — Vol. 1: Summary, 1962. (United Nations Document,
    A/AC. 82/G/L. 792. 1962.)
141. Kuroda, P.K. et al. — Science, Vol. 32:742. 1960.
142. Kuroda, P.K. et al. — Science, Vol. 137:15. 1962.
143. Kruger, P. and A. Miller. — J. Geophys. Res., Vol. 69:1469. 1964.

144. Lindell, B. A Review of Measurement of Radioactivity in Food, Especially Dairy Milk, and a Presentation of the 1963 Data on Cs^{137} and Sr^{90}. — United Nations Document A/AC. 82/G/L. 934. 1964.
145. Machta, L. A Survey of Information on Meteorological Aspects of World-Wide Fallout. — Ibid., A/AC. 82/G/R, 81. 1959.
146. Mamuro, T. et al. — Nature, Vol. 197:964. 1959.
147. Martell, E. A. — Science, Vol. 129:1197, 1956.
148. Miyake, Y. et al. — Papers on Meteorology and Geophysics, Vol. 11:151. 1960.
149. Miyake et al. Deposition of Sr-90 and Cs-137 in Tokio. — United Nations Document, A/AC. 82/G/L. 688. 1961.
150. Nagayama, Sh. — Nucl. Sci. Abstr. of Japan, Vol. 4:90. 1965.
151. Ohta, S. — Papers of Meteorology and Geophysics, Vol. 11:6. 1960.
152. Pasak, V. Niederschläge in den Waldständen und ihre Radioaktivität. Biometeorology. Oxford — London — New York — Paris, p. 582. 1962.
153. Pasák, V. — Sb. ČSAZV. Lesn., Vol. 7:849. 1961.
154. Parker, R.P. and J.O. Crokkal. — Nature, Vol. 190:574. 1961.
155. Pavlotskaya, F.I. et al. Radioecological Concentration Processes. — Proceedings of the International Symposium Held in Stockholm 25—29 April 1966. Edited by B. Aberg and F. P. Hungate. London, Pergamon Press, p. 25. 1967.
156. Physical Aspects of the Radioactive Fallout over Scandinavia Especially During the Period October 1958 — October 1959. — United Nations Document, A/AC. 82/G/L. 343. 1960.
157. Prawitz, J. A Fallout Model. II. Some Quantitative Properties. — Ibid., A/AC. 82/G/L. 760. 1962.
158. Radioactivity Survey Data in Japan. — National Institute of Radiological Science. Chiba, Japan, No. 1. 1963.
159. Radioactivity Survey Data in Japan. — Ibid., No. 5. 1964.
160. Radioactivity Survey Data in Japan. — Ibid., No. 9. 1965.
161. Radioactivity Survey Data in Japan. — Ibid., No. 10. 1966.
162. Reintschler, W. et al. — Atomkernenergie, Vol. 9:297. 1964.
163. Reintschler, A. and H. Schreiber. — Atomkernenergie, Vol. 7:325. 1962.
164. Report of the Federal Radiation Council. Revised Fallout Estimations for 1964—1965 and Verification of the 1963 Predictions. Report No. 6. 1964.
165. Salo, A. — Nature, Vol. 212:61. 1966.
166. Schumann, G. — Naturwissenschaften, Vol. 54:6. 1967.
167. Schumann, G. and G. Eulitz. — Naturwissenschaften, Vol. 47:13. 1960.
168. Squire, H.M. — Agricultural Research Council Radiobiological Laboratory. Annual Report 1964—1965. ARCRL, No. 14, p. 64. 1965.
169. Stewart, N.G. et al. — Atomic Energy Research Establishment, AERE-HR/R 2354. 1958.

170. Stewart, N.G.   The Deposition of Long-Lived Products
     from Nuclear Test Explosions: Results to the Middle of 1958. — Ibid.,
     AERE-HP/R 2790. 1959.
171. Szepke, R. — Atompraxis, Vol. 11:7, 391. 1965.
172. Thornthwaite, C.W. et al. — Science, Vol. 131:1015. 1960.
173. Tyuryukanova, E.B. et al. — Proceedings of the International
     Symposium Held in Stockholm 25 — 29 April 1966. Edited by
     B. Aberg and F.P. Hungate. London. Pergamon Press, p. 33. 1967.
174. United Nations Scientific Committee on the Effects of Atomic Radiation.
     Report of the United Nations Scientific Committee on the Effects
     of Atomic Radiation, Annex. F., Part 2. General Assembly,
     Seventeenth Session. Suppl. No. 16 (A/5216), N.Y. 1962.
175. Ibid., General Assembly, Nineteenth Session. Suppl. No. 14 (A/5814),
     N.Y. 1964.
176. Volchok, H.L. — Science, Vol. 156:1, 1487. 1967.
177. Volchok, H.L. World-Wide Deposition of Sr[90] Through 1968. —Health
     and Safety Laboratory. Fallout Program Quarterly Summary
     Report. USAEC Report, HASL-183, p. 2. 1967.
178. Walton, A. — J. Geophys. Res., Vol. 68:1485. 1963.
179. Welford, J.A. and W.R. Collins. — Science, Vol. 131:1711. 1960.
180. Wilson, D.W. et al. — Radiol. Health Data, Vol. 6:675. 1967.
181. Wykes, E.R. — Radiol. Health Data, Vol. 7:545. 1966.

Chapter 3

LANDSCAPE-GEOCHEMICAL ASPECTS OF
STRONTIUM-90 MIGRATION

Studies of radioactive fission products in the biosphere make use of landscape-geochemical methods, which examine different landscape constituents (soil, rock, water, vegetation) on genetically strictly defined areas differing from adjacent areas in age and composition of soils and parent rocks, relief, water regime, and plant cover. In accordance with the principles of landscape geochemistry concerning the role played by runoff in the migration of chemical elements, the samples for radiochemical investigations should be taken according to the profile method, consisting in studies of soil profiles on major landscape elements linked with runoff. A typical example is provided by a series of soil profiles along a line traversing water divides, ancient terraces and floodplains of individual rivers.

In forecasting the radiological situation on a given territory, special attention must be paid to the geochemistry of typomorphic elements of every landscape and elements acting as carriers of radioactive fission products. Polynov /19/ and Perel'man /18/ defined typomorphic elements in a landscape as macroelements that are present in appreciable concentrations and belong to the most mobile elements under the given geochemical situation. In the geochemistry of bog landscapes in a forest zone, Fe is a typomorphic element along with hydrogen; the typomorphic element for steppe landscapes is Ca, and for dry-steppe and coastal marine landscapes, Na. Being present in macroconcentrations and being most responsive to changes in the redox and alkali-acid conditions, typomorphic elements largely control the migration of trace elements and radionuclides. The distribution of ^{90}Sr in the bogged soils of Polysye landscapes in the forest zone has been found to correlate with the distribution of Fe, while in the steppe zone it is correlated with the distribution of Ca and stable Sr.

Consider the influence of various environmental factors on the migration of ^{90}Sr in a landscape. Geomorphological conditions controlling the magnitude and direction of surface and subsurface runoff affect the removal of radionuclides and formation of their secondary accumulation foci. A buildup of ^{90}Sr occurs in soils of depressions and in diluvial deposits /16, 27, 29, 30, 43 —45, 49, 52/. Not all depressions, however, become accumulation sites of radionuclides. An important role is played by runoff direction, shape of depression, exposure, amount of turf on slopes, etc. As a rule, ^{90}Sr accumulates in closed depressions without outflow and with plowed-up slopes. Secondary accumulation of radionuclides often occurs in the lower third of slopes amply irrigated by subsurface runoff water and continuously supplied with nutrients. In the forest zone, depressions are usually occupied by bogs. Accumulation of ^{90}Sr in bogs depends on the latter's type. Highest

TABLE 3.1. Strontium-90 distribution in soils as a function of geomorphological conditions

Site	^{90}Sr concentration in 0−60 cm soil layer, ncuries/m^2	Site	^{90}Sr concentration in 0−60 cm soil layer, ncuries/m^2
Profile 1 (Ryazan Region), 1967		Profile 4 (Ryazan Region), 1967	
Flat summit of a hill; C l a d i n u m moss forest (Pinetum cladinosum)	57	Summit of a hill; pine-oak forest	45
		Slope; oak forest	30
Gentle forested slope	30	Alder bog	112
Lower third of a slope; whortleberry pine forest (Pinetum myrtillosum) with smallreed	103	Profile 5 (Ryazan Region), 1967	
		Summit of a hill; whortleberry spruce forest (pinum myrtillosum)	78
Sedge bog	35		
Profile 2 (Ryazan Region), 1967		Unforested slope	36
Summit of a hill; green moss pine forest (Pinetum hylocomiosum)	56	Profile 6 (Zaporozhe Region), 1966	
		Water divide; tilled chernozem	161
Gentle slope; birch-oak forest	32	Slope; gray forest soil; "dubrava"-type oak forest	43
Birch-sphagnum bog	28		
Profile 3 (Ryazan Region), 1967		Terrace under "bor"-type pine forest; low-humus sandy soil	11
Summit of a hill; green moss pine forest (Pinetum hylocomiosum)	56		
		Hollow; meadow solonchak	44
Slope; birch-smallreed forest	39	Depression; wet-solonchak	8
Bentgrass-sphagnum bog	54		

^{90}Sr concentrations are found in dark-colored soils of depressions occupied by alder bogs with sluggishly flowing waters; ^{90}Sr concentrations are low in sedge and green moss bogs. The bogs can be arranged in the following order with respect to ^{90}Sr concentration in their soil: lowland alder bogs with dark-colored soils > sphagnum transition bogs with peat soils > > sphagnum lowland bogs with peat soils > sedge lowland bogs with sodgley soils. In the steppe zone, ^{90}Sr concentrations are lower in shallow hollows serving for the runoff of spring waters and occupied by hydromorphic soils than on water divides (Table 3.1).

Vegetation, responsible for biogenic migration and accumulation of chemical elements, has a strong effect on the behavior of radionuclides in the biosphere /7/. Land vegetation is the first screen retaining radionuclides deposited from the atmosphere. The forest displays an especially important part in this respect /8, 34−38, 51/. During intensive fallout, soils under mixed forests had a higher content of ^{90}Sr than soils of coniferous forests (Table 3.2), owing to the higher capacity of coniferous species for retaining fallout in their crowns. The ^{90}Sr concentration in soils under coniferous forests increases over subsequent years on account of additional influx with the needles shed later on, as well as the lower decomposition capacity of coniferous litter.

TABLE 3.2. Strontium-90 concentration in podzolic and sandy brown forest soils in the 0–30 cm layer (Ryazan Region), ncuries/m^2

Forest	Observation year		
	1963	1964	1967
Pine	$\dfrac{15-20}{17}$	$\dfrac{20-44}{40}$	$\dfrac{20-100}{60}$
Spruce-pine-birch	$\dfrac{23-30}{25}$	$\dfrac{50-143}{55}$	$\dfrac{30-40}{33}$
Pine-oak	$\dfrac{37-50}{44}$	$\dfrac{50-60}{56}$	$\dfrac{30-50}{40}$

N o t e : Numerators and denominators respectively indicate the variation ranges and mean concentrations.

TABLE 3.3. Mean ^{90}Sr concentration in forest litter on podzolic and sandy brown forest soils in 1967 (Ryazan Region)

Forest	A_0'		A_0''	
	ncuries/kg	pcuries/cm^3	ncuries/kg	pcuries/cm^3
Oak	5.0	0.4	1.9	1.0
Deciduous, small-leaved	1.6	0.8	2.0	0.3
Pine	2.9	0.5	7.5	2.3

Migration of ^{90}Sr in sod-podzolic and brown forest soils is largely controlled by the composition and degree of decomposition of forest litter (Table 3.3). The buildup of ^{90}Sr is especially pronounced in the lower part of the semidecomposed, raw-humus coniferous litter. The buildup of ^{90}Sr is directly proportional to the humus content in the humified litter and depends on the composition of the humus.

The distribution of radionuclides in a geochemical landscape is strongly affected not only by arboreal but also by herbaceous and moss vegetation. Strontium-90 concentration varies widely in different kinds of land vegetation, depending on the fallout level of nuclear explosion products and the particular properties of plant species (Table 3.4). In years with comparatively intensive influx of ^{90}Sr with the fallout (1964, 1965) its concentration in herbaceous and moss vegetation was considerably higher than in years with only small influx of ^{90}Sr from the atmosphere (1967, 1968). As a rule, moss vegetation has a higher buildup of ^{90}Sr than herbaceous vegetation, on account of the capacity of mosses, especially sphagnums, to retain atmospheric precipitation. Sphagnum has high ^{90}Sr concentrations even in years with comparatively low fallout levels.

Distribution of ^{90}Sr in root systems and aerial parts of plants depends on the specific properties of the plants. The root system of certain plant species, mainly gramineous grasses (foxtail, smallreed, feathergrass), act as a barrier retaining considerable amounts of ^{90}Sr (Table 3.5).

TABLE 3.4. Strontium-90 concentration in land vegetation, ncuries/m^2

Sampling site	Vegetation	1964,1965	1967,1968
Ryazan Region	Green moss (Pleurozium schreberi)	20	6
	Lichen	15	3
	Sphagnum	25	17
	Whortleberry	0.5	0.2
	Lilly of the valley	—	0.4
	Smallreed	—	0.2
	Foxtail	0.3	0.3
	Sedge	0.7	0.1
	Hairgrass	0.6	0.2
	Mixed forbs	0.5	0.3
Mountain-Badakhshan Autonomous Region, Tadzhik SSR	Winterfat	0.2	—
	Wormwood	0.2	—
	Leaf cushion	0.4	—

TABLE 3.5. Strontium-90 concentration in herbaceous vegetation in 1965, pcuries/kg

Sampling site	Species	Aerial portion	Root, cm 0 — 20	Root, cm 20 — 50
Ryazan Region	Foxtail (Alopecurus pratensis)	930	770	—
	Meadow forbs	2,200	210	130
	Sedge (Carex inflata)	3,700	430	160
	Smallreed (Calamgrostis epigeios)	2,500	6,250	—
Mountain-Badakhshan Autonomous Region, Tadzhik SSR	Winterfat (Eurotia ceratoides)	2,200	330	—
	Wormwood (Artemisia rhodontha)	2,000	820	—
	Feathergrass (Stipa glareosa)	1,100	3,000	—

The nonuniformity of land vegetation and soil cover produces a mosaic distribution of ^{90}Sr on territories with smooth relief. For instance, ^{90}Sr concentration in soils under forest in sandur landscapes in the forest zone on level territories vary from 20 to 120 ncuries/m^2.

The physicochemical properties of soils play a significant role in the migration and redistribution of ^{90}Sr /1, 6, 9, 11, 12, 21—24, 31—33, 39—42, 46—52/. The distribution and migration of radionuclides in every landscape proceed under the combined effects of all natural factors. Therefore, landscape-geochemical and comparative-geographical methods should be used in evaluating the radiological situation on a given territory, with simultaneous analysis of soils, vegetation, waters and parent rocks. This fairly extensively employed method, elaborated by Dokuchaev /10/ at the end of the 19th century, permits the characterization of the soil-radiogeochemical situation and the main trends of its variation on considerable territories with minimal expenditure (analysis of a small number of expertly taken samples) and a high degree of reliability.

The comparative-geographic method was used in soil-radiogeochemical investigations of landscapes on the Russian plain /3, 4, 15—16, 25—30, 52/, which revealed zonality in the concentration and distribution of ^{90}Sr in soils, in agreement with the zonality of soil-forming and weathering processes /13, 17/.

TABLE 3.6. Strontium-90 concentration in soils (0—30 cm), ncuries/m^2

Zone	Observation year						
	1960	1961	1962	1963	1964	1965	1966
Forest	$\dfrac{8-30}{16}$	$\dfrac{8-95}{15}$	$\dfrac{8-41}{21}$	$\dfrac{15-50}{25}$	$\dfrac{5-50}{44}$	$\dfrac{17-134}{62}$	$\dfrac{37-76}{53}$
Forest-steppe	—	$\dfrac{16-40}{25}$	—	$\dfrac{11-60}{33}$	—	—	$\dfrac{43-161}{100}$
Steppe	—	$\dfrac{20-37}{29}$	—	$\dfrac{30-50}{44}$	$\dfrac{89-175}{125}$	—	$\dfrac{8-294}{154}$

Note: Numerators and denominators respectively indicate variation ranges and mean values.

The analytical results listed in Table 3.6 show that ^{90}Sr concentrations are lower in the soils of humid regions (forest zone) than in those of arid territories (steppe zone), on account of the latitudinal effect of global fallout and partly the leaching water regime of soils in humid regions. Moreover, the extensive production of living matter in the humid zone causes the presence of large quantities of organic matter and carbon dioxide in the soil waters that are conducive to "acid" leaching, as a result of which the soils become depleted in mobile elements (Ca, Mg, Na, Sr, etc.) and also ^{90}Sr. The light atmospheric precipitation and intensive evaporation in arid landscapes (steppes, deserts) produce a nonleaching water regime of soils and are conducive to the accumulation of elements in the surface horizons owing to evaporative concentration /5/. This applies especially to mobile salts, and consequently also to the bulk of radioactive fission products deposited with fallout occurring in a mobile form /2, 14/.

TABLE 3.7. Strontium-90 concentration in soil (0—60 cm, Ryazan Region), ncuries/m^2

Soil	Observation year					
	1960	1961	1962	1963	1964	1965
Sod-podzolic, sandy	$\dfrac{9-15}{12}$	$\dfrac{8-26}{13}$	$\dfrac{29-41}{35}$	$\dfrac{24-50}{37}$	$\dfrac{18-59}{47}$	$\dfrac{55-83}{70}$
Sod-gley, sandy-loamy, floodplain	—	$\dfrac{75-94}{85}$	$\dfrac{8-13}{10}$	$\dfrac{21-26}{24}$	$\dfrac{30-470}{88}$	$\dfrac{80-134}{107}$

Note: Notation is the same as in Table 3.6.

Humid and arid regions differ not only in the concentration of ^{90}Sr in their soils, but also in its distribution in geochemical landscapes linked with runoff. In humid regions, where water divides are dominated by sod-podzolic soils, the soils of accumulation floodplains have been found to contain secondary accumulation of ^{90}Sr removed from the water divides (Table 3.7). In arid regions, where the water divides are dominated by chernozem and chestnut soils, secondary accumulation of ^{90}Sr and hydromorphic soils does not occur and its maximum concentration occurs in chernozem soils on water divides.

Floodplain landscapes of humid regions are distinguished by high variability of ^{90}Sr concentrations. The distribution and migration of ^{90}Sr in the soils of floodplains is largely controlled by the relief and vegetation. For instance, ^{90}Sr concentrations in floodplain soils in the middle reaches of the Oka River in 1968 vary from 30 to 90 ncuries/m^2. The highest ^{90}Sr concentrations were determined in sod-meadow soils of shallow hollows characterized by high content of humus and exchange bases. Sod-gley soils of floodplain bogs and underdeveloped sod soils of near-channel "griva" ridges are distinguished by low ^{90}Sr concentrations.

In humid regions, ^{90}Sr migrates fairly deeply into the soil profiles and often reaches the water tables (Table 3.8).

TABLE 3.8. Mean ^{90}Sr concentration in water (Ryazan Region), pcuries/liter

Water	Year of observation				
	1962	1963	1964	1967	1968
Groundwater on water divides	4.5	1.0	2.8	0.2	0.3
Groundwater in floodplains	4.0	2.3	1.1	–	1.5
River water	3.7	4.6	2.0	1.4	3.1

For a quantitative evaluation of the water migration of elements, Polynov /20/ suggested calculation of water migration coefficients based on comparison of the chemical composition of rocks or soils with the composition of waters draining them.

Starting from Polynov's ideas, Perel'man suggested that in order to evaluate the migration capacity of elements, the water migration coefficient K_x should be computed. The latter quantity is defined as the ratio of the concentration of a given element x in the mineral residue of natural water to its concentration in rocks or soils drained by these waters. Since the concentration m_x of an element in water is usually stated in g/liter, while their concentration n_x in rocks and soils is stated in percentages, the calculation formula for K_x has the following form:

$$K_x = \frac{m_x}{an_x}\ 100,$$

where a (g/liter) is the sum total of mineral substances dissolved in the water.

The formula for the determination of the migration capacity of radio-nuclides in different landscapes is

$$K_x = \frac{m_x}{an_x},$$

where m_x (curies/liter) is the concentration of the radionuclide in the water, while n_x (curies/g) is its concentration in the soil from which the water is drained.

Computed water migration coefficients for ^{90}Sr (K = 90 to 100) testify to the high mobility of this radionuclide in the landscapes of humid regions.

Thus, the behavior in the biosphere of radionuclides deposited from the atmosphere is affected by the entirety of natural conditions of every specific landscape. Since it is often very difficult to isolate and estimate the effect of every individual factor, and in view of certain features in the behavior of radionuclides due to their distinctive source, the form of their deposition on the earth's surface and the time of their interaction with soil, landscape-geochemical investigations must be accompanied by experiments to determine the specific influence of individual environmental factors on the behavior of radioactive fission products.

References

1. Aleksakhin, R.M. Radioaktivnoe zagryaznenie pochvy i rastenii (Radioactive Contamination of Soil and Plants). Moskva, Izdatel'-stvo Akademii Nauk SSSR. 1963.
2. Baranov, V.I. and V.D. Vilenskii. — Radiokhimiya, Vol. 4:486. 1962.
3. Baranov, V.I. et al. — Atomnaya Energiya, Vol. 18:246. 1965.
4. Baranov, V.I. et al. — In Sbornik: "Problemy geokhimii," p. 556. Moskva, "Nauka." 1965.
5. Batulin, S.G. — Voprosy Geografii, No. 59:202. 1962.
6. Bochkarev, V.N. et al. — Pochvovedenie, No. 9:56. 1964.
7. Vernadskii, V.I. Biosfera (The Biosphere). Moskva-Leningrad. Nauchno-khimiko-tekhnicheskoe otdelenie VSNKh. 1926.
8. Demkin, O.T. Thesis. Kiev. 1967.
9. Dibobes, I.K. et al. Global'nye vypadeniya Sr^{90} na territorii Urala v period 1961 — 1966 gg. (Total Fallout of ^{90}Sr Over the Urals During 1961 — 1966). Moskva, Atomizdat. 1967.
10. Dokuchaev, V.V. Uchenie o zonakh prirody (Theory of Natural Zones). Sankt-Peterburg. 1899.
11. Kvaratskhelia, N.T. and G.G. Glonti. — Pochvovedenie, No. 10:64. 1965.
12. Klechkovskii, V.M. and I.V. Gulyakin. — Pochvovedenie, No. 3:1. 1958.
13. Kovda, V.A. and E.M. Samoilova. — Pochvovedenie, No. 9:1. 1966.
14. Pavlotskaya, F.I. and L.N. Zatsepina. K voprosu ob izuchenii form postupleniya nekotorykh produktov deleniya na zemnuyu

poverkhnost' (Forms of Influx of Certain Fission Products to the
Earth's Surface). Moskva, Atomizdat. 1965.

15. P a v l o t s k a y a, F. I. et al. — In Sbornik: "Radioaktivnost' pochv i
metody ee opredeleniya," p. 20. Moskva, "Nauka." 1966.

16. P a v l o t s k a y a, F. I. and E. B. T y u r y u k a n o v a. — Atomnaya
Energiya, Vol. 23:229. 1967.

17. P e r e l' m a n, A. I. — Trudy Instituta Rudnykh Mestorozhdenii, Petro-
grafii, Mineralogii i Geokhimii Akademii Nauk SSSR, No. 99:114.
1963.

18. P e r e l' m a n, A. I., Geokhimiya landshafta (Landscape Geochemistry).
Moskva, "Vysshaya shkola." 1966.

19. P o l y n o v, B. B. Geokhimicheskie landshafty. Izbrannye Trudy
(Geochemical Landscapes. Selected Works). Moskva, Izdatel'stvo
Akademii Nauk SSSR. 1956.

20. P o l y n o v, B. B. Uchenie o landshaftakh. Izbrannye Trudy (Theory
of Landscapes. Selected Works). Moskva, Izdatel'stvo Akademii
Nauk SSSR. 1956.

21. P o l y a k o v, Yu. A. — Pochvovedenie, No. 11:45. 1962.

22. R o s y a n o v, S. P. et al. Raspredelenie Sr^{90} i Cs^{137} po profilyu pochv
v prirodnykh usloviyakh v 1964 g. (Distribution of ^{90}Sr and ^{137}Cs
Over Soil Profiles Under Natural Conditions in 1964). Moskva,
Atomizdat. 1967.

23. T i m o f e e v a, N. A. — DAN SSSR, Vol. 131:488. 1960.

24. T i m o f e e v - R e s o v s k i i, N. V. et al. — In Sbornik: "Radioaktivnost'
pochv i metody ee opredeleniya," p. 46. Moskva, "Nauka," 1966.

25. T y u r y u k a n o v a, E. B. — Vestnik MGU. Seriya Biologicheskaya,
No. 4:115. 1957.

26. T y u r y u k a n o v a, E. B. — Nauchnye Doklady Vysshei Shkoly. Biologi-
cheskie Nauki, No. 12:131. 1967.

27. T y u r y u k a n o v a, E. B. et al. — Pochvovedenie, No. 10:66. 1964.

28. T y u r y u k a n o v a, E. B. et al. — Pochvovedenie, No. 8:88. 1964.

29. T y u r y u k a n o v a, E. B. et al. — Informatsionnyi Byulleten' Radiobiologiya, No. 9:9. 1966.

30. T y u r y u k a n o v a, E. B. and F. I. P a v l o t s k a y a. — Agrokhimiya,
No. 9:50. 1967.

31. T y u r y u k a n o v a, E. B. O metodike issledovaniya povedeniya radioaktivnogo strontsiya v pochvakh razlichnykh geokhimicheskikh
landshaftov (Investigations into the Behavior of Radioactive
Strontium in Soils of Different Geochemical Landscapes). Moskva,
Atomizdat. 1968.

32. C h u l k o v, P. M. et al. — Pochvovedenie, No. 4:28. 1957.

33. Y u d i n t s e v a, E. V. — Agrokhimiya, No. 6:82. 1967.

34. A u e r b a c h, S. I. et al. Trends of Cs^{137} in the Liriodendron Forest
Vegetation, p. 58. ORNL-4007 (Oak Ridge National Laboratory).
1966.

35. A u e r b a c h, S. I. et al. Summary of Radiocesium Movement in a
Forest Landscape, p. 79. ORNL-4168 (Oak Ridge National
Laboratory). 1967.

36. A u e r b a c h, J. I. et al. Losses of Cs^{137} and Co^{60} from Leaf Litter,
p. 73. ORNL-4007 (Oak Ridge National Laboratory). 1966.

37. Auerbach, J. Movement of Cs^{137} in Pine Forest Floor Subsystems,
      p. 74. ORNL-4007 (Oak Ridge National Laboratory). 1966.
38. Franklin, R. et al. — Soil Sci. Soc. Amer. Proc., Vol. 31:39. 1967.
39. Hardy, E.P. Vertical Penetration of Sr^{90} in Sandy Soil from Alapaha,
      Georgia. 1959—1960. — Fallout Program Quarterly Summary
      Report, HASL-113. 1961.
40. Hardy, E.P. Vertical Penetration of Sr^{90} in Three Soils Sampled in
      March 1963. — Fallout Program Quarterly Summary Report,
      HASL-138, p. 249. 1963.
41. Knoop, E. and D. Schroeder. — Naturwissenschaften, Vol. 45:436.
      1958.
42. Kulp, J.S. Radionuclides in Man from Nuclear Tests. — In: Radio-
      active Fallout, Soils, Plant, Foods, Man, p. 247. Amsterdam —
      London — New York, Elsevier Publishing Company. 1965.
43. Kwaratskhelia, N. et al. The Influence of Some Natural Factors
      on the Behaviour of Radioactive Strontium in Soils. Radioecologi-
      cal Concentration Processes. — Proceedings of an International
      Symposium Held in Stockholm. April 1966. London, Pergamon
      Press, p. 19. 1967.
44. Menzel, G.M. — Science, Vol. 131:499. 1960.
45. Milbourn, G.M. and R. Taylor. Initial Retention and Subsequent
      Loss of Fission Products in Edible Tissues of Pastures. — Annual
      Report ARCRL, No. 10:43. 1962—1963.
46. Miller, R. and R.F. Reitmeier. — Soil Sci. Soc. Amer. Proc.,
      Vol. 27:141. 1963.
47. Mortensen, L. and E. Marcusin. — Soil Sci. Soc. Amer. Proc.,
      Vol. 27:653. 1963.
48. Reissig, H. — Kernenergie, Vol. 7:117. 1966.
49. Reissig, H. — Chem. Erde, Vol. 25:204. 1966.
50. Schilling, G. and D. Richter. Über das Verhalten von Sr^{90} und
      Y^{90} in mitteldeutschen Böden. — Albrecht-Ihaer-Arch., Vol. 8:85.
      1964.
51. Szepke, R. — Atompraxis, Vol. 11:391. 1965.
52. Tyuryukanova, E.B. et al. Peculiarities of Radiostrontium Dis-
      tribution in Some Landscapes of the Southern Taiga. Radioecolog-
      ical Concentration Processes. — Proceedings of an International
      Symposium Held in Stockholm, April 1966, p. 33. London,
      Pergamon Press. 1967.

Chapter 4

BEHAVIOR OF STRONTIUM-90 AND CESIUM-137 IN SOILS

Introduction

Of the large number of radioactive fission products fromed by the fission of heavy nuclei, an important role in the biotic cycle is played by ^{90}Sr and ^{137}Cs. The latter are long-lived and are respectively chemical analogs of calcium, which is an essential component of skeleton, and potassium, which is an essential component of intercellular metabolism /4, 5, 7/.

The physicochemical properties of ^{90}Sr and ^{137}Cs and the behavior of these radionuclides in the biosphere have been the subject of numerous investigations /1, 6, 9, 23, 36, 39, 43, 48, 49/. Nevertheless, many features of their behavior in soils are still rather obscure.

At present, the laws governing the migration of substances and ion exchange principles can yield, in many cases (but not always), both qualitative and quantitative results which, in addition to their theoretical significance, can be used to predict the intake of ^{90}Sr and ^{137}Cs from the environment by the human body /40, 50/. Moreover, the behavior of ^{90}Sr and ^{137}Cs in soils is of importance for devising protective measures, since any techniques for restoring contaminated soils to their original status must be based on data concerning the behavior of these isotopes in soils.

The behavior of ^{90}Sr and ^{137}Cs in soils raises several questions of general theoretical interest /4/, including the interaction of trace amounts of ^{90}Sr and ^{137}Cs and their nonisotope analogs (Ca, Mg, K, Na) with soil colloids (organic matter and anions of the crystal lattice of clay minerals), as well as exchange reactions in soil with the participation of these elements. There is no doubt that these reactions control the specific behavior of ^{90}Sr and ^{137}Cs and their nonisotope analogs in soil and this specifically effects the intake of the radionuclides by plants from the soil.

This chapter deals with the forms of ^{90}Sr and ^{137}Cs compounds in soils and the impact on the behavior of these nuclides of the water factor, absorption capacity, complex formers, diffusion and ion exchange. It is also concerned with energy characteristics of ion exchange reactions of the analog element, of importance when studying the phenomenon of ^{90}Sr and ^{137}Cs discrimination in different stages of the ecological cycle /4, 26, 44, 45/.

Forms of ^{90}Sr compounds

The radioactive fission product ^{90}Sr initially exists as strontium oxide /43/, which subsequently combines with atmospheric water vapor to form

strontium hydroxide and then with atmospheric carbon dioxide to form
strontium carbonate and bicarbonate.

After deposition on the soil surface, some of the ^{90}Sr atoms are captured
by anions of the crystal lattice of clay minerals, by exchange sorption and
other reactions. Exchange of ^{90}Sr is a most common and relatively better
studied reaction. Exchange strontium atoms are distributed over capillary
walls and also over the surface of colloid particles in soil aggregates /24,
25/. They are in an adsorbed state and are readily desorbed by neutral
salts /59/; their behavior in soil obeys the laws of ion-exchange equilibria
/14—16, 25/. Strontium-90 may also be present in the form of sparingly
soluble compounds that are not leached by water or solutions of neutral
salts. This form of strontium compounds has been less thoroughly studied
than the preceding one. Finally, a fraction of ^{90}Sr exists in ionic form which
is readily soluble in water and possesses the highest mobility and chemical
activity.

The relative amounts of water-soluble, exchange and nonexchange forms
of ^{90}Sr in soil depend on several factors and vary widely. For instance,
water-soluble and exchange ^{90}Sr fractions amounted to 80—90% on certain
soils (in cases of global fallout) /8, 33/. The forms of ^{90}Sr compounds
were studied by us in the latitudinal zone of Darvinskii Reservation with a
global soil contamination level (Table 4.1)

TABLE 4.1. Contents of different forms of ^{90}Sr compounds in sod-podzolic soils (September 1967), pcuries/kg

Horizons, cm	Water-soluble		Exchange		Nonexchange		Total
	absolute content	%	absolute content	%	absolute content	%	absolute content
0 — 2	27.0	3.3	520	63.1	277	33.6	824
2 — 4	11.9	0.5	1,200	61.2	750	38.3	1,961.9
4 — 6	0	0	185	80.0	46.5	20.0	231.5
6 — 8	0	0	123	92.0	10.8	8.0	133.8
8 — 10	0	0	58	81.2	13.4	18.8	71.4

From Table 4.1 it is seen that water soluble forms of ^{90}Sr compounds
were detected only in the uppermost soil horizons (0—4 cm), the amount of
water-soluble ^{90}Sr varying from 0.5 to 3.3%. The presence of exchangeable ^{90}Sr was detected in all the soil profile horizons. Characteristically,
the amount of exchange ^{90}Sr was considerably higher than that of other ^{90}Sr
forms, varying from 61 to 92%. Nonexchange ^{90}Sr was concentrated mainly
in the humus soil horizons, where it amounted to 34—38%. The amount of
nonexchange ^{90}Sr in the lower horizons varied from 8 to 19%. The methods
for determining different forms of ^{90}Sr compounds in soil are based on extraction, but identification of these forms involves certain difficulties: it is
impossible to compose an extractant solution that would extract only the
^{90}Sr atoms belonging to a single specified form of this nuclide /19/.

Significance of the water factor

The water factor plays an important role in the behavior of ^{90}Sr and ^{137}Cs in soils and also in the intake of these nuclides by the human body from the environment. Importance is assumed by the rate of migration of these nuclides in soils, when the latter are being flushed with distilled water or electrolyte solutions. This point was examined by means of the following experiments. Samples of three soil varieties (chernozem, gray forest and podzolic soils) with different initial ^{90}Sr contents were placed in columns and leached with 0.1 N CaCl$_2$. Experiments on leaching ^{90}Sr with distilled water were conducted in parallel.

The eluate was studied by experimental determinations of the volume V of every fraction, time t required for the passage of a given volume eluate through the column, and ^{90}Sr concentration Z .

Experimental data and chromatographic principles /21, 22/ were used to construct two series of curves $Z=f(V)$, displaying ^{90}Sr concentrations in the successive portions of eluate as a function of the volume of outflowing solution; and $VZ=f(V)$, displaying the variation in the total ^{90}Sr content in the eluate during its desorption, expressed as a percentage of the total amount of desorbed ^{90}Sr (Figure 4.1).

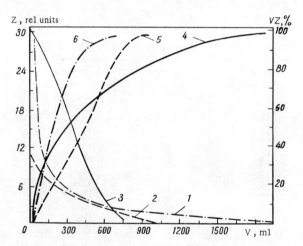

FIGURE 4.1. Curves $Z=f(V)$, displaying ^{90}Sr concentration as a function of the volume (V) of outflowing solution $(1-3)$ and total ^{90}Sr content in the eluate $(4-6)$ when leaching with 0.1 N CaCl$_2$:

1, 4 — loamy podzolic soil; 2, 5 — gray forest soil; 3, 6 — chernozem.

The curve $Z=f(V)$ is exponential for podzolic type soils, linear for light-textured chernozems, and near-linear for gray forest soils. Variation of ^{90}Sr concentration in the eluate is largely dependent on the soil properties: those with higher filtration coefficients and lighter texture give up their ^{90}Sr when treated with solutions of electrolytes much more readily than heavier soils with lower filtration coefficients. Desorption of ^{90}Sr proceeds at different rates in different soils, most rapidly on chernozem-type soils, less so on gray forest soils, and slowest on podzolic soils.

It is of interest to study the direct effect of water free from electrolytes and organic substances on ^{90}Sr migration over the soil profile. During 1954 — 1958 in the New York area, ^{90}Sr migration under the influence of atmospheric precipitation proceeded very slowly and at least about 95% of ^{90}Sr was concentrated in the surface horizon (0 — 5 cm) /54/. Mathematical models were proposed for forecasting further migration of ^{90}Sr over the soil profile /56, 57/; however, owing to the extreme variety of soils differing very significantly in their physicochemical properties, each particular case must be studied by laboratory simulation and field observations of ^{90}Sr migrations. Laboratory experiments with different soils possessing textures varying from sands to loams showed that after the soil samples had been washed with twice-distilled water (by sprinkling), the surface soil layer (0 — 5 cm) retained 62 — 92% of the originally introduced ^{90}Sr. In the presence of vanishingly small admixtures of electrolytes (0.0005 N NaCl) the same soils retained only 31 — 82% of the originally introduced ^{90}Sr /46/.

The phenomena under consideration were studied by means of laboratory and field investigations. Columns shaped as parallelepipeds, 35 cm tall and 20 cm^2 cross-section, were packed with soils of different textures. The height of the soil columns in every experiment was 34.5 cm. A layer of soil, 0.5 cm thick with known content of ^{90}Sr, was spread over the top surface of the soil column, covered with a perforated plate (to avoid washout) and washed with distilled water. After passing a quantity of water equivalent to 2,000 m^3/ha through the column, the latter was dismantled, samples were

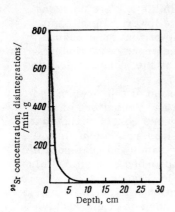

FIGURE 4.2. Strontium-90 distribution over the soil column profile after washing with distilled water

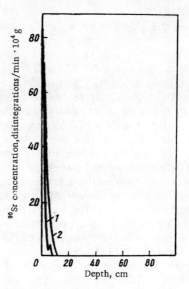

FIGURE 4.3. Strontium-90 distribution over the profile of chernozem soil at the start of experiments (1) and five years later (2)

taken at 5 mm intervals and analyzed for ^{90}Sr. Studies of the curve representing the distribution of ^{90}Sr over the soil profile revealed that the ^{90}Sr
leached by the washing of soil did not exceed 1% of the amount introduced
(Figure 4.2). The migration of ^{90}Sr may not have been due to leaching, but
to the mechanical effect of gravity water during washing of the soil.

This curve may be regarded as very typical of soil varieties with slightly
alkaline, slightly acid or neutral reaction and free from readily soluble
salts in their surface horizon. Observations showed that curves of this
shape are obtained for certain varieties of sod-podzolic, gray forest and
chernozem soils, provided their physicochemical parameters satisfy the
above conditions.

The effect of the water factor on the behavior of ^{90}Sr in soil was also
studied in the field. At the start of the experiments, ^{90}Sr was introduced
locally on the surface of chernozem soil and was fully confined to the top
soil layer. No significant changes in the nature of the ^{90}Sr distribution were
observed five years after the start of the experiments, although the total
atmospheric precipitation in the zone of observations totaled at least
2,000 mm over the elapsed period (Figure 4.3).

However, analogous experiments on sandy soils and sands containing
trace amounts of readily soluble salts revealed considerably more rapid
leaching processes, resulting in a gradual redistribution of ^{90}Sr over the
soil profile. Between these extremes, there also exist intermediate
regimes, in which the role played by the water factor in the leaching processes and behavior of ^{90}Sr in soils may vary quite widely.

Effect of the absorption capacity of soils

An important part is played by the absorption capacity of soils in the
migration processes of ^{90}Sr and ^{137}Cs in soil /2, 6/. For the same radioactive fallout density, ^{90}Sr and ^{137}Cs concentration in the surface horizons
of podzolic-type sandy soils has been reported to be lower than in loamy
podzolic soils, because sandy varieties possess a lower cation-exchange
capacity than loamy and clayey varieties /31/. The organic fraction has
been reported to sorb Ca stronger than ^{90}Sr, whereas the inorganic fraction
is distinguished by preferential sorption of ^{90}Sr /37/.

It was therefore of interest to study the distribution of ^{90}Sr over the soil
profile in relation to the amount of sorbed calcium and the pattern of its
distribution among the genetic soil horizons, taking into consideration that
this element is one of the major components of the absorption complex.
For this purpose we selected weakly podzolic silty-sandy and peat-podzolic
soils, differing not only in the absolute content of absorbed calcium but also
in the pattern of its distribution over the soil profile (Figure 4.4).

Figure 4.5 shows the variation of ^{90}Sr concentration with depth in the
same soils. The ^{90}Sr concentration in the surface horizon of the weakly
podzolic silty-sandy soil proved to be approximately one-third of its concentration in the peat-podzolic variety. The weakly podzolic silty-sandy
soil possesses a significantly lower cation-exchange capacity than the peat-

podzolic soil, and therefore it may be concluded that a decrease in the cation-exchange capacity of soils of podzolic type is accompanied by a decrease of ^{90}Sr concentration in the surface horizons.

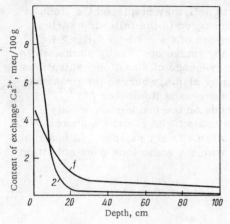

FIGURE 4.4. Distribution of absorbed Ca^{2+} over the profile of a weakly podzolic silty-sandy soil (1) and a peat-podzolic soil (2)

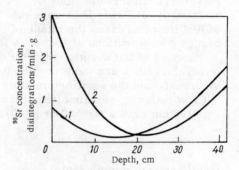

FIGURE 4.5. Strontium-90 distribution over the profile of a weakly podzolic silty-sandy soil (1) and a peat-podzolic soil (2) (Darvinskii Reservation)

This conclusion is further supported by the absence of any reason to believe that other physical factors, such as soil consistence or permeability, are responsible for this effect. On the contrary, the effect of these factors should enhance ^{90}Sr retention in the surface horizons of silty-podzolic soils, possessing a lower permeability and denser consistence than the peat-podzolic soil.

The behavior of ^{90}Sr is more complex in deeper horizons (Figure 4.5) on account of the mineralogical composition and permeability of underlying rocks as well as the presence of geochemical barriers in the form of illuvial horizons. These horizons usually occur at depths of $40-60$ cm in the group of soils under consideration. Since these horizons proved to be practically impermeable to ^{90}Sr ions, it is also possible to explain the absence of ^{90}Sr from groundwaters. Illuvial horizons also play the role of barriers in the presence of foci of high ^{90}Sr concentrations on the surface of the soils in question.

Effect of various cations

The behavior of ^{90}Sr and ^{137}Cs in soils is significantly affected by various chemical substances. The presence in soils of the cations NH_4^+, K^+, Ca^{2+}, H^+ and complexers in soils enhances the mobility of ^{90}Sr and ^{137}Cs in soil, while the presence of SO_4^{2-} and PO_4^{3-} lowers their mobility (owing to the formation of sparingly soluble sulfates and phosphates) /29/.

Studies of ^{90}Sr and ^{137}Cs, as well as Ca, in soils should take account of the degree of hydration of these ions. Sr^{2+} ions are known to be less hydrated than Ca^{2+} ions, and are therefore more strongly sorbed by clay minerals. Different degrees of hydration may also affect the rate of intake of these ions by plants.

When NH_4^+, K^+ and Rb^+ were added to soil, they displaced Cs^+ ions at different rates. These cations can be arranged in the following series with respect to their effect on the desorption of Cs^+: $Rb^+ > NH_4^+ > K^+$. However, addition of K^+ ions decreases the intake of ^{137}Cs by plants, while addition of NH_4^+ ions increases it. The presence of small concentrations of Rb^+ ions increases the intake of ^{137}Cs by plants, whereas the presence of high concentrations of the same ions decreases the intake /47/.

Kinetics of the sorption of ^{137}Cs depends on the properties of clay minerals. Illite and montmorillonite saturated with K and Ca absorb ^{137}Cs vigorously and the equilibrium state is attained very rapidly. In contrast, ^{137}Cs sorption by vermiculite saturated with the same ions does not attain equilibrium even after 500 hrs /30/.

Significance of complexants

Complexants are of interest in studies of ^{90}Sr and ^{137}Cs in soil, including their vertical migration /53/. The addition of a complexant is known to raise ^{90}Sr concentration in the liquid phase, the total quantity of free ions in the liquid phase being equal to the sum total of the two ingredients $Sr^{2+} + Sr\gamma^-$, where γ^- denotes the complexant.

Polyakov /12/ has shown that the coefficient of distribution of ^{90}Sr between the solid and liquid phases in the presence of a complexant K'_d can be determined from the equation

$$K'_d = \frac{K_d}{K(\gamma^{----})}, \qquad (4.1)$$

where K_d is the coefficient of distribution of ^{90}Sr in the absence of complexants, K is the constant of exchange between free Sr^{2+} ions and the complexant (association constant), and γ^{----} is the complexant concentration.

The association constant for rare-earth elements varies from 10^{14} to 10^{19}, depending on the atomic number /58/. For elements of the second group, this quantity is $6-8$ orders of magnitude smaller than the indicated values.

The association constant of strontium and EDTA is $K = 6.3 \cdot 10^8$ l/mole /12/. On the other hand, for Sr^{2+} in sandy-loamy soils $K_d = 0.1 - 0.6$ l/g /32/. Taking the mean value $K_d = 0.3$ l/g, we find that for a complexant concentration equal to 10^{-4} mole/g equation (4.1) yields

$$K'_d = \frac{0.3 \, \text{l/g}}{6.3 \cdot 10^8 \, \text{l/mole} \cdot 10^{-4} \, \text{mole/l}} = 4.7 \cdot 10^{-6} \, \text{l/g}.$$

Even for relatively low complexant concentrations the product $K(\gamma^{----})$ is a fairly large quantity. It follows that the presence of a complexant diminishes the coefficient of distribution of ^{90}Sr between the solid and liquid phases. This effect becomes stronger for higher complexant concentrations and larger association constants.

Equation (4.1) limits the possible use of complexants for deactivation of soils. Since K_d in (4.1) is a constant, the distribution coefficient K_d' may be regarded as a function of variables γ^{----} and K. Clearly $K_d' = 0$ only when $K \cdot \gamma^{----}$ is infinite. However, this condition is practically unrealizable, and therefore it is impossible to imagine a system in which $K_d' = 0$. Therefore, ^{90}Sr cannot be completely removed from soils with complexants alone, as reported in /53/, although their use may considerably lower the ^{90}Sr concentration in soil.

In deriving equation (4.1), the soil in its initial state was assumed to be saturated by only one cation. Under natural conditions one deals with much more complex systems characterized, among other things, by their absorption centers being mainly saturated by calcium ions.

Consider the case of two cations, Sr^{2+} and Ca^{2+}, in the solid phase of soil, instead of a single cation. Suppose their constants of association with EDTA are K_{Sr} and K_{Ca}, respectively. Then for the first system, with Sr^{2+} as one of its ingredients,

$$K_d' = \frac{K_{dSr}}{(\gamma^{----}) K_{Sr}},\qquad (4.2)$$

where K_d' and K_{dSr} are the respective distribution coefficients of Sr^{2+} in the presence and absence of a complexant, while K_{Sr} is the association constant of Sr^{2+} and EDTA.

For the second system, with Ca^{2+},

$$K_d'' = \frac{K_{dCa}}{(\gamma^{----}) K_{Ca}},\qquad (4.3)$$

where K_d'', K_{dCa}, K_{Ca} have the same physical meaning as in equation (4.2).

Eventually K_d''', the coefficient of distribution of Sr^{2+} and Ca^{2+} between the solid and liquid phases in the presence of a complexant, will be numerically equal to the ratio K_d'/K_d''. Here

$$K_d''' = \frac{K_d'}{K_d''} = \frac{\dfrac{K_{dSr}}{(\gamma^{----}) K_{Sr}}}{\dfrac{K_{dCa}}{(\gamma^{----}) K_{Ca}}} = \frac{K_{dSr} \cdot K_{Ca}}{K_{dCa} \cdot K_{Sr}}.\qquad (4.4)$$

These data suggest that the presence of natural complexants in soil, as well as the addition of artificial complexants, such as EDTA and citrates, have a considerable effect on the behavior of ^{90}Sr in soil. Experiments in columns showed that in sandy soils (or soils mixed with sand) the presence of complexants, especially citrates, accelerates the leaching of ^{90}Sr from soils. On the other hand, the effect is scarcely observed on chernozem-type soils containing a large content of organic matter and possessing a heavy texture.

Strontium forms more stable compounds with complexants than does calcium. In light-textured soils and in the presence of complexants, strontium exhibits a more marked tendency toward transition from the solid to the liquid phases than calcium. In this case addition of complexants becomes a very efficient technique for leaching trace amounts of ^{90}Sr from soils.

Significance of **diffusion phenomena**. Diffusion characteristics have recently been employed frequently when studying ^{90}Sr and ^{137}Cs in soils. Diffusion of ^{90}Sr and ^{137}Sc in soils and its role in the behavior of these nuclides in soil has become the subject of an extensive literature, describing all aspects of the diffusion problem. Without dwelling on the theory of diffusion and the mechanism of diffusion phenomena, and also various techniques for determining diffusion characteristics and various devices used in such determinations, we only note that the processes of ^{90}Sr and ^{137}Cs diffusion in soil can be represented as particular cases of a more general law of nature according to which all natural systems tend to become homogeneous systems /28/. This definition is satisfied by any system in which one substance diffuses into another.

Application of the Fick law to studies of the diffusion of ^{90}Sr and ^{137}Cs ions in soils

In the absence of groundwater motion in soils and underlying rocks, the movement of ions is mainly due to diffusion. This migration of ions can be described by means of the Fick equation

$$\frac{\partial c}{\partial t} = D \frac{\partial^2 c}{\partial x^2} , \tag{4.5}$$

where D is the diffusion coefficient, c is the concentration of the diffusing substance, x is the distance measured along a normal to the interface, and t is the diffusion time. Qualitatively, the diffusion coefficient can be regarded as the effect of soil resistance to the movement of ^{90}Sr and ^{137}Cs ions. On the whole, if the resistance increases, the velocity of ions decreases, with a corresponding decrease in the diffusion coefficient. However, if the resistance decreases, the velocity of ions increases, with a corresponding increase in the diffusion coefficient.

The Fick equation theoretically allows one to find the diffusion coefficient in any system, given the variation in the concentration of the diffusing substance as a function of time and space. However, application of the second-order Fick equation involves certain difficulties and therefore in practice the diffusion coefficient is determined by means of linear equations, derived from the Fick equation by means of certain transformations.

The Fick equation can be transformed in a variety of ways, depending on the purpose of the investigation and the experimental conditions. We /11/ and others /38/ used the equation

$$c = At^{-\frac{1}{2}} e^{-\frac{x^2}{4Dt}} , \tag{4.6}$$

where A is a constant, while c, D, t, x have already been defined. Note that equation (4.6) does not involve any simplifying assumptions.

Equation (4.6) can be converted to linear form /11/ and D calculated by the following formula:

$$D = -\frac{0.1086}{\tan \alpha t},$$ (4.7)

where α is the slope of the straight line $\log c = f(x^2)$ with respect to the x-axis.

Approximate values of D can be determined more simply by making use of numerical values of c_{x_1} and c_{x_2} determined at two arbitrary points x_1 and x_2:

$$D = \frac{b \, (cm^2)}{\ln \varepsilon \, 4t \, (sec)},$$ (4.8)

where

$$b = x_2^2 - x_1^2; \quad \varepsilon = \frac{c_{x_1}}{c_{x_2}}.$$

In computing diffusion characteristics use was also made of the equation

$$D = \frac{x^2 \, (cm^2)}{4ut \, (sec)},$$ (4.9)

where u is a dimensionless quantity (the meanings of the other symbols are as above).

From this equation D can be readily calculated provided parameters t, x, u are known. The values of the first two parameters are indicated above; they were determined experimentally. The value of the first parameter and also the methods of its determination were described elsewhere /11/.

Field determination of diffusion coefficients

Diffusion can be observed in the field under a great variety of boundary conditions. However, in practice the most important case is that in which the diffusing substance is located at a known depth from the surface in the initial period of time. The following method was used for these boundary conditions.

Soil sampling pits, 1 m deep and 1 m^2 cross-section, were dug on selected plots. Every genetic horizon was removed separately, placed on a tarpaulin and thoroughly mixed. When the pit was dug to its full depth, its bottom was marked with a thin layer of sand (or another readily distinguishable material), and then the soil layers previously extracted from the pit were returned in the same order as that occupied by them under natural conditions. The pit was thus filled to a total depth of 50—55 cm. Then the open surface of soil in the pit was covered with a thin layer of comminuted soil containing the desired amount of ^{90}Sr or ^{137}Cs. It must be emphasized that the inter-layer of diffusing substances must be strictly horizontal and the initial con-centration of the substance must be as uniform as possible over its hori-zontal plane. The pit was then filled with the remaining soil layers in their natural sequence.

Six years later, soil samples to be tested for ^{90}Sr contents were taken with an auger. It was found that $x = 12$ cm and $u = 3.06$.

The diffusion coefficient D in the investigated soil was determined by the method of curve intersections (see equation (4.9)) and found to be

$$D = \frac{x^2 (\text{cm}^2)}{4ut\,(\text{sec})} = \frac{(12)^2\,\text{cm}^2}{4 \cdot 3.06 \cdot 1.9 \cdot 10^8\,\text{sec}} = 0.6 \cdot 10^{-7}\ \text{cm}^2/\text{sec}.$$

The numerical values of D for other soil varieties were calculated in a similar manner.

The diffusion coefficient varies from $0.6 - 8.1 \cdot 10^{-7}\ \text{cm}^2/\text{sec}$ in cher-nozems to $14 - 20 \cdot 10^{-7}\ \text{cm}^2/\text{sec}$ in alluvial soil.

Laboratory determination of the diffusion coefficient of ^{137}Cs

The diffusion coefficient of ^{137}Cs was determined in several varieties of podzolic soils. In accordance with the adopted method the samples were comminuted, sifted through a sieve with 1 mm meshes, moistened uniformly and placed in two parallelepipeds. A membrane with activity 10^7 disin-tegrations/min $\cdot$ g was placed at the interface of these parallelepipeds, the two parallelepipeds were joined into one, and the joint was paraffined for insulation from external influences. The combined parallelepiped was placed in a thermostat at 25°C for the entire observation period. After time $t = 1.04 \cdot 10^7$ sec one wall of the parallelepiped was removed and several layers cut away along a normal to the x-axis at different distances from the interface.

Quantity D was determined by equation (4.8) using the numerical values of c_{x_1} and c_{x_2} determined for the given pair of layers. For two layers with parameters $c_{x_1} = 3.37 \cdot 10^6$ and $x_1 = 0.45$; $c_{x_2} = 1.23 \cdot 10^6$ and $x_2 = 1.0$,

$$D = \frac{b\,(\text{cm}^2)}{\ln e 4t\,(\text{sec})} = \frac{1 - 0.45^2\,(\text{cm}^2)}{\ln \dfrac{3.37 \cdot 10^6}{1.23 \cdot 10^6} \cdot 4 \cdot 1.04 \cdot 10^7\,(\text{sec})} = 1.9 \cdot 10^{-8}\ \text{cm}^2/\text{sec}.$$

For different layers the value of D varied (owing to the nonuniformity of the diffusion medium) from $0.45 \cdot 10^{-8}$ to $1.9 \cdot 10^{-8}\ \text{cm}^2/\text{sec}$.

Energy of the bond of ^{90}Sr and ^{137}Cs with soil colloids

Thermodynamic investigation methods are of considerable interest for a quantitative estimation of the energy of the bond of ^{90}Sr and ^{137}Cs with

soils. Thermodynamically, the sorption of, for instance, ^{90}Sr and ^{137}Cs ions
by soils and migration of these ions from the solid into the liquid phase
can be described uniquely by the change in free energy ΔF^0. The value of ΔF^0
shows the amount of energy that must be expended for desorption of 1 mole
of a given element from the solid soil phase into the soil solution. This
parameter can thus be used also to estimate the rate of intake of elements
from soil solution by plants, since the existing relationship between the
nutrition of plants and exchange energy is of a universal nature and can be
applied to all soils. The thermodynamic characteristics are also important
when forecasting the behavior of radionuclides in soil, and therefore their
determination, in addition to its theoretical interest, also possesses considerable practical importance.

Determination of thermodynamic characteristics

Thermodynamic characteristics are determined by a variety of methods
which can be grouped, somewhat arbitrarily, into three groups. The first
group includes methods based on a determination of the numerical value
of the exchange constant and its temperature dependence /16, 51/. The
second group comprises calorimetric methods based on measuring the
thermal effect of reaction (determination of enthalpy ΔH^0), while the third
group consists of techniques utilizing the electrochemical properties of
mineral membranes for calculations of ΔF^0 and other thermodynamic
functions /42, 55/. The actual choice of method depends on the aim of the
investigation and also on the properties of the systems for which these
characteristics are being determined. Nevertheless, it should be noted
that direct determinations of the heat content of a system ΔH^0 do not produce sufficiently reliable results /42, 52/, because of the usually small
size of samples, containing mg-equivalents of exchange ions, used in studies
of exchange reactions. If quantity ΔH^0 is referred to 1 mg-eq instead of
1 mole, then the change of heat content in the systems under investigation
must be calculated from very small thermal effects that are commensurable
with the almost inevitable experimental errors. Therefore more reliable
results are obtained with techniques belonging to the first group, making
use of numerical values of the exchange constant and its temperature dependence when determining thermodynamic characteristics.

Theory of the method. Techniques to be described in detail further on
were used to determine experimentally the equilibrium constant at different
temperatures. The change in the heat content of the system ΔH^0 was determined by the Van't Hoff equation

$$\frac{d \ln K}{dT} = \frac{\Delta H^0}{RT^2}.$$

(4.10)

In solving this equation one must start from the relationship between the
equilibrium constant and temperature. The following two cases are possible:
1) the equilibrium constant is independent of temperature:

$$\frac{d \ln K}{dT} = 0;$$ (4.11)

2) the equilibrium constant varies with temperature, in which case equation (4.10), after integration and suitable transformations, can be expressed in the form

$$\log \frac{K_2}{K_1} = -\frac{\Delta H^0}{4.57}\left(\frac{1}{T_2}-\frac{1}{T_1}\right).$$ (4.12)

The change in enthalpy in the first case is clearly zero, as is revealed by comparing equations (4.10) and (4.11):

$$\Delta H^0 = 0.$$ (4.13)

The change of heat content in the second case is

$$\Delta H^0 = -4.57\left[\log\frac{\left(\frac{K_2}{K_1}\right)}{\left(\frac{1}{T_2}-\frac{1}{T_1}\right)}\right].$$ (4.14)

The numerical values of other thermodynamic characteristics can be determined in accordance with the general thermodynamic principles with the aid of the following equations:

$$\Delta F^0 = -RT \ln K;$$ (4.15)
$$T\Delta S^0 = \Delta H^0 - \Delta F^0.$$ (4.16)

Determination of the thermodynamic characteristics ΔH^0, ΔF^0 and ΔS^0 for exchange and absorption of Cs, K, Sr and Ca in soils

During the migration of Sr and Cs from soils into plants a significant role is played by the energy of their displacement from the solid into the liquid phase. However, available experiment data are as yet limited. The displacement energy of Sr and Cs is equivalent to the change in the free energy ΔF^0, which in turn is related to the change in enthalpy ΔH^0 and entropy ΔS^0 /34, 50/. The change in ΔF^0, ΔH^0 and ΔS^0 in soil systems was studied for several reactions /13, 18/, The change in the free energy involved in the formation of salts of metal cations on synthetic resins was also studied /27/, and the thermodynamic functions of ion-exchange reactions on clays were computed /41/. Experimental data were used to calculate the equilibrium criteria and the values of ΔH^0, ΔS^0 and ΔF^0 for exchange sorption of several pairs of ions on soils and resins: Cs^+, Na^+; Cs^+, K^+; Sr^{2+}, Ca^{2+}; Sr^{2+}, Cs^+. The first three pairs of ions can be regarded as three systems of analog elements, especially interesting in that one of them is present in soils in trace amounts, and the other in macro amounts. The specific features of the interaction of analog elements with each other and with other elements present in the solid phase of soils and in soil solution are of practical interest because studies of these aspects may lead to measures for controlling the migration of ^{90}Sr and ^{137}Cs from soils into food chains /4/.

Exchange sorption of Cs^+ and Na^+ ions in soil

The equilibrium criteria for exchange of Cs^+, Na^+ ions were determined by several methods.

Dynamic (Nikol'skii's) method.[*] Soil samples were saturated with hydrogen ions (by treatment with 0.025 N HCl) dried on a water bath, triturated and sifted through a sieve with 1 mm meshes and washed with buffer solutions of sodium and cesium chlorides taken in the ratios 90:10, 70:30, 50:50, 30:70 and 10:90, at constant ionic solution strength $\mu = 0.1$. Sample washing was terminated when the pH of the solution emerging from the funnel became equal to its value in the original solution (pH = 6.5).

Sorbed Cs^+ and Na^+ ions were displaced with 1 N CH_3COONH_4 (500 ml). Equilibrium amounts of cesium and sodium ions proved in this experiment to be respectively $X_{Cs} = 15.16$ meq/100 g, $X_{Na} = 5.59$ meq/100 g in the solid phase, $c_{Cs} = 50$ meq/ml, $c_{Na} = 50$ meq/ml in the liquid phase.

By means of Nikol'skii's equation, it is found that

$$K_{Cs, Na} = \frac{X_{Cs}\, c_{Na}}{X_{Na}\, c_{Cs}} = 2.7.$$

Static method[**]. The soil sample was saturated with sodium ions (according to Gedroits), brought to air-dry condition, comminuted, sifted through a sieve ($d = 0.25$ mm), then treated with 0.1 N CsCl + NaCl.

The pre-experimentally determined quantities were the equilibrium concentrations of sodium ions c_{Na}, cesium ions c_{Cs}, and absorption capacity Q. The other quantities required in the determination of the exchange constant were found by calculation. The results were as follows: $X_{Cs} = 0.3$ meq/g; $X_{Na} = 0.02$ meq/g; $c_{Cs} = 2.2$ meq/ml; $c_{Na} = 0.3$ meq/ml.

Proceeding as in the foregoing case, it is found that

$$K_{Cs, Na} = \frac{0.3 \cdot 0.3}{0.02 \cdot 2.2} \approx 2.0.$$

Larger values of the constant were found in other experimental variants (static conditions), but none exceeded $K_{Cs, Na} = 2.5$.

Comparison of the numerical values of the constant determined under dynamic and static conditions reveals that the difference between them does not exceed 26%. Since the constants were determined by a variety of techniques, these differences are quite normal and readily explicable.

[*] The experimental work was conducted by Ryzhova.
[**] The experiments were performed by Rutkovskii.

Determination of ΔH^0. Experiments /3, 20/ have shown that the exchange of ions of the same valence in soils is practically independent of temperature (at least within $20-70°C$). Consequently

$$\Delta H^0 = 0, \qquad (4.17)$$

i. e., the exchange reaction of Na^+, Cs^+ ions in soils proceeds without any appreciable thermal effect.

Determination of ΔF^0. Determination of the change in free energy is of the most significance, since its numerical value characterizes not only the trend of exchange sorption of Cs^+ and Na^+ ions under natural conditions, but also the thermodynamic stability of the compounds of these elements with soil. The change in free energy was determined by equation (4.15), assuming $K_{Cs, Na} = 2.5$ as the most probable value of the constant. Under these conditions, $\Delta F^0 = -1,359 \cdot \log 2.5 = -540$ cal/mole.

Determination of ΔS^0. This was conducted by equation (4.16), assuming $\Delta H^0 = 0$:

$$\Delta S^0 = \frac{540}{298} = 1.8 \text{ cal/deg.}$$

The fairly similar values of $K_{Cs, Na}$ obtained by the two independent methods testify that the exchange sorption of Cs, Na in the given soil can be regarded as a process that is very close to reversible. The value of ΔF^0 shows that this process proceeds preferentially toward the formation of X_{Cs}. Consequently, the compound Cs-soil is thermodynamically more stable than the compound Na-soil. The change in entropy is very small, and therefore it seems quite probable that the exchange of Cs^+ and Na^+ ions in the investigated soil is not complicated by irreversible absorption of Cs atoms.

Exchange of Cs^+ and K^+ in soils

The equilibrium constant of this system can be calculated from data yielded by the determination of the equilibrium criteria of intermediate reactions. The exchange constant $K_{NH_4, K} = 0.58$ /10/, and $K_{NH_4, Cs} = 0.48$ /12/. Quantity $K_{K, Cs}$ was determined in the following manner. The exchange constant $K_{NH_4, K}$ is given by

$$K_{NH_4, K} = \frac{NH_4 - soil \cdot K^+}{K - soil \cdot NH_4^+}.$$

This formula corresponds to the exchange reaction between K^+ and NH_4^+ ions, the following stoichiometric equation being adopted for its description:

$$K - soil + NH_4^+ \rightleftarrows NH_4 - soil + K^+.$$

Similarly

$$K_{NH_4, Cs} = \frac{NH_4 - soil \cdot Cs+}{Cs - soil \cdot NH_4^+}.$$

Division of $K_{NH_4, Cs}$ by $K_{NH_4, K}$ yields

$$\frac{K_{NH_4, Cs}}{K_{NH_4, K}} = \frac{\dfrac{NH_4 - soil \cdot Cs+}{Cs - soil \cdot NH_4^+}}{\dfrac{NH_4 - soil \cdot K+}{K - soil \cdot NH_4^+}} = \frac{K - soil \cdot Cs+}{Cs - soil \cdot K+}.$$

The last term in this equation represents quantity $K_{K, Cs}$. Consequently

$$K_{K, Cs} = \frac{K_{NH_4, Cs}}{K_{NH_4, K}} = \frac{0.48}{0.58} = 0.83.$$

Hence

$$\Delta F^0 = -1359 \log 0.83 \approx 110 \text{ cal/mole};$$
$$\Delta S^0 = -\frac{\Delta F^0}{T} = -\frac{110}{298} = -0.37 \text{ cal/deg.}$$

The small numerical values of ΔF^0 and ΔS^0 (and their signs) show that the exchange reaction of Cs^+ and K^+ ions in soils is very dynamic. It can readily be reversed with small changes in the concentrations of Cs and K. Nevertheless, the bond between K atoms and soil colloids is less strong than the bond of Cs atoms, and therefore exchange sorption of K^+, Cs^+ in soils proceeds preferentially toward the formation of the compound Cs-soil. The thermodynamic characteristics cannot always be determined with the degree of accuracy with which they are determined for other systems /29, 41/, owing to the well-known anomalous behavior of Cs^+ ions in the course of their absorption and exchange in soils and certain differences in ion radii (Cs = 1.69 Å; K = 0.92 Å) and the degree of their hydration. However, the cited data were obtained from a fairly large amount of experimental data and can be regarded as fully reliable.

Exchange of Sr and Ca ions in soils and clays

The exchange reaction of Sr and Ca ions on soils has been treated in the literature. The equilibrium constant of this reaction depends on the properties of soils, some soils sorbing Ca in preference to Sr, others displaying a reversed preference. Thus, in the determination of equilibrium constant $K_{Sr, Ca}$ on Indian soils this quantity varied from 0.6 to 1.5, depending on the properties of soils /30/. Similar phenomena were also observed by others, according to whom the constant varied from 1.0 to 1.4 /35/, from 1.1 to 1.2 /30/ and from 0.36 to 3.3 /30/. The inconstancy of numerical values of the equilibrium constant is explained by different affinities of Sr and Ca ions toward the mineral and organic soil fractions, considerable variation in the absorption capacity and composition of absorbed bases, different mineralogical compositions, as well as the possibility of the formation of insoluble compounds with soil solution anions. It was pointed out that the presence of organic matter reduces the equilibrium constant as a result of the stronger retention of Ca than Sr atoms by the organic soil fraction. The value of $K_{Sr. Ca}$ has also been reported to increase as the exchange capacity of soils decreases /30/.

Turning to the experimental data, attention is drawn to the fact that the determination of the equilibrium constant in the experiments described below was performed when the solid phase initially contained only Ca atoms (i. e., the soils underwent preliminary saturation with Ca) while their liquid phase contained Sr atoms or a mixture of Ca +Sr atoms. In the latter case the reaction of calcium substitution by strontium proceeded in the presence of free calcium atoms. Under these conditions the exchange reaction of Ca and Sr ions on soils satisfies the following stoichiometric equation:

$$Ca - soil + Sr^{2+} \rightleftarrows Sr - soil + Ca^{2+}, \qquad (4.18)$$

where Ca-soil and Sr-soil respectively signify the exchange amounts of calcium and strontium in the solid phase, while Ca^{2+}, Sr^{2+} signify equilibrium concentration of these ions in the liquid phase (meq).

The equilibrium constant $K_{Sr. Ca}$ for the reaction under consideration can be determined from Nikol'skii's equation

$$K_{Sr. Ca} = \frac{(Sr)\,Ca^{2+}}{(Ca)\,Sr^{2+}}. \qquad (4.19)$$

Direct determination of the equilibrium constant $K_{Sr. Ca}$ from equation (4.19) involves certain difficulties on account of the chemical similarity of strontium and calcium and the presence of stable strontium in soils /32/. Therefore the value of $K_{Sr. Ca}$ was determined by calculations from the constant K' and K'', respectively, corresponding to the following two intermediate reactions:

$$NH_4 - soil + \frac{1}{2} Ca^{2+} \rightleftarrows \frac{1}{2} Ca - soil + NH_4^+; \qquad (4.20)$$

$$NH_4 - soil + \frac{1}{2} Sr^{2+} \rightleftarrows \frac{1}{2} Sr - soil + NH_4^+. \qquad (4.21)$$

The equilibrium constant of the first reaction is

$$K' = \frac{\sqrt{(Ca)} \cdot NH_4^+}{(NH_4) \cdot \sqrt{Ca^{2+}}}, \qquad (4.22)$$

determined as 5.8 /17/.

The equilibrium constant K'' of the second reaction was determined by two independent methods, by means of ^{89}Sr and in the usual manner, from analytical data on the contents of Sr^{2+} and NH_4^+ ions in the liquid and solid phases. The most probable value, derived from several determinations of the quantity by the two independent methods, is

$$K'' = \frac{\sqrt{(Sr)} \cdot NH_4^+}{(NH_4) \cdot \sqrt{Sr^{2+}}} = 6.5. \tag{4.23}$$

Hence

$$K_{Sr, Ca} = \left(\frac{K''}{K'}\right)^2 \approx 1.2.$$

In determining ΔH^0 the equilibrium constant was assumed to be independent of temperature in the exchange of ions of identical valence. Therefore, in the case under consideration, $\Delta H^0 = 0$, i.e., the exchange reaction of Sr^{2+} and Ca^{2+} ions on soils proceeds without any appreciable thermal effect.

In accordance with equation (4.15) the change in free energy (for $K_{Sr, Ca} = 1.2$, $t = 25°C$) proved to be -107.5 cal/mole.

Quantity ΔS^0 was determined from equation (4.16) assuming $\Delta H^0 = 0$:

$$\Delta S^0 = \frac{107.5}{298} = 0.3 \text{ cal/deg}.$$

Thus, the entropy of Sr-soil compound is very small and consequently the exchange of calcium and strontium ions in soils can be regarded as close to isoenergetic. However, the positive sign of ΔS^6 indicates that the substitution of calcium by strontium ions in soils proceeds with an increase in entropy. Therefore the adsorption compound Sr-soil is thermodynamically more stable than the compound Ca-soil, and under natural conditions the process will preferentially proceed toward the formation of the compound Sr-soil.

Exchange of Ca and Sr ions in soils under natural conditions

The equilibrium constant of the exchange of Sr, Ca cations at ordinary temperatures is constant and its most probable value is $K_{Sr, Ca} = 1.2$.

However, this value was determined under laboratory conditions on systems with more or less stable values of the ionic strength of solution and with only Ca atoms present in the solid phase at the initial moment. It is not yet sufficiently clear whether these data can aid in describing the phenomena of Ca, Sr exchange under natural conditions. Below, results are discussed of determining equilibrium criteria for Ca^{2+}, Sr^{2+} ion exchange on soil samples for a ratio of equilibrium concentrations and exchange masses of these ions that became established as the result of multisecular soil-formation processes. The experiments were conducted on a chernozem soil, the physicochemical properties of which are listed in Table 4.2.

The soil sample was placed in a flask with addition of 1 N CH_3COONH_4 (solid:liquid = 1:10), shaken for 2 hrs, transferred to a filter and filtered. The filtrate was analyzed for Sr and Ca (Table 4.3)

TABLE 4.2. Physicochemical properties of soil

Horizon	Depth, cm	Humus according to Tyurin, %	pH	Absorption capacity, meq/100 g	Absorbed bases, meq/100 g		
					Ca^{2+}	Mg^{2+}	K^+
A_1	0 − 10	7.8	6.5	44.9	36.4	7.0	0.7
A_1	10 − 20	7.4	6.6	44.6	36.5	7.0	0.6
$A_1 B_1$	20 − 30	4.0	6.6	37.4	29.2	6.5	0.9
B_1	35 − 60	1.8	6.6	22.4	17.1	4.2	1.1
B_2	60 − 90	0.6	6.8	20.6	14.0	4.8	1.2

TABLE 4.3. Contents of exchangeable and water-soluble Ca and Sr and numerical values of the equilibrium constant

Depth, cm	Composition of the liquid phase, meq/100 g		Composition of the solid phase, meq/100 g		$K_{Ca, Sr}$
	Ca^{2+}	$Sr^{2+}, 10^{-4}$	Ca	Sr	
0 − 10	0.342	6.5	22.1	0.048	0.88
10 − 20	0.237	3.6	30.0	0.057	0.80
20 − 30	0.198	2.7	25.0	0.043	0.79
30 − 40	0.307	5.7	23.5	0.046	0.95
40 − 50	0.274	5.7	25.0	0.050	1.04
50 − 60	0.229	3.8	22.0	0.046	0.79
60 − 70	0.283	3.6	22.7	0.032	0.90
70 − 80	0.593	5.3	43.2	0.055	0.70
80 − 90	0.860	8.7	30.0	0.100	0.91
90 − 100	0.595	7.0	180.0	0.252	0.84
100 − 110	0.623	6.1	210.0	0.252	0.82
110 − 130	0.500	5.1	170.0	0.228	0.76
130 − 150	0.500	5.1	160.0	0.228	0.71
Mean	−	−	−	−	0.84

During determination of the equilibrium constant it was assumed that the exchange reaction of Ca^{2+} and Sr^{2+} ions in natural systems satisfied the following stoichiometric equation:

$$Sr - soil + Ca^{2+} \rightleftarrows Ca - soil + Sr^{2+}, \qquad (4.24)$$

where $Sr - soil$, $Ca - soil$, Sr^{2+}, Ca^{2+} are explained above.

Proceeding as in the foregoing case it is found that the equilibrium constant of this reaction can be calculated from the equation

$$K_{Ca, Sr} = \frac{(Ca) \cdot Sr^{2+}}{(Sr) \cdot Ca^{2+}} . \qquad (4.25)$$

Comparison of equations (4.18) and (4.24), and also (4.19) and (4.25), shows that the exchange of Sr and Ca atoms corresponding to equation (4.18) is a direct reaction of the formation of the absorption compound Sr-soil, while the exchange of the same ions corresponding to equation (4.24) is the reverse reaction. Under these conditions the product of the constants should be unity: $K_{Sr,Ca} \cdot K_{Ca,Sr} = 1$. It follows that if $K_{Sr,Ca} = 1.2$ (as already established), then its reciprocal $K_{Ca,Sr}$ in the case of reversibility of the reaction under consideration should be $K_{Ca,Sr} = \dfrac{1}{K_{Sr,Ca}} = 0.83$. Thus, in determining the equilibrium constant $K_{Ca,Sr}$ by equation (4.25) it could theoretically be expected that $K_{Ca,Sr} = 0.83$.

From Table 4.3 it follows that the value of $K_{Ca,Sr}$ varies within a relatively narrow interval, and the value $K_{Ca,Sr} = 0.83$ can be accepted as most probable.

The thermodynamic characteristics of the reaction in question retain the same absolute values as those found for the direct course of this reaction, with only a change in their signs: $\Delta F^0 = 107.5$ cal/mole, $\Delta S^0 = -0.3$ cal/deg.

It is important to note that the data on equilibrium criteria and thermodynamic characteristics were obtained in studies of exchange sorption of Ca^{2+}, Sr^{2+} cations in different samples taken from different depths in the soil profile (down to 1.5 m). The values of equilibrium criteria and also the thermodynamic characteristics of the direct and reverse reactions of Ca^{2+}, Sr^{2+} exchange in soils proved to be close to the theoretically forecast values. Thus, data obtained in laboratory studies of the exchange sorption of Ca, Sr atoms may be extrapolated for exchange processes under natural conditions.

Exchange sorption of Cs^+ and Sr^{2+} ions on resins

Studies into the exchange sorption of Cs, Sr were conducted on Dowex-50W resin saturated with strontium ions at two ion strengths of solutions (μ = 0.1 and 0.05) and two temperatures ($t_1 = 25°C$ and $t_2 = 50°C$). The equilibrium concentrations of Cs^+ and Sr^{2+} ions and the absorption capacity of resin were determined experimentally. All the other quantities required to describe the reaction in question were calculated.

In accordance with the experimental conditions the exchange sorption of Cs and Sr atoms on resins can be described by the following stoichiometric equation:

$$\frac{1}{2} Sr\,x + Cs^+ \rightleftarrows Cs\,x + \frac{1}{2} Sr^{2+}, \qquad (4.26)$$

where $Sr\,x$ and $Cs\,x$ signify exchange masses of strontium and cesium ions, while Sr^{2+}, Cs^+ signify their concentration in equilibrium solutions. The exchange constant of this reaction was determined by Nikol'skii's equation

$$K_{Cs,Sr} = \frac{Cs\,x \sqrt{Sr^{2+}}}{\sqrt{Sr\,x}\ Cs^+}. \qquad (4.27)$$

From Table 4.4 it is seen that in the first series of experiments (μ_1 = = 0.1, $t_1 = 25°C$) the value of the equilibrium constant was $K_1 = 0.24$. In the second series of experiments, in which the ion strength of solution was

TABLE 4.4. Equilibrium constant of exchange sorption of Cs and Sr on Dowex-50W resin for different ionic strengths and temperatures

Percentage ratio in the initial solutions		pH of solution		Equilibrium concentration		Exchange mass		Ratio of equilibrium concentrations $\dfrac{Cs^+}{\sqrt{Sr^{2+}}}$	Ratio of exchange masses $\dfrac{(Cs)}{\sqrt{(Sr)}}$	$K_{Cs,\,Sr}$
Sr	Cs	initial	equilibrium	$\sqrt{Sr^{2+}}$	Cs^+	$\sqrt{(Sr)}$	(Cs)			
					$\mu = 0.1$, $t = 25\,°C$					
80	20	6.20	5.40	1.628	0.680	1.734	0.233	0.418	0.125	0.30
70	30	6.20	5.55	1.564	1.019	1.724	0.335	0.652	0.194	0.30
50	50	6.20	5.45	1.422	1.799	1.688	0.458	1.265	0.272	0.21
30	70	6.25	5.40	1.264	2.549	1.641	0.610	2.017	0.372	0.18
20	80	6.30	5.45	1.182	2.800	1.580	0.811	2.369	0.513	0.22
										0.24
					$\mu = 0.05$, $t = 25\,°C$					
80	20	6.25	5.45	1.193	0.320	1.738	0.140	0.268	0.080	0.30
70	30	6.30	5.25	1.150	0.470	1.726	0.220	0.409	0.127	0.31
60	40	6.30	5.20	1.118	0.620	1.710	0.300	0.555	0.175	0.31
50	50	6.25	5.20	1.072	0.800	1.694	0.350	0.746	0.207	0.28
40	60	6.30	5.15	1.025	0.940	1.682	0.440	0.917	0.262	0.29
30	70	6.30	5.20	0.974	1.120	1.670	0.490	1.150	0.293	0.25
										0.29
					$\mu = 0.05$, $t = 50\,°C$					
80	20	6.25	5.30	1.265	0.359	1.688	0.100	0.284	0.059	0.21
70	30	6.30	5.25	1.218	0.530	1.682	0.160	0.437	0.095	0.22
60	40	6.30	5.10	1.162	0.700	1.677	0.220	0.602	0.131	0.22
50	50	6.25	5.10	1.116	0.860	1.664	0.290	0.771	0.175	0.23
40	60	6.30	5.10	1.061	1.000	1.659	0.380	0.943	0.229	0.24
30	70	6.30	5.10	1.000	1.160	1.655	0.450	1.160	0.272	0.23
										0.22

only one-half that in the first series ($\mu_2 = 0.05$), the constant was similar to the above, $K_2 = 0.29$. Thus, a decrease in the ion strength of solution did not appreciably affect the equilibrium constant, fluctuations of individual values of constant $K_{Cs, Sr}$ being insignificant. This circumstance is very important, revealing that the exchange sorption of Cs and Sr atoms on resins may be regarded as an ordinary chemical reaction obeying the law of mass action.

The effect of temperature on the equilibrium of the reaction under consideration can be established from Table 4.4. When $t_1 = 25°C$, $K_1 = 0.29$, and when $t_2 = 50°C$, $K_2 = 0.22$. This shows that the equilibrium constant of exchange sorption of Cs^+ and Sr^{2+} ions on resins decreases at higher temperatures. True, the decrease is very slight, but it appears to be quite significant.

According to Table 4.4, $K_{T_1} = 0.29$ and $K_{T_2} = 0.22$. The change in the heat content of the system under consideration can then be calculated from equation (4.14):

$$\Delta H^0 = -4.57 \left[\frac{\log \frac{0.29}{0.22}}{\left(\frac{1}{298} - \frac{1}{323} \right)} \right] = -2110 \text{ cal/mole.}$$

Substitution of the numerical values of K and T in (4.15) yields

$$\Delta F^0 = -1.98 \cdot 298 \cdot 2.303 \cdot \log 0.29 = 729 \text{ cal/mole.}$$

When determining ΔS^0 it is sufficient to substitute the numerical values of ΔH^0 and ΔF^0 in equation (4.16):

$$\Delta S^0 = -\frac{2110 + 729}{298} = -9.5 \text{ cal/deg.}$$

The numerical values of thermodynamic characteristics and their signs show that the formation reaction of compounds of type Cs-resin corresponding to equation (4.26) is endothermic and the change in its heat content is equivalent to 2,110 cal/mole.

The absolute value of the change in its free energy is small ($\Delta F^0 = 729$ cal/mole), but quite significant. In view of the stoichiometric characteristics of the reaction under consideration and the positive value of ΔF^0, it is seen that the compound Cs-resin is thermodynamically less stable than the compound Sr-resin. A similar effect was observed also in the exchange of this pair of ions on soils /12/. It is of interest that the numerical values of ΔF^0 determined on resins were almost identical to the values determined on soils (the value of ΔF^0 on soils for the reverse reaction was found to be −666 cal/mole). Thus, the use of different and independent methods for determining ΔF^0 yielded fairly similar results, testifying not only to the high efficiency of thermodynamic investigation methods but also to the adequate reliability of data characterizing the energy of the bond of Cs and Sr with soil colloids.

Equation (4.27) shows that

$$K_{Cs,\,Sr} = f\left(\frac{\dfrac{Cs\,x}{\sqrt{Sr\,x}}}{\dfrac{Cs+}{\sqrt{Sr^{2+}}}}\right).$$

Figure 4.6 reveals a linear relationship between $Cs_x/\sqrt{Sr_x}$ and $Cs+/\sqrt{Sr^{2+}}$ and practical constancy of $K_{Cs,\,Sr}$ for the experimental conditions in question.

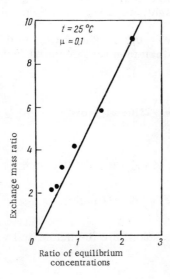

FIGURE 4.6. Isotherm of exchange sorption of Cs, Sr on Dowex-50W resin

The above data as well as those reported in the literature suggest that the behavior of ^{90}Sr and ^{137}Cs in soil depends on the physicochemical properties of these nuclides, the absorption capacity of soils, artificial and natural complexers, as well as ion exchange and diffusion phenomena.

All these factors, along with the pH of soil, play an important role in controlling the concentrations of ^{90}Sr and ^{137}Cs in soil solution.

References

1. Voprosy radioekologii (Problems of Radioecology). Moskva, Atomiz-
        dat. 1968.
2. Gulyakin, I.V. and E.V.Yudintsev. Radioaktivnye produkty
        deleniya v pochve i rasteniyakh (Radioactive Fission Products in
        Soil and Plants). Moskva, Gosatomizdat. 1962.
3. Ionnyi obmen (Ion Exchange). Moskva, Izdatel'stvo inostrannoi litera-
        tury. 1951.

4. Klechkovskii, V. M. — Vestnik Akademii Nauk SSSR, No. 5:93. 1966.
5. Klechkovskii, V. M. — In Sbornik: "Trudy nauchno-metodicheskogo
    soveshchaniya uchenykh sotsialisticheskikh stran po primeneniyu
    izotopov i izluchenii v sel'skom khozyaistve," p. 54. Moskva,
    Sel'khozgiz. 1961.
6. Klechkovskii, V. M. et al. Povedenie radioaktivnykh produktov
    deleniya v pochvakh, ikh postuplenie v rasteniya i nakoplenie v
    urozhae (Behavior of Radioactive Fission Products in Soils, Their
    Intake by Plants and Buildup in Crops). Moskva, Izdatel'stvo
    Akademii Nauk SSSR. 1956.
7. Leipunskii, O. I. — Atomnaya Energiya, Vol. 2:278. 1957.
8. Pavlotskaya, F. I. et al. — Informatsionnyi Byulleten' Radiobio-
    logiya, No. 9:3. 1961.
9. Pavlotskaya, F. I. et al. — In Sbornik: "Radioaktivnost' pochv i
    metody ee opredeleniya," p. 174. Moskva, "Nauka." 1966.
10. Polyakov, Yu. A. — Trudy Vsesoyuznogo Nauchno-Issledovatel'skogo
    Instituta Udobrenii, Agrotekhniki i Agropochvovedeniya imeni
    K. K. Gedroitsa," No. 29:182. 1949.
11. Polyakov, Yu. A. — In Sbornik: "Fiziko-khimicheskie metody issle-
    dovaniya pochv," p. 32. Moskva, "Nauka." 1968.
12. Polyakov, Yu. A. — In Sbornik: "Radioaktivnost' pochv i metody ee
    opredeleniya," p. 81. Moskva, "Nauka." 1966.
13. Polyakov, Yu. A. — Informatsionnyi Byulleten' Radiobiologiya,
    No. 9:33. 1966.
14. Polyakov, Yu. A. — Kolloidnyi Zhurnal, Vol. 21:221. 1959.
15. Polyakov, Yu. A. — Kolloidnyi Zhurnal, Vol. 26:705. 1964.
16. Polyakov, Yu. A. — In Sbornik: "Doklady sovetskikh uchenykh k
    VII Mezhdunarodnomu kongressu pochvovedov v SShA," p. 100.
    Moskva, Izdatel'stvo Akademii Nauk SSSR. 1960.
17. Polyakov, Yu. A. — Pochvovedenie, No. 7:59. 1955.
18. Polyakov, Yu. A. — In Sbornik: "Fiziko-khimicheskie metody issle-
    dovaniya pochv," p. 57. Moskva, "Nauka." 1966.
19. Polyakov, Yu. A. — In Sbornik: "Radioaktivnost' pochv i metody ee
    izmereniya," p. 5. Moskva, "Nauka," 1966.
20. Polyakov, Yu. A. — In Sbornik: "Doklady VI Mezhdunarodnomu kon-
    gressu pochvovedov, 2-ya komissiya. Khimiya pochv," p. 187.
    Moskva, Izdatel'stvo Akademii Nauk SSSR. 1956.
21. Rachinskii, V. V. — Izvestiya Moskovskoi Sel'skokhozyaistvennoi
    Akademii imeni Timiryazeva (TSKhA), No. 3:161. 1954.
22. Rachinskii, V. V. — Ibid., No. 5:184. 1960.
23. Yudintseva, E. V. and I. V. Gulyakin. Agrokhimiya radioaktiv-
    nykh izotopov strontsiya i tseziya (Agrochemistry of Radioactive
    Isotopes of Strontium and Cesium). Moskva, Atomizdat. 1968.
24. Alexander, L. T. et al. Radioisotopes in the Biosphere, p. 3.
    Univ. Minnesota 1959. Minneapolis, Minn. 1960.
25. Amphlett, C. B. — Research, Vol. 8:335. 1955.
26. Bowen, H. J. M. and J. A. Dymond. — J. Exptl. Bot., Vol. 7:264.
    1956.
27. Boyd, G. E. et al. — J. Amer. Chem. Soc., Vol. 69:2818. 1947.

28. C r a n k, J. The Mathematics of Diffusion. Oxford, Clarendon Press. 1956.
29. G i l b e r t, M. and H. L a u d e l o t. — Soil Sci., Vol. 100:157. 1965.
30. H a l s t e a d, E. H. et al. — Soil Sci. Soc. Amer. Proc., Vol. 32:69. 1968.
31. H e a l d, W. R. — Soil Sci. Soc. Amer. Proc., Vol. 24:103. 1960.
32. H e n r y, R. M. — Soil Sci. Soc. Amer. Proc., Vol. 22(6):514. 1958.
33. H o w a r d, R., Jr., and R. G. M e n z e l. — In: Radioactive Fallout, Soils, Plants, Foods, Man. Amsterdam — London — New York, Elsevier Publishing Company. 1965.
34. J a c k s o n, R. D. — Soil Sci. Soc. Amer. Proc., Vol. 29:144. 1965.
35. J a c k s o n, W. A. et al. — Soil Sci., Vol. 99:345. 1965.
36. K a h n, H. On Thermonuclear War. Princeton, 1961.
37. K h a s a n e n, F. F. et al. — Soil Sci. Soc. Amer. Proc., Vol. 32:209. 1968.
38. L a i, T. M. and M. M. M o r t l a n d. — Soil Sci. Soc. Amer. Proc., Vol. 25:353. 1961.
39. L a n g h a m, W. H. — In: Radioactive Fallout, Soils, Plants, Foods, Man. Amsterdam — London — New York, Elsevier Publishing Company. 1965.
40. L a u r e n c e, I. et al. — Science, Vol. 136:19. 1962.
41. M a r s h a l l, C. E. and G. G a r c i a. — J. Phys. Chem., Vol. 63:1663. 1959.
42. M a r s h a l l, C. E. and R. S. G u n t a. — J. Soc. Chem. Ind., Vol. 52:483. 1933.
43. M a r t e l l, E. A. — Science, Vol. 129:1197. 1959.
44. M e n z e l, R. G. and W. R. H e a l d. — Soil Sci., Vol. 80:287. 1955.
45. M e n z e l, R. G. and W. R. H e a l d. — Soil Sci. Soc. Amer. Proc., Vol. 23:110. 1959.
46. M i l l e r, J. R. and R. E. R e i t e m e i e r. Rate of Leaching of Radio-strontium Through Soils by Simulated Rain and Irrigation Water. — U. S. Department of Agriculture. ARS Research Rept., 300. 1957.
47. O l i v e r, S. and S. A. B a r b e r. — Soil Sci. Soc. Amer. Proc., Vol. Vol. 30:82. 1966.
48. Radioactive Fallout, Soils, Plants, Foods, Man. Amsterdam — London — New York, Elsevier Publishing Company. 1965.
49. R u s s e l l, R. S. (editor). Radioactivity and Human Diet. New York, Pergamon Press. 1966.
50. R e d i s k e, J. H. and A. A. S e l d e r s. — Plant Physiol., Vol. 28:594. 1953.
51. R u s s e l l, R. S. et al. The Availability to Plants of Divalent Cations in the Soil. — Proc. Soc. Intern. Conf. Peaceful Uses Atomic Energy, Pergamon Press. 1958.
52. S c h r o e d e r, D. van et al. — Landwirtsch. Forsch., Vol. 15:48. 1962.
53. S c h u f l e, G. A. — Texas J. Sci., Vol. 133. 1961.
54. S c h u l e r t, A. et al. Distribution of Nuclear Fallout. — Annual Report to AEC under contract A. T. (30 — 1), 1956. 1 October 1959.
55. S c h u l z, R. K. et al. — Hilgardia, Vol. 27:333. 1958.
56. T h o r n t h w a i t e, G. W. et al. — Science, Vol. 131:1015. 1960.
57. W a l t o n, A. — J. Geophys. Res., Vol. 68:1485. 1963.
58. W h e e t w i r g h t, E. I. et al. — J. Amer. Chem. Soc., Vol. 75:4196. 1953.
59. W i k l a n d e r, L. — Soil Sci., Vol. 97:168. 1964.

Chapter 5

ION DIFFUSION IN SOILS AND ITS ROLE IN
RADIONUCLIDE MIGRATION

Introduction

In a radioactively polluted biosphere artificial radionuclides migrate along the food chains, their ultimate link being the human organism. Soil constitutes an important link in many of these chains, and therefore the attention paid by researchers to migration of radionuclides in soils is readily understood /1/. Migration in soils is a complex phenomenon, the motivating forces directly responsible for it being extremely varied. They include a variety of biological transport processes, climatic factors and human economic activities.

Nevertheless, in spite of this variety, the causes of migration can be divided into the following two main groups: 1) external natural or human factors; 2) differences in the equilibrium concentration of a nuclide at different points within the soil. As a result of the chaotic thermal movement of molecules and ions, a difference in concentrations "spontaneously" gives rise to a diffusional flux of a nuclide.

A distinctive feature of diffusion processes due to their ionic-molecular nature is their uninterrupted continuity for prolonged periods of time. Unlike other transport processes, the diffusion of nuclides in soil occurs continuously throughout the year (even at subzero soil temperatures). Therefore, in spite of the comparatively low velocities of diffusion processes, their consequences are liable to exceed those of stronger, but transient effects.

For diffusion to take place it is sufficient that the soil contains just one nuclide concentration gradient, and therefore diffusion may be regarded as the most widespread process of nuclide migration. On the other hand, proper studies of this phenomenon have been conducted for only about 10 years. In migration studies under natural conditions, in the presence of a multitude of significant factors, it is difficult to isolate an effect caused by any single process, including diffusion. Therefore, field observations must necessarily be supplemented with laboratory experiments conducted under controlled conditions.

Radionuclide diffusion in soil

Diffusion of radionuclides in soil possesses certain special features distinguishing it from other diffusion processes.

1.   Soil is an adsorbent medium with respect to the majority of nuclides and therefore the diffusion process is complicated by simultaneous adsorption (desorption).

2.   Soil is a polyfunctional adsorbent, and therefore the same nuclide may be present in several adsorbed states differing in the strength and nature of the bond with the solid phase of soil.

3.   The isotherm of nuclide adsorption in soil can be regarded as linear in the majority of cases, the amount of nuclides being negligibly small in comparison to the adsorption capacity of soil.

4.   Owing to the inhomogeneity of soil, the diffusion rate of radionuclides in it generally depends on the coordinate along which the diffusion takes place.

5.   The diffusion rate in soil varies with time, because of variations in soil humidity and temperature.   (Items 4 and 5 refer to natural conditions.)

These features, save for the linearity of the isotherm, complicate examination of the diffusion process, its basic laws having been derived for diffusion in a homogeneous medium not interacting with the diffusing substance. Theoretical analysis yielded differential equations describing the diffusion of a substance in an adsorbent disperse medium, of more complex form than the conventional diffusion equation /9/.   It has been demonstrated, however, that when certain conditions are fulfilled, the diffusion of radionuclides in soil obeys the conventional differential equation of diffusion, i. e., the Fick law

$$\frac{\partial c}{\partial t} = D \frac{\partial^2 c}{\partial x^2} , \qquad\qquad (5.1)$$

where c is the concentration of diffusing substance, t is the diffusion time, x is the coordinate in the direction of diffusion, and D is the diffusion coefficient.   Physically, the diffusion coefficient describes the amount of substance passing through unit surface per unit time with unit substance concentration gradient.   The diffusion coefficient has dimensions cm^2/sec.

Conditions governing the applicability of the Fick law to diffusion in a disperse adsorbent medium are as follows /9, 12/: 1) adsorption must be reversible; 2) adsorption equilibrium between different states of the substance in any plane perpendicular to the direction of diffusion must become established more rapidly than an appreciable variation in the concentration of the substance in the same plane due to diffusion; 3) the diffusion medium should be homogeneous when considered on a scale considerably larger than the size of its constituent particles; 4) the actual diffusion coefficient in the medium should be independent of the concentration of diffusing substance (as with linear adsorption isotherms).   These conditions are fulfilled with respect to diffusion of the majority of nuclides (^{137}Cs being one possible exception), and therefore the diffusion rate can be described by the actual diffusion coefficient D alone, which will be referred to simply as the diffusion coefficient.

Solutions of equation (5.1) related to certain, practically important cases will be given below.

The inhomogeneous properties of soil raise serious difficulties in attempts at a quantitative description of the diffusional migration of nuclides. In laboratory experiments it is not difficult to achieve a high degree of homogeneity of the soil material, but field studies call for various

simplifying assumptions.   The more accurately that these assumptions
describe the real conditions, the more labor-consuming becomes the
calculation.

The time-dependence of the diffusion rate presents no difficulties,
provided one knows the time-variation of the diffusion coefficient.   In this
case it is possible to use the mean diffusion coefficient $\bar{D}$ defined by the
formula

$$\bar{D} = \frac{D_1 t_1 + D_2 t_2 + \ldots + D_n t_n}{\tau} \qquad (5.2)$$

or

$$\bar{D} = \frac{1}{\tau} \int_0^\tau D(t)\, dt,$$

where D_1, D_2, ..., D_n are the diffusion coefficients in time intervals
t_1, t_2, ..., t_n, respectively; $\tau = t_1 + t_2 + \ldots + t_n$ is the total diffusion time.

An estimation of the contribution made by diffusion to different phenomena is impossible without a knowledge of numerical values of the diffusion
coefficient and its dependence on the different properties of soil.   An
extensive body of experimental data has already been accumulated on
diffusion rate measurements in soils and subsoils for ions of different
elements, including artificial radionuclides.   Although we are mainly
interested in data on the diffusion of ^{90}Sr, ^{137}Cs, ^{144}Ce and other fission
products, in many cases we shall refer to results related to diffusion of
other nuclides, provided this information reveals general regularities.

From the standpoint of the possibility of diffusion in it, soil is a system
consisting of three phases: soil solution, solid phase (mainly soil minerals
of crystalline structure), and soil air.   The highest mobility is possessed
by ions and molecules in the soil solution.   At 25°C the diffusion coefficient
in water is $7.93 \cdot 10^{-6}$ cm^2/sec for Sr^{2+} ions, $2.06 \cdot 10^{-5}$ cm^2/sec for Cs^+
ions, $2.05 \cdot 10^{-5}$ cm^2/sec for I^- ions, and $7.92 \cdot 10^{-6}$ cm^2/sec for Ca^{2+} ions.

The solution layer adjacent to the surface of soil particles possesses
somewhat different properties from the rest of the solution and can be
conditionally regarded as a separate phase.   This layer constitutes the
outer plate of the double electrical layer surrounding negatively charged
soil particles.   This is the layer containing the exchange-adsorbed ions in
soil, i.e., 90—99% of the entire ^{90}Sr present in soil in cation form.   The
surface layer thickness is roughly estimated at 10—20 Å for nonsaline soils.
The mobility of adsorbed ions is less than in solution.   For instance, the
mobility (and the diffusion coefficient, which is proportional to it) of adsorbed Na^+ ions in clay pastes is 30—40% of the mobility of the corresponding ions in free solution /29/, while the mobility of adsorbed Ca^{2+}
ions is only 8% of the mobility of its ions in free solution /36/.   The
mobility of trace amounts of Sr^{2+} ions adsorbed by moist soils turned out
to be 1—8% of their mobility in soil solution /9/.

In spite of their lesser mobility, adsorbed cations play an important role
in the overall diffusion flux.   According to /29, 36/, the diffusion of
adsorbed Na^+ is responsible for about 60% of the total flux, and the diffusion of adsorbed Ca^{2+} for 10—50%.   In the case of trace amounts of Sr^{2+}

this percentage may be even greater; for instance, it amounted to 80—90% in chernozem and chestnut soil, where 98—99% of the radionuclides were in an adsorbed state /9/.

Many ions are capable of diffusion in the solid phase of soil. However, the diffusion rate in the crystal lattice of minerals is usually many orders of magnitude less than in liquid. For instance, the diffusion coefficient of ^{45}Ca in CaCO$_3$ is 10^{-19} cm^2/sec /67/. Nevertheless, diffusion in the solid phase plays an important part in fixation of ^{137}Cs and ^{90}Sr by clay minerals of soil.

The factors controlling the diffusion rate of radionuclide ions in soil may be divided into two groups: properties of the ion itself and those of soil.

The most important property of an ion is the sign of its charge, nuclides in anion form usually diffusing much more rapidly than cations. As a rule, the magnitude of charge has an unambiguous effect on the diffusion rate, namely, the larger the charge of a cation, the lower is the diffusion rate. A possible exception is ^{137}Cs, possessing an anomalously low diffusion rate in certain clays. Ions can be arranged in the following series with respect to their diffusion coefficients:

$$K^+ > Na^+ > Sr^{2+} > Ca^{2+} > Ba^{2+} \gg Ce^{3+}, Y^{3+}$$

(bentonite paste; the ion was added as hydroxide) /41/;

$$Na^+ > Cs^+ > Ba^{2+} > Sr^{2+} > Ca^{2+} \gg Ce^{3+}, Y^{3+}$$

(bentonite paste; the ion was added as chloride) /41/;

$$SO_4^{2-} > Na^+ > Rb^+ > Ca^{2+} > Cs^+$$

(bentonite paste) /40/;

$$Sr^{2+} \geqslant Cs^+ > Ce^{3+}$$

(light-loamy sod-podzolic soil) /7, 10, 18/.

The following properties of soil exert the strongest influence on the diffusion of nuclides: moisture content, density, concentration of salts in the soil solution, humus contents, ion composition of absorption complex and soil solution.

An increase in moisture content and density is accompanied by an increase in the relative volume occupied by the soil solution, the diffusion paths are straightened and shortened, improving the opportunities for migration of ions. According to experiments, the diffusion coefficients of ^{89}Sr, ^{90}Sr, ^{137}Cs and ^{144}Ce increase with increasing volumetric moisture content /7, 10, 18, 45, 56, 65, 69/. The diffusion coefficient of ^{90}Sr in the range of moisture contents that is characteristic of natural conditions is most commonly of the order of 10^{-8} cm^2/sec, although values of the order of 10^{-7} cm^2/sec have been reported for some soils.

With increasing concentration of salts in the soil, the solution increases the fraction of cations in free solution, where they possess higher mobility than in adsorbed state. In this case, the mean diffusion rate increases /15,

17, 57, 69/. For instance, the diffusion coefficient of ^{90}Sr is liable to increase tenfold and more with increasing concentration of salts in soil /15/. At very high concentrations (comparable with the soil's exchange capacity) further increase in the quantity of salts may cease to affect the diffusion rate, or the latter may even decrease.

Due to the large absorption capacity of humus with respect to ^{90}Sr, ^{137}Cs and ^{144}Ce /22, 76/, it is natural to expect that the diffusion rate in high-humus soils should be lower than in low-humus soils. In model experiments on quartz sand with different additions of humic acids the diffusion coefficient of ^{90}Sr decreased from $1.1 \cdot 10^{-6}$ (pure sand) to $1 \cdot 10^{-8}$ cm^2/sec (10% humic acids), i. e., it attained numerical values characteristic of strongly sorbing soils /16/.

The diffusion rate of nuclides in soils also depends on the ion composition of the absorption complex. For instance, the values of the diffusion coefficient of ^{90}Sr in clay form the following series with respect to the different ion forms of clay /45/:

$$(H + Al) > Ca > K > Na,$$

which, not accidently, coincides with the sequence for the ion adsorption energies; the more difficult the displacement of adsorbed ions by Sr^{2+} ions, the higher is the diffusion rate of ^{90}Sr.

The diffusion rate of ^{90}Sr may vary in a complex manner when several properties of soil are varied simultaneously, as, for instance, when soil is limed /14/.

The diffusion rate of ions in liquids and solids increases at higher temperatures. The quantitative characteristic of the temperature-dependence of diffusion coefficients is usually provided by activation energy E from equation

$$D = D_0 e^{-E/RT}, \qquad (5.3)$$

where D_0 is a constant, R is the gas constant, while T is the absolute temperature. For diffusion of Cs^+ in bentonite $E = 6.3 - 7.6$ kcal/mole /44/. There are practically no available data on activation energy for diffusion of radioactive fission products in soil.

Comparison and analysis of published results of laboratory investigations and field measurements delineate four groups of phenomena in which diffusion processes play an appreciable, and occasionally the main, part: 1) vertical migration of radionuclides over the soil profile; 2) fixation of radionuclides by the soil's clay minerals; 3) adsorption and desorption of radionuclides by the bottom of water bodies; 4) intake of radionuclides from soil by plants through their roots.

Migration of nuclides over the soil profile

Factual data on the distribution of radionuclides over soil profiles are very extensive. They refer to different types of soils, climatic zones,

isotope composition of nuclides, conditions of their deposition on soil surface, and migration time.   However, the bulk of this information is of a
descriptive nature, only a few papers being concerned with the quantitative
laws governing the migration of radionuclides over the soil profile.

The major physicochemical processes responsible for vertical migration
of radionuclides in soils are convective transport by descending or ascending flow of soil solution and ionic or molecular diffusion.   Both these processes are accompanied by ion exchange, since soil possesses an exchange
capacity with respect to cations.   It should be noted that ion exchange in
itself is not a cause for the migration of nuclides, but it modifies the relative
amounts of nuclides in the adsorbed and unadsorbed states.

Essentially the convective-diffusion transport in the soil profile is
analogous to the migration of a substance in a chromatographic column
filled with an ion-exchange resin.   The theories of chromatographic transport have been adequately elaborated and are often used to describe the
migration of salts and radionuclides in soil /4, 5, 23, 38, 74/.   A review of
the literature on the subject can be found in /39/.

The transport of nuclides over the soil profile possesses distinctive
features in comparison to other chromatographic processes.   One of them
is the very low mean migration rate of soil solution.   Thus, with annual
atmospheric precipitation of, say, 1,000 mm (assuming the entire precipitation to percolate into the soil, thereby making the results certainly too high),
the mean filtration velocity is $3 \cdot 10^{-6}$ cm/sec, which is several orders of
magnitude less than the ordinary filtration velocity in ion-exchange columns.
In soils with exudational instead of leaching water regime the migration rate
of soil solution is likewise low.   As follows from the ion exchange dynamics
theory, diffusion of the substance along the column begins to play a significant and occasionally even the leading role in its migration at low filtration
velocity /46/.   With respect to the migration of radioactive isotopes of
strontium and calcium in soils it is pointed out that lengthwise diffusion
becomes the decisive factor in the washout of radioactive zones at soil
solution migration velocity < 3,000 mm/yr /40/.

Another peculiar aspect of the conditions under which vertical migration
of nuclides takes place, namely, the nonuniformity of soil properties over
the profile, creates certain mathematical difficulties in attempts at applying the theory of chromatography to nuclide migration under natural conditions.   Moreover, the nature of vertical variation of soil properties differs
for soil types, and it is scarcely feasible to develop separate mathematical
expressions for every type of soil.   These difficulties can be avoided in two
ways, either by using equations based on various simplifying assumptions
or by resorting to numerical solution of chromatographic equations with
electronic computers.   Either choice depends on the migration conditions
and the investigator's facilities.

One possible assumption, for instance, is uniformity of soil over its
profile, in the first approximation.   This assumption becomes more accept-
able as the depth down to which nuclides penetrate is less and as the thickness of the upper soil horizon is greater.   Another assumption is that the
total results of all the vertical migration processes are regarded as the
results only of diffusion characterized by an apparent diffusion coefficient.

The author used both these assumptions in proposing a simple calculation method for radionuclide migration over a soil profile /8, 9/. It consists in finding the percentage content of nuclide in the upper soil layer of any given thickness at any given time by a selected apparent diffusion coefficient, using the provided calculation curves. The curves were calculated by formulas relating to the following three different cases of nuclide deposition on the soil surface:

1) a single (one-time) nuclide deposition on the surface,

$$q = \Phi(y);$$

2) constant concentration of nuclides on the surface,

$$q = 1 - \exp(-y^2) + \sqrt{\pi}\; y [1 - \Phi(y)];$$

3) nuclide concentration on the surface increasing linearly with time,

$$q = \frac{1}{2}\left\{ 3\sqrt{\pi}\; [1 - \Phi(y)]\left(1 + \frac{2}{3}y^2\right) y + \right.$$
$$\left. + 3\left[1 - \frac{2}{3}\exp(-y^2)(1 + y^2)\right] - 1\right\},$$

where q is the fraction of radionuclide, with respect to its total content in soil, in the upper layer of thickness l; $y = l/2\sqrt{D_a t}$ is a dimensionless parameter; D_a is the apparent diffusion coefficient; t is the diffusion time; $\Phi(y) = \frac{2}{\sqrt{\pi}}\int\limits_0^y e^{-u^2}du$ is the error function (a tabulated function).

The method is also applicable to the inverse problem of finding the numerical value of an apparent diffusion coefficient from the measured content of a nuclide in the upper soil layer.

Analysis of numerous data on the distribution of ^{90}Sr under natural conditions (results from several dozen soil profiles) revealed that values of the apparent diffusion coefficient for ^{90}Sr lie between $2 \cdot 10^{-8}$ and $2.5 \cdot 10^{-7} cm^2/$/sec. Yet, values of the diffusion coefficient of ^{90}Sr measured in the laboratory, where all the other migration causes were excluded, are likewise of the order of magnitude $10^{-8} - 10^{-7}$ cm^2/sec. The similarity of the apparent and true diffusion coefficients shows that in many cases diffusion constitutes the major process responsible for migration, and that assumption pertaining to the diffusional nature of all migration processes largely fits the nature of the phenomenon. For an accurate determination of the fraction of nuclides transported to a given soil layer by the diffusion process, it is necessary, starting from the mean value of the apparent diffusion coefficient, to calculate the vertical nuclide distribution profile and to compare it with the experimental profile.

The conclusion concerning the importance of diffusion in vertical migration was subsequently confirmed by Makhon'ko /4, 5/, who considered the migration of ^{90}Sr, ^{137}Cs, ^{144}Ce and ^{106}Ru as combined convective-diffusional transport in a medium, where the diffusion coefficient increases linearly with increasing depth of soil. Makhon'ko studied the extensive literature referring to global and also local radioactive soil pollution. He represented the diffusional penetration rates of the nuclide by the magnitude of the effective diffusion coefficient $\bar{D}$ averaged over the $0-20$ cm layer and

also by the rate of diffusional penetration v_D, numerically equal to the
effective diffusion coefficient of the nuclide at 1 cm depths but with dimen-
sion cm/sec. The velocity of downward movements v served as an overall characteristic of the intensity of the remaining processes, which included transport by infiltration, silting and buildup in plants.

For all four nuclides, the mean value $\bar{D}$ was of the order of 10^{-7} cm^2/sec,
v_D was of order of magnitude 1 cm/yr, and v was of order of magnitude
0.1 —1 cm/yr. As a rule, the diffusional and convective transport rates
were the same for light-textured soils, but in the case of heavy soils the
diffusional transport rate for nuclides was tenfold and more the rate of
convective transport.

The relative importance of diffusion among the physicochemical vertical
migration processes clearly differs under different soil conditions and
with different radionuclides. Basing ourselves on general considerations,
it may be asserted that diffusion should play a more important role in the
following cases: 1) in clayey soils, in which the filtration rate is low;
2) in high-humus soils, possessing high adsorption capacity with respect to
many nuclides; 3) in bog soils with high and relatively permanent groundwater tables; 4) in soils with a water balance in which both the descending
and ascending moisture fluxes are small. In the case of soils in which diffusion plays a predominant part, the use of purely diffusional equations does
not introduce a large error in the calculation results and at the same time
renders the calculations much less labor-consuming.

Fixation of nuclides by clay minerals

Fixation of ions by soils and clays is defined as absorption which is not
of the usual ion-exchange nature and causes the ions to pass into a strongly
combined state. The radioecological significance of this phenomenon is that,
as a result of fixation, a part of the nuclide becomes exclusive, as it were,
from exchange reactions in soil and, as long as the nuclide is in the fixed
state, it is not taken up by plants. Fixation of K^+ and NH_4^+ by soils is well
known. Cesium and rubidium, being chemical analogs of potassium and
possessing very similar ion dimensions, are likewise capable of being fixed
by soils. Fixation of Sr^{2+} and Zn^{2+} has likewise been observed, but much
less thoroughly studied.

The fixation is caused by interaction of ions with the crystal lattice of
certain clay minerals. As yet, there is no concensus of opinion concerning
the exact lattice structures in which different fixed ions are localized /3/.
Apparently, different ions are absorbed through different mechanisms.
With some ions, fixation means their interstitial penetration between the
lattice layers, with others, migration through internodes, with still others,
fixation in special lattice cavities ("traps"). However, in any case, the
conversion of ions to a fixed state involves a stage of migration toward the
fixation loci, carried out by diffusion within the crystal lattice. In the
opinion of many investigators, this stage controls the fixation rate in the
majority of cases.

Though dealing mainly with fixation of ^{137}Cs, we shall occasionally refer to results obtained in studies of the fixation of potassium, such references being rendered legitimate by the similar behavior of these elements.

In his studies of the kinetics of the adsorption of ^{137}Cs by different minerals, Sawhney /70/ found that adsorption equilibrium of ^{137}Cs with illite and montmorillonite suspensions was established in less than 2 hours, but in the case of vermiculite suspensions equilibrium was not attained even after 500 hours. The difference is due to the fact that ^{137}Cs adsorption by vermiculite is the result of two processes, rapid ion exchange with cations located on external surfaces of mineral particles and slow diffusion into the interlayer space. In order to examine the part played by the second process, it suffices to say that the interlayer space contains over 90% of all the minerals' possible sorption loci. The behavior of K^+ ions in illite and vermiculite is quite analogous /30, 33/.

These results mean that the idea of rapidly established adsorption equilibrium in soil, allowing application of static ion exchange equations to the solid soil phase — soil solution system, is often inapplicable. Under these conditions, along with the usual factors controlling ion-exchange adsorption (such as the concentration and ion composition of soil solution), there are also other active factors, including the mineralogical composition of soil, structural peculiarities of soil clay minerals, the shape and size of the particles of these minerals, their percentage content in soils. The part played by some of these factors manifests itself in the numerical values of the diffusion coefficient. For instance, the diffusion rate of K^+ in illites is very low, the D values being of the order of 10^{-23} cm^2/sec /30/ and 10^{-20} cm^2/sec /32/. In contrast, the diffusion rate of K^+ in micas (including biotite) is very much larger: $D \approx 10^{-10}$ cm^2/sec /32/. An intermediate position is occupied by clinoptilolite, the diffusion coefficient of K^+ in this mineral being $0.4 - 8.3 \cdot 10^{-12}$ cm^2/sec, depending on the degree of saturation of the mineral with K^+ ions /34/. Obviously, fixation of K^+ will proceed more rapidly and to a higher degree in soils where the clay fraction is dominated by biotite than in soils with a predominance of illite.

In speaking of fixation, one must not imagine it as an irreversible process. Ions having diffused into the crystal lattice are capable of leaving it and migrating to the surface, once the conditions have become altered. Although the diffusion in the reverse direction proceeds as slowly, the defixed (released) quantities of ions may add up to an appreciable quantity over a sufficiently long time. For instance, plants have been found to absorb a much larger amount of potassium over their growth period than the content of exchange potassium in soil, the difference being due to the diffusional conversion of fixed potassium to its exchangeable form, which is available to plants /52/.

In this connection, Mortland /52/ noted that measurement of the ratio of the amounts of potassium in the exchangeable, acid-soluble and other forms, though yielding definite information concerning the status of potassium in soil, imparts no knowledge concerning the rate of its conversion from one form to another. Yet, in many cases it is just the conversion rate that is liable to prove the principal factor controlling the availability of potassium to plants. This largely applies also to ^{137}Cs.

Until now we have been dealing with fixation of Cs^+ and K^+ ions. However, more than ten years ago Libby /48/ suggested that, similar to ^{137}Cs, ^{90}Sr penetrating into the soil may in the course of time be converted into a nonexchangeable form becoming less available to plants. It should be specified that the expression "nonexchangeable form," similar to the case of ^{137}Cs, should not be interpreted as the impossibility of its conversion to other forms (say, an exchangeable form), but in the sense that this form does not directly pass into solution with the usual techniques for the determination of exchange cations.

Libby's suggestion was soon confirmed. In experiments on five different soils, Squire /73/ determined the degree of extraction of ^{90}Sr introduced onto the soil under conditions approximating the natural conditions, 3.5 years after its introduction. Squire found that some of the nuclide (0.3 —1.2%) could not be extracted from soil even by repeated extractions with boiling 8 M HNO_3.

Schulz and Riedel /71/, in a similar study, likewise discovered that the content of nonexchangeable ^{90}Sr 3.5 years after its introduction (2.5 years in the case of one of their soils) was much higher than directly on intro- duction. It reached 6 —7% of the total ^{90}Sr content in soil, and there were no grounds to assume that this content could not increase with time. In order to explain the presence of an appreciable amount of nonexchangeable ^{90}Sr in soil, these authors suggested that Sr^{2+} ions were capable of migrat- ing into the lattice of calcium carbonate in the soil. However, the fixation was observed also in the soil that was devoid of carbonate, suggesting the existence of two other fixation mechanisms. According to the authors, such mechanisms were diffusion into insoluble calcium and strontium phos- phates or other insoluble solid phases containing these cations. However, it had been experimentally demonstrated even before, in the instance of vermiculite, that Sr^{2+} is capable of diffusing into a phase that was initially devoid of either calcium or strontium /75, 76/.

More specific results were obtained by Roberts and Menzel /68/, who found that of several soils investigated by them fixation of strontium was strongest in soils with kaolinite coated with iron and aluminum hydroxides as their predominant clay mineral. Their studies were continued under laboratory conditions by Frere and Champion /37/, who investigated the kinetics of ^{85}Sr fixation by suspensions of soils and mixtures of kaolinite with sesquioxide gels. In all the investigated systems, the content of fixed ^{85}Sr increased with time while tending to a certain equilibrium value (3 — 15% of the entire adsorbed quantity), which was achieved 20 —40 days after addition of the nuclide to the suspension. The conditions of exchange between the phases in suspensions were evidently incomparably better than in natural soils, owing to the better contact between liquid and solid phases and the large quantity of liquids. Therefore equilibrium in these experiments was obtained much more rapidly than it would have been in the same soils and mixtures at a moisture content that is characteristic under natural conditions. Nevertheless, in our opinion the equilibrium value must be the same in both cases.

As in many other investigations of fixation, Frere and Champion based their interpretation of results on the assumption that the process was of a diffusional nature. This concept provides the investigators with a

quantitative basis for calculating the fixation (or defixation) rate. The
formula proposed by Mortland et al. /53, 55/ may be regarded as the
first attempt in this direction:

$$r = B(c' - c),$$ (5.4)

where r is the defixation rate of potassium, c' is the concentration of
fixed potassium, c is the concentration of potassium in solution, and B is
the diffusion constant. When $c' > c$ potassium is defixed (released) from
the mineral, and when $c' < c$ it is fixed from solution, If $c' = c$, there is
equilibrium between the mineral and solution. If quantities B and c are
known, it is possible to use expression (5.4) to determine not only the rate
but also the trend of the process (whether fixation or defixation) depending
on the potassium concentration in the soil solution. It follows from (5.4)
that the fixation rate r is linearly related to the starting concentration of
fixed potassium in clay c'. This law was confirmed by experiments using
clay fractions of six different soils /52/.

Frere and Champion /37/ proposed the following formula describing
the fixation kinetics:

$$Y = \alpha(1 - e^{-\beta t}),$$ (5.5)

where Y is the quantity of fixed strontium, α is the capacity of soil with
respect to fixation of strontium, i. e., the maximum amount of strontium
that can be fixed by 1 g soil, β is a constant involving the diffusion coefficient and cross-section open to diffusion, and t is time. Having processed
20 experimental curves, the authors found that the numerical values of β
for the systems investigated by them lay within $0.2 - 1.0$ day^{-1}. These
data made possible an immediate estimation of the time in which fixation
reaches a given fraction of the equilibrium value. From (5.5) it follows that

$$t = -\frac{1}{0.434\beta} \log\left(1 - \frac{Y}{\alpha}\right).$$ (5.6)

Substituting the degree of attainment of equilibrium in which we are
interested, Y/α, and with known constant β, the corresponding time t can
readily be determined. It should be borne in mind that these data were
obtained for suspensions and their applicability to soils in their natural
state has not yet been confirmed.

Unfortunately the above papers /37, 53 — 55/ lack derivation of the latter
formulas or even a statement of the main assumptions, so making it impossible to discuss the reliability of the formulas and the limits of their
applicability. Nevertheless, the diffusional nature of the process of fixation
allows one to resort to already known diffusional kinetics equations /21/
referring to different cases.

In the simplest case, the mineral particles capable of fixing the nuclide
can be regarded as spheres of uniform radius, while the concentration of
nuclide on the surface of particles may be assumed to be constant throughout the diffusion time. Let the particles be free from nuclide at the initial
moment. Then the time-dependence of saturation of the particles with the

nuclide is described by the equation

$$\frac{a}{a_{\infty}} = 1 - \frac{6}{\pi^2} \sum_{n=1}^{\infty} \frac{e^{-n^2\pi^2\tau}}{n^2}, \tag{5.7}$$

where a and a_{∞} are the respective amounts of nuclide fixed by the mineral at a given moment of time and after attainment of equilibrium; $\tau = Dt/r^2$; D is the diffusion coefficient of the nuclide in the mineral particle; r is the particle radius; t is time. If equation (5.7) is limited to the first term of the series (which is permissible for large values of τ) the formula then reduces to that proposed by Frere and Champion /37/. The curve calculated from (5.7), representing ratio a/a_{∞} as a function of τ, was provided in /21, p.95/. By means of this curve it is possible to solve a variety of problems: 1) calculation of the degree of saturation of the mineral with a nuclide for a given time from given D and r; 2) calculation of the time in which a given degree of saturation will be attained, from given D and r; 3) calculation of quantity D from given a/a_{∞} and r, and so on.

For example, if it is assumed that $r=10^{-4}$ cm, then, according to calculation, the degree of saturation $a/a_{\infty}=0.90$ is attained in 30 min for $D= 10^{-12}$ cm^2/sec, in 2 days for $D=10^{-14}$, in 7 months for $D=10^{-16}$, in 58 years for $D=10^{-18}$ and in 5,820 years for $D=10^{-20}$ cm^2/sec. The range of these values emphasizes the role of the diffusion coefficient in fixation kinetics.

When applied to natural conditions, the assumption of constant concentration on the surface of a particle does not exactly reflect the nature of the phenomenon. It is more correct to assume that the diffusion of a nuclide into a particle proceeds from a certain limited volume of surface layer in contact with the solid phase. The amount of nuclide in it decreases with time as the nuclide penetrates into the particle. The shape of clay particles should be regarded as cylindrical (flat disk) rather than spherical. In this case, the solution of the diffusion equation assumes the form /21, p. 98/

$$\frac{a}{a_{\infty}} = 1 - \sum_{n=1}^{\infty} \frac{4\lambda(1+\lambda)}{4(1+\lambda)+\lambda^2 q_n^2} e^{-q_n^2\tau}, \tag{5.8}$$

where $\lambda = 2V/(1+\Gamma)\pi r^2 l$; V is the volume of the surface layer; Γ is the ratio of equilibrium concentrations in the particle and surface layer; r is the disk radius; l is the thickness; $\tau = Dt/r^2$; q_n are roots of the characteristic equation $2I_1(q)+\lambda q I_0(q)=0$ (their numerical values are provided in /21/); I_1 and I_0 are Bessel functions of first and zero orders, respectively.

The applicability of equations (5.7) and (5.8) to a quantitative description of the phenomenon of nuclide fixation under natural conditions can readily be tested by using suitable experimental data. The time-independence of the diffusion coefficient calculated from experimental data may serve as the applicability criterion.

Adsorption and desorption of radionuclides
by the bottom of water bodies

Radioactive fission products deposited on the earth's surface as global or
local fallout often become dissolved in the water of ponds, lakes and other
nonflowing water bodies. In the course of time and in the absence of further
influx of radionuclides their content in the water diminishes due to their
absorption by the bottom of the water body and also (but to a considerably
lesser degree) by the biomass of the water body. The converse phenomenon
is likewise possible, namely, an increase in the concentration of radionuclides in water after a change of water in such a water body.

In all these cases it is very desirable to be able to forecast the concentration of radionuclides, especially ^{90}Sr, in the water of lakes and ponds for
more or less prolonged periods. Such a possibility is provided by the concept of the diffusional nature of ion-exchange adsorption of radionuclides
in water bodies /11, 12/.

Ion-exchange adsorption of ^{90}Sr by bottom sediments is not confined to
exchange between the bottom surface and water, but is accompanied by
penetration of the nuclides into the bottom sediments. If the bottom surface
is defined as a thin layer (say, 1 mm thick), then adsorptional equilibrium
between water and surface must be established very rapidly, since the exchange conditions in this case are approximately the same as in kinetic
experiments on ^{90}Sr adsorption by soil suspensions, in which equilibrium
was attained after about 1 hour /2, 58/. However, a decline in the concentration of nuclides in natural water bodies takes months and years, because
the process of adsorption of a nuclide is not terminated by the establishment of equilibrium between water and the surface layer; there is continuous ion exchange between the surface layer and the underlying layer, the
latter and its underlying layer, and so on.

As a result, it is an experimental fact that the contents of a nuclide
differ in different vertical layers of bottom sediments. To put it differently,
there is a migration of radionuclides into the bottom sediments. If filtration of water in the bottom sediments does not take place or if it proceeds
slowly, then diffusion becomes the principal migration mechanism. It also
controls the kinetics of nuclide adsorption by the bottom from water for
fairly long periods following the deposition of a nuclide in the water body
(weeks, months and years).

Prokhorov /11, 12/ used corresponding solutions of diffusion equation
(5.1) to derive formulas describing the time-variation of radionuclide concentration in water, assuming that natural mixing of water ensures uniform
concentration of the nuclide in water at any given time and at any distance
from the bottom surface.

In the case of a single (one-time) deposition of a radionuclide in a water
body, the adsorption kinetics equation assumes the form

$$\frac{u}{u_0} = e^{y^2}[1 - \Phi(y)], \qquad (5.9)$$

where

$$y^2 = \frac{K_d^2 Dt}{h^2}; \qquad (5.10)$$

u_0 and u are the nuclide concentrations in water at the initial moment of time (immediately after its deposition) and at time t, respectively; $\Phi(y)$ is the error function; K_d is the coefficient of nuclide distribution in the bottom sediments —water system; D is the actual diffusion coefficient; h is the mean depth of the water body.

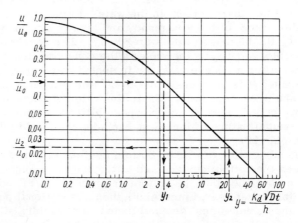

FIGURE 5.1. Relative radionuclide concentration u/u_0 in the water of a nonflowing water body as a function of $y = \dfrac{K_d \sqrt{Dt}}{h}$

Figure 5.1 shows a plot of functions (5.9). All the quantities entering dimensionless parameter y^2 possess a clear physical meaning and can be determined experimentally by independent measurements; its form indicates how the adsorption rate is influenced by its elements. For instance, an n-fold increase in the mean depth of the water body or distribution coefficient has the same effect on the concentration of nuclide in water, other conditions being equal, as an n^2-fold decrease in the diffusion coefficient or time of residence of the nuclide in the water body.

It is important that the distribution and diffusion coefficients enter parameter y^2 as the product $K_d^2 D$, because in practical calculations quantity $K_d^2 D/h^2$ can be regarded as a single coefficient characterizing a given water body.

By making use of Figure 5.1 and with known concentration u_1 at time t_1, it is possible to calculate concentration u_2 for any other time t_2. For this purpose, the graph is used to find the value y_1^2 corresponding to the experimentally determined quantity u_1/u_0, and this is used to determine the numerical value of $K_d^2 D/h^2 = y_1^2/t_1$. Then y_2^2 for time t_2 is determined by substituting $K_d^2 D/h^2$ in (5.10), and the corresponding value of u_2/u_0 is found from the graph.

For large values of y (corresponding to large values of t), the calculations may be based on the following simplified formula (5.9):

$$\frac{u}{u_0} \approx \frac{h}{K_d \sqrt{\pi D t}} .$$

(5.11)

The error of formula (5.11) in comparison to (5.9) is less than 5% when $y^2 \geqslant 10.5$.

If it is necessary to introduce a correction for radioactive decay of the nuclide during its adsorption, the right-hand side of (5.9) or (5.11) is multiplied by $e^{-\lambda t}$, where λ is the decay constant.

Applications of formulas (5.9) and (5.11) are illustrated by the calculation curves for the relative concentration of ^{90}Sr as a function of time /13/ for four water bodies, the characteristics of which are presented in /20/.

For the continuous (constant rate) influx of a radionuclide into a water body, the equation of diffusional adsorption kinetics is

$$\frac{uh}{qt} = \frac{1}{y^2} \left\{ \frac{2y}{\sqrt{\pi}} - 1 + e^{y^2}[1 - \Phi(y)] \right\}, \qquad (5.12)$$

where $q =$ const signifies the amount of nuclide deposited in the water body from external sources per unit time through unit area*.

The diffusional concept also allows one to compute the radionuclide migration from bottom sediments into the water of a water body after the contaminated water has been replaced with clean water. A somewhat more complex problem (though, in principle, solved similarly) is the calculation of radionuclide adsorption and desorption in a flowing water body such as a river, a pond with flowing water, etc.

The diffusional approach to the bottom adsorption phenomenon yields a mathematical description which avoids the use of purely empirical quantities and unsubstantiated initial assumptions (such as the possibility of establishing adsorption equilibrium between the water and the bottom of the water body).

Root nutrition of plants

We have dealt with the roles played by diffusion processes in the migration of radionuclides within a single link in the food chain (soil or subsoil). Now we consider the role played by diffusion in migration directly from one link to another (soil to plant). All the experimental data were obtained for ordinary mineral nutrients of plants (including trace elements). However, although there are still no data available on the contribution of diffusion to the intake of ^{90}Sr, ^{137}Cs and other radionuclides by plants, there is no doubt that suitable investigations will be undertaken in the near future. The mechanisms of adsorption of radionuclide ions by plants in no way differ from the mechanisms of adsorption of ions of nutrients. Therefore, the methods and concepts briefly described below are applicable to the root intake of radionuclides by plants.

The root system of plants does not absorb ions or molecules of a substance unless they are located directly at the surface of roots or a root hair. From there they are adsorbed by the surface layer of cells and are

---

* In this case, contrary to the foregoing case, the radioactive decay of the nuclide cannot be taken into account by simply multiplying (5.12) by $e^{-\lambda t}$, as stated erroneously in /11/.

then included in metabolic processes. However, at any given moment no more than 3% of the total content of nutrients in the root zone are in contact with the entire surface of all the roots of a plant /28/. Even smaller values have been reported, from $2 \cdot 10^{-3}$ to 0.2% of the total content of nutrients /64/. These amounts are obviously inadequate for satisfying the plants' needs in many elements. The reserve of nutrients in contact with the root is rapidly exhausted and the plant would suffer from nutrient deficiency but for their replenishment at the root surface. Here and below we refer to plants growing in soil, not in nutrient solution. Only the physiological patterns of nutrition are operative in the latter case.

Recent investigations /25, 27, 28, 72/ have demonstrated that the depletion of ions and molecules of nutrients at the root surface is replenished by the following three processes: 1) penetration of roots into microareas of the root zone, where the reserves of nutrients have not yet been exhausted (this process is known as "root interception"); 2) migration of ions and molecules in the soil solution moving due to absorption of water by the root (convective transport); 3) diffusion of ions and molecules toward the root through soil solution and over the surface of soil particles. The convective transport and diffusion of cations (as well as some anions) are complicated by simultaneous adsorption and desorption of ions.

In order to compare the relative contribution of these three processes Barber et al. /28/ determined a given element in the plant and the separate contributions of its root interception and convective transport with the flow of solution. The contribution made by diffusion is calculated as the difference between the content of the element in the plant and these two quantities.

The determination of the contribution made by individual processes merits a somewhat more detailed examination. As the roots grow they occupy the space in soil that was originally occupied by soil particles carrying adsorbed ions as well as the space originally occupied by soil solution containing these ions. It is assumed that the entire quantity of ions originally present in such a space is absorbed by the roots. Once the total volume of root system and the mean volume concentration of an element in soil are known, the amount of element absorbed in the process can readily be discerned. It should be noted that this quantity represents the maximum possible absorbed amount, i.e., it is somewhat larger than the actual absorption.

Absorption of water by the root system generates an influx of the soil solution toward the root surface. As a result the elements in the solution are transported to the roots, where they can be assimilated by the plant. The amount of element thus absorbed is found by multipliyng its mean concentration in the soil solution by the total amount of moisture transpired by the plant. It is however improbable that the amount of an element arriving at the root surface would be exactly identical with the amount assimilated by the root. Unless the first two processes ensure a rate of influx of the element that is equal to the rate of its absorption, the soil layer adjacent to the root is depleted in the given element (depletion zone). These conditions generate a concentration gradient of this element and consequently a diffusion flux toward the root surface (Figure 5.2, a). This flux is responsible for the remainder of the element assimilated by the plant. If the element's influx rate exceeds the plant's needs, the element is accumulated in the soil layer adjacent to the root (accumulation zone). The concentration gradient generated under these conditions actuates a diffusion flux in the reverse direction to the convective flux (see Figure 5.2, b). In this case the calculation based on the mean concentration of the soil solution will produce a somewhat high result for the contribution of the convective flux, since the amount of element brought by the flux to the root surface is not entirely absorbed by the latter.

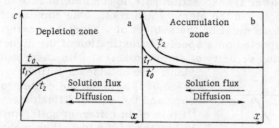

FIGURE 5.2. Concentration profile of an element at a root surface in convective-diffusional transport:

a — rate of absorption by the surface exceeds the supply rate of the element; b — rate of absorption by the surface is less than the supply rate of the element.

Thus, diffusion may play a dual role, either supplying the lacking amounts of element to the surface or removing its excess, depending on the conditions (transpiration rate, soil solution concentration, degree of adsorption of the element, rate of its absorption by the root surface) governing whether the element's concentration at the root surface is higher or lower than its concentration at some distance from the latter; in both cases diffusion has a beneficial effect on the functioning of the root system.

The interaction between the convection and diffusion fluxes was experimentally confirmed by autoradiography both for cations and anions /26/. Corn plants were grown in boxes (with a side wall of thin transparent plastic) filled with initially uniformly tagged soil (^{86}Rb, ^{32}P, ^{35}S). Autoradiographs were made of the soil together with the plant roots. The autoradiographs displayed distinct depletion (^{86}Rb, ^{32}P) and accumulation (^{35}SO$_4$) zones near the roots. The presence of these zones provided convincing proof of the possibility of both direct and reverse diffusional fluxes.

As already stated, the contribution of diffusion to the supply of elements to a plant is determined as the difference between the total amount of elements in the plant and the contribution of the other two processes. The results for the latter are too high and consequently the contribution of diffusion calculated in this way is too low. Nevertheless, it is not directly determined at present owing to difficulties involved in measuring the total surface area of the root system of plants grown in soil. Knowledge of this quantity and also of the diffusion coefficient of an element and its concentration in soil will permit direct calculations of the contribution made by the diffusion process.

Barber et al. /26, 28/ studied the relative importance of individual mechanisms in supplying plants with different nutrients. By investigating more than 100 soils they established that diffusion is usually the principal source of supply of potassium and phosphorus to the plant (as much as 90% and more of the total supply). Moreover, depending on the conditions, it in itself is capable of controlling the intake by the plants of many trace elements, including manganese, iron and zinc /61/.

In general, the relative importance of diffusion is greater at higher soil moisture contents /49/, lower concentrations of the element in question in the soil solution, lower transpiration rate, lower growth rate of roots and shorter distance between the element and root. The comparative importance of these factors must be the subject of a special investigation.

Barber /26/ reported on a special investigation of the direct relationship between the diffusion velocity of an ion in soil and its absorption by the plant. He obtained a satisfactory correlation ($r^2 = 0.939$) between the content of ^{86}Rb in corn and the actual diffusion coefficient of ^{86}Rb in eight different soils, while the contributions of the two other mechanisms were minimized as far as possible (Figure 5.3). In our opinion, Barber's work is very important, since it approaches an estimation of the intake of elements by plants starting from the soil characteristics that are directly related to the intake mechanism, instead of being based on arbitrarily selected soil properties.

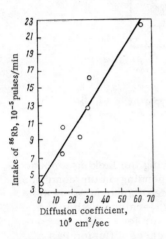

FIGURE 5.3. Relationship between actual diffusion coefficient in soil and intake of ^{86}Rb by corn plants

Although experimental data on the supply mechanisms of ^{90}Sr and ^{137}Cs to roots of plants are not yet available, some of the above-enumerated irregularities suggest that diffusion plays a fairly important role. Indeed, the absolute concentrations of these nuclides in the soil solution are infinitesimal; as a rule they are located near the soil surface, i. e., in the zone with the highest density of roots. Finally, in the case of ^{137}Cs the predominant role of diffusion is confirmed also by the chemical similarity between cesium and potassium.

The concept of the diffusional nature of the supply of nutrients to the root systems of plants provided the basis for several mathematical models of the process of ion intake by plants.

One of the earliest publications in this field was by Bouldin /31/, who noted that the transport of ions toward roots involves two components (diffusional and convective). Bouldin demonstrated by means of simple calculations that in the case of phosphorus intake convective transport does not play the dominant role. A plant absorbs 500 —1,000 g water per gram dry mass. (Still smaller values were reported by others, e. g., 250 —300 g per 1 g dry mass in the case of gramineous plants /63/.) One gram dry mass usually contains at least 2 mg phosphorus. If all the phosphorus is transported to the root surface in the flux of water, its concentration in the soil solution should exceed $2 \cdot 10^{-4}\%$, which is more than the actual values. Bouldin considers only the diffusional components in his model in order to estimate its role in the transport of phosphorus.

The root of a plant was regarded as a cylindrical drain of radius R located in an infinite medium with initial concentration c_0 of the element under consideration. The rate of absorption of the element by the root sur- face F was assumed to be proportional to the element concentration at this surface. In this case the diffusion equation and boundary conditions assume the form

$$\frac{\partial c}{\partial t} = D \left(\frac{\partial^2 c}{\partial r^2} + \frac{1}{r} \frac{\partial c}{\partial r} \right), \qquad (5.13)$$

$$c = c_0, \quad R < r < \infty, \quad t = 0, \qquad (5.14)$$

$$F = D \frac{\partial c}{\partial r} = Mc, \quad r = R, \qquad (5.15)$$

where c is the element concentration; t is time; r is the distance to the root axis; D is the diffusion coefficient; M is a proportionality factor. Bouldin made use of the well-known solution of the diffusion equation for these conditions in the form

$$b = \frac{F}{Mc_0} = I(N, A), \qquad (5.16)$$

where b is the ratio of the rates of absorption from soil F and from thoroughly mixed solution Mc_0; $I(N, A)$ is the function described in /31/; $N = RM/D$ and $A = Dt/R^2$ are dimensionless parameters. The paper provides computational graphs for b as a function of time (0 to 48 hr), diffusion coefficient ($10^{-5} - 10^{-9}$ cm^2/sec), radius of roots ($5 \cdot 10^{-2}$ and $7.5 \cdot 10^{-4}$cm), and coefficient M.

The calculation showed that the part played by root hairs ($R = 7.5 \cdot 10^{-4}$ cm) in the absorption of phosphorus was much more important than the part played by the roots proper ($R = 5 \cdot 10^{-2}$ cm), on account of their larger sur- face area and also their higher efficiency (as much as tenfold) per unit area. It should be emphasized that this is not a physiological but purely a geo- metrical property of diffusional absorption. At diffusion coefficients equal to 10^{-7} cm^2/sec and less the absorption of phosphorus is almost entirely dependent on its assimilation by the root hairs.

By using data on phosphorus absorption from solutions of different con- centrations, Bouldin calculated coefficient M, equal to the ratio of absorption rate per unit area to concentration at the surface. Coefficient M is essen- tially the linear velocity of the phosphorus during its absorption by the root; its values vary between $2 \cdot 10^{-5}$ and $18 \cdot 10^{-5}$ cm/sec, depending on the con- centration and pH.

Other diffusional mathematical models of the absorption of phosphorus and potassium, differing from Bouldin's model mainly in the boundary con- ditions adopted for the root surface, were used by Olsen et al. /63/, Miles /51/, Lewis and Quirk /47/, Graham-Bryce /42/, Jenny /43/, and Nye /59/.

Nye and Spiers /60/ also proposed a more general model involving combined convective-diffusional transport.

A recently suggested model takes into account all the three processes replenishing the loss of elements at the root surface /13/, namely, root interception, convective transport, and diffusion.

We shall briefly consider two investigations which demonstrated the possibilities offered by studying the intake of individual elements through the roots.

Olsen and Watanade /62/ performed independent experimental measurements of the mean absorption rate of phosphorus per cm^2 root surface, the diffusion coefficient of phosphorus in soil, its concentration in soil, its concentration in the soil solution and the soil's capacity for absorbing phosphorus. They used the diffusion equation to calculate that the depleting (with respect to phosphorus) influence of a corn root on the surrounding soil manifests itself after 24 hr contact at distances up to 0.04 cm from the root surface, over one-half of the total amount of elements being absorbed from the 0.01-cm-thick layer. Once the total absorbing surface of the root system is known, it is not difficult to calculate the effective soil volume from which the element is absorbed by the plant, and then to compare it with the total value of the root zone. Lewis and Quirk /47/ thus found that in the case of wheat plants grown in vegetative pots each of capacity 600 cm^3, the effective volume of soil from which phosphorus is absorbed by roots is only 110 cm^3 per plant. The difference is liable to be even more pronounced under field conditions, since a plant seems to utilize only an insignificant fraction of the root zone for its nutrition. If this is so, then any irregularities in the distribution of an element in this layer combined with individual differences in the degrees of development of the root system may result in discrepancies between the contents of the elements in individual plants. This may be one cause of the usual scatter of the results of field observations. For successful utilization of diffusional and convective-diffusional models information is required concerning the structure and dimensions of the root systems (including the effective surface area capable of absorbing radionuclides).

There is still another process of a diffusional nature allied to the root intake of radionuclides by plants, namely, the penetration of radionuclides into tubers of plants from soil directly through the tuber surface. In experiments on potatoes grown in soil with different locations of the ^{90}Sr-tagged horizon, an overwhelming part of the radionuclide accumulated in the tubers toward the end of the growing season had been absorbed by direct diffusion from soil in contact with the tubers /6, 50/. In our opinion this process, as yet little studied, warrants further research.

Quite possibly, phenomena in which the diffusion of radionuclides in soils is capable of playing an appreciable role are not exhausted by those treated above, in which the presence of diffusional contributions may be regarded as proved. In any case, when considering nuclide migration in soil one must always remember the inevitable existence of diffusion, in order to attempt a quantitative estimation of its results in each particular case.

Research into the role played by diffusion in radionuclide migration is now proceeding in three directions. First, the diffusion velocity of different radionuclides (or nutrient ions) in soil is being measured under conditions approximating natural conditions, as well as the dependence of this rate on different factors. Second, the actual behavior of radionuclides under natural conditions is being studied, including their distribution over the soil profile in relation to the latter's properties: rate of fixation, rate of adsorption from the water in water bodies, their contents and the rate of their intake by plants with simultaneous investigation of the peculiarities of the root system of plants. Finally, mathematical models are being constructed in order to combine experimental data obtained by research proceeding along the first two directions. At present, these three trends have not yet completely merged, but interest in this field of research is steadily growing, as evidenced by the growing number of relevant publications.

The majority of diffusional problems are rather trivial from the purely mathematical standpoint, solutions of many analogous problems having long been in use in heat conduction theory. Therefore these problems do not generally attract the interest of experts in the fields of mathematics and mathematical physics. However, radioecologists are inadequately acquainted with the theory of diffusional processes and often underestimate the possibilities offered by it or cannot utilize them owing to their inadequate mastery of the mathematical apparatus. As a result, one occasionally encounters a definite discrepancy between the large volume of accumulated factual information concerning radionuclide migration in soil and the degree of theoretical processing of this information. One purpose of this chapter was to narrow this gap by systematizing the range of migration phenomena to which diffusion theory and the information on the diffusion of nuclides in soils are applicable.

There is no doubt that systematization of this kind can also be carried out with respect to transport processes other than diffusion, such as convective transport, the theory of which has likewise been fairly well examined. Understanding of the quantitative theoretical laws governing the migration of radionuclides in soil, including their diffusional migration, will enable their intake by plants to be forecast and will help in the preparation of measures aimed at regulating this intake.

References

1. Aleksakhin, R. M. Radioaktivnoe zagryaznenie pochv i rastenii (Radioactive Pollution of Soils and Plants). Moskva, Izdatel'stvo Akademii Nauk SSSR. 1963.
2. Gromov, V. V. — Zhurnal Fizicheskoi Khimii, Vol. 34:1357. 1960.
3. Kokotov, Yu. A. and S. R. Vil'ken. — In Sbornik: "Radioaktivnye izotopy v pochvakh i rasteniyakh," Trudy po Agronomicheskoi Fizike, No. 18:35. Moskva — Leningrad, "Kolos." 1969.
4. Makhon'ko, K. P. — Ibid., p. 48.
5. Makhon'ko, K. P. and V. B. Chunichev. — Ibid., p. 57.
6. Mel'nikova, M. K. and Z. A. Baranova. — Ibid., p. 116.
7. Prokhorov, V. M. — Kolloidnyi Zhurnal, Vol. 21:60. 1963.
8. Prokhorov, V. M. — Atomnaya Energiya, Vol. 18:631. 1965.

9. Prokhorov, V. M. Issledovanie diffuzii strontsiya-90 v pochve (Investigation of the Diffusion of Strontium-90 in Soil). Thesis, Leningrad, LGU. 1965.
10. Prokhorov, V. M. — Pochvovedenie, No. 10:61. 1965.
11. Prokhorov, V. M. — Atomnaya Energiya, Vol. 20:448. 1966.
12. Prokhorov, V. M. — Radiokhimiya, Vol. 11:217. 1969.
13. Prokhorov, V. M. — Agrokhimiya, No. 7:126. 1970.
14. Prokhorov, V. M. et al. — Agrokhimiya, No. 2:40. 1970.
15. Prokhorov, V. M. and A. S. Frid. — Radiokhimiya, Vol. 7:496. 1965.
16. Prokhorov, V. M. and A. S. Frid. — Pochvovedenie, No. 3:68. 1966.
17. Prokhorov, V. M. and A. S. Frid. — Radiokhimiya, Vol. 8:705. 1966.
18. Prokhorov, V. M. and Chai Tien-ying. — Radiokhimiya, Vol. 5:639. 1963.
19. Prokhorov, V. M. and Chai Tien-ying. — Pochvovedenie, No. 7:107. 1963.
20. Rovinskii, F. Ya. — Radiokhimiya, Vol. 9:80. 1967.
21. Timofeev, D. P. Kinetika adsorbtsii (The Kinetics of Adsorption). Moskva, Izdatel'stvo Akademii Nauk SSSR. 1962.
22. Chuveleva, E. A. et al. — Zhurnal Fizicheskoi Khimii, Vol. 36:1378. 1962.
23. Anochin, V. L. et al. — In: Radioecological Concentration Processes, p. 409. Oxford, Pergamon Press. 1966.
24. Van der Molen, N. H. — Soil Sci., Vol. 81:19. 1956.
25. Barber, S. A. — Soil Sci., Vol. 93:39. 1962.
26. Barber, S. A. Limiting Steps in Ion Uptake by Plants from Soil. — Techn. Rep. Ser. N 65:39. Vienna, IAEA. 1966.
27. Barber, S. A. et al. — Trans. Comm. IV and V, Int. Soil Conf., New Zealand, Vol. 3:121. 1962.
28. Barber, S. A. et al. — J. Agr. Food Chem., Vol. 11:204. 1963.
29. Bloksma, A. H. — J. Coll. Sci., Vol. 12:40. 1957.
30. Bolt, G. H. et al. — Soil Sci. Soc. Amer. Proc., Vol. 27:294. 1963.
31. Bouldin, D. R. — Soil Soil Sci. Amer. Proc., Vol. 25:476. 1961.
32. Chute, J. H. and J. P. Quirk. — Nature, Vol. 213:1156. 1967.
33. De Haan, F. A. M. et al. — Soil Sci. Soc. Amer. Proc., Vol. 29:528. 1965.
34. Deist, J. and O. J. Talibudeen. — Soil Sci., Vol. 18:138. 1967.
35. Drew, M. C. et al. — Plant and Soil, Vol. 30:252. 1969.
36. Fletcher, G. and W. H. Slabaugh. — J. Coll. Sci., Vol. 15:485. 1960.
37. Frere, M. H. and D. F. Champion. — Soil Sci. Soc. Amer. Proc., Vol. 31:188. 1967.
38. Frissel, M. J. and P. Poelstra. — Soil Sci., Vol. 98:274. 1964.
39. Frissel, M. J. and P. Poelstra. — Plant and Soil, Vol. 26:285. 1967.
40. Frissel, M. J. and P. Poelstra. — Plant and Soil, Vol. 27:20. 1967.
41. Gast, R. G. — J. Coll. Sci., Vol. 17:492. 1962.
42. Graham-Bryce, I. J. Plant Nutrient Supply and Movement. — Techn. Rep. Ser., No. 48:42. Vienna, IAEA. 1965.
43. Jenny, H. — Plant and Soil. Vol. 25:265. 1966.
44. Lai, T. M. and M. M. Mortland. — Soil Sci. Soc. Amer. Proc., Vol. 25:353. 1961.

45. Lai, T. M. et al. — Proc. 4th Conf. Radioisotopes, p. 1127.   Tokyo,
          Oct. 1961. Tokyo, Japan Atomic Ind. Forum. 1962.
46. Lapidus, L. and N. R. Amundson. — J. Phys. Chem., Vol. 56:984.
          1952.
47. Lewis, D. G. and J. P. Quirk. Plant Nutrient Supply and Movement.—
          Techn. Rep. Ser., No. 48:71. Vienna, IAEA. 1965.
48. Libby, W. F. — Science, Vol. 128:1134. 1958.
49. Limiting Steps in Ion Uptake by Plants from Soil. — Techn. Rep. Ser.,
          No. 65:147. Vienna, IAEA. 1966.
50. Melnikova, M. and Z. Baranova. — In: Radioecological Concen-
          tration Processes, p. 409. Oxford, Pergamon Press. 1966.
51. Miles, J. W. — Proc. Roy. Soc., Ser. A., Vol. 284:137. 1965.
52. Mortland, M. M. — Soil Sci., Vol. 91:11. 1961.
53. Mortland, M. M. et al. — Soil Sci. Soc. Amer. Proc., Vol. 20:476. 1956.
54. Mortland, M. M. et al. — Soil Sci. Soc. Amer. Proc., Vol. 22:503. 1958.
55. Mortland, M. M. et al. — Soil Sci. Soc. Amer. Proc., Vol. 23:363. 1959.
56. Newbould, P. and C. H. S. Walter.— Agric. Res. Council Radiobiol.
          Lab.   Annual Report, 1964 — 65. ARCRL, Vol. 14:49. 1965.
57. Newbould, P. and C. H. S. Walter.— Agric. Res. Council Radiobiol.
          Lab.   Annual Report, 1965 — 66, ARCRL, Vol. 16:51. 1966.
58. Nishita, H. and P. Taylor.— Soil Sci., Vol. 98:181. 1964.
59. Nye, P. H. Limiting Steps in Ion Uptake by Plants from Soil. —
          Techn. Rep. Ser., No. 65:66. Vienna, IAEA. 1966.
60. Nye, P. H. and J. A. Spiers.—Trans. 8th Intern. Congress Soil Sci.,
          Bucharest, Vol. 3:535. 1964.
61. Oliver, S. and S. A. Barber. — Soil Sci. Soc. Amer. Proc.,
          Vol. 30:468. 1966.
62. Olsen, S. R. and F. S. Watanabe. — Soil Sci. Soc. Amer. Proc.,
          Vol. 30:598. 1966.
63. Olsen, S. R. et al. — Soil Sci. Soc. Amer. Proc., Vol. 26:222. 1962.
64. Rassioura, J. B. Limiting Steps in Ion Uptake by Plants from Soil. —
          Techn. Rep. Ser., No. 65:82. Vienna, IAEA, 1965.
65. Phillips, R. E. and D. A. Brown. — J. Soil Sci., Vol. 17:200. 1966.
66. Reiniger, P. et al. — In: Actes du Symposium International de Radio-
          ecologie, p. 589. C. E. N. Cadarache. 1969.
67. Reiniger, P. and N. Lahav. Plant Nutrient Supply and Movement. —
          Techn. Rep. Ser., No. 48:85. Vienna, IAEA. 1965.
68. Roberts, H. and R. G. Menzel. — J. Agr. Food Chem., Vol. 9:95.
          1961.
69. Rowell, D. L. et al. — J. Soil Sci., Vol. 18:204. 1967.
70. Sawhney, B. L. — Soil Sci. Soc. Amer. Proc., Vol. 30:565. 1966.
71. Schulz, R. K. and H. H. Riedel. — Soil Sci., Vol. 91:262. 1961.
72. Shapiro, R. E. et al. — Soil Sci. Soc. Amer. Proc., Vol. 24:161. 1960.
73. Squire, H. M.. — Nature, Vol. 188:108. 1960.
74. Thorntwaite, C. W. et al. — Science, Vol. 131:1015. 1960.
75. Walker, G. F. — Nature, Vol. 184:1392. 1959.
76. Winkler, R. and E. Leibnitz. — Kernenergie, Vol. 3:992. 1960.

Chapter 6

RADIONUCLIDE MIGRATION IN FOREST BIOGEOCENOSES

Introduction

Investigation of radionuclide migration in forest stands is one of the major tasks of forest radiation biocenology. This task has acquired special practical and theoretical urgency owing to the spreading of artificial radioactive substances in the biosphere. There are many reasons for treating the migration of radionuclides in forests as an independent problem of forest radiation biogeocenology.

The distribution, redistribution and circulation of radionuclides in forest cenoses possess specific features in comparison to migration of radioactive substances in other types of natural cenoses. In view of the economic importance of forest products the distribution of radionuclides in a forest cenosis is important from the standpoint of examining possible economic uses for contaminated forest products. Forests, being independent natural formations occupying a considerable proportion of land, exert an appreciable influence on the global migration of radioactive substances. Forest stands accumulate radionuclides and prevent their spread over the earth's surface. Information on the distribution and migration of radionuclides provides the basis for reliable forestry measures on radioactively contaminated territories.

Finally, it should be emphasized that radionuclides in the biosphere and forest biogeocenoses act as carriers of ionizing radiation — an important ecological factor, the effect of which is studied on a cenological level by radiation biogeocenology or radioecology. Radioactive fallout fission products produced by nuclear explosions and radionuclides entering the biosphere as a result of emergencies, as well as radioactive wastes in the environment are of special interest from their practical standpoint as a source of forest cenosis irradiation. In analyzing the possible consequences one must bear in mind that forests as a type of biogeocenosis are extremely sensitive to the effects of ionizing radiation in comparison to other types of biogeocenoses. Therefore studies of radionuclide migration as a source of radiation are of interest from the standpoint of predicting possible radioecological consequences of radioactive pollution of forests. Accumulation of quantitative information on the circulation of radionuclides in forest stands is a necessary and sufficient condition of calculating irradiation doses of different components in a forest biogeocenosis.

The migration and circulation of radionuclides in different landscapes is, generally speaking, a subject dealt with not only by radioecology and radiation biogeocenology, but also by biogeochemistry, which treats the biogeochemical migration cycles of natural stable and radioactive nuclides. The specific features of the radioecological approach is that it deals primarily with the behavior of artificial radionuclides, the migration of which in natural landscapes has a markedly unsteady character. This differs from natural stable and radioactive nuclides, the distribution of which among living matter and dead environment is generally close to steady-state and varies only slowly with time.

Biogeochemical migration of natural stable and radioactive nuclides is also a steady-state quasi-equilibrium process occurring at an approximately constant rate and in many cases over a closed cycle. In contrast to natural stable and radioactive nuclides, as soon as artificial radionuclides penetrate the biosphere they are distinguished by a distribution that is spatially nonuniform and varies rapidly with time. Radionuclide migration under these conditions is distinguished by an irreversible initial stage, since the distribution of artificial radionuclides among external environmental components in the subsequent period does not, generally speaking, return to its initial state. In the course of time the distribution of artificial radionuclides among the components of biogeocenoses gradually approaches the distribution of stable nuclides of the same chemical elements. However, the attainment of ecological equilibrium state between artificial radionuclides and stable nuclides of the same elements usually requires considerable periods of time, significantly longer than the half-life of long-lived artificial radionuclides.

On the whole, migration of radionuclides in forest stands has been less thoroughly investigated than the migration of radioactive substances on other farmlands. One case of this relative inadequacy lies in methodological difficulties involved in studying the migration of radioactive substances in forests, especially under conditions of aerial contamination. These difficulties are primarily due to the necessity of introducing radioactive substances over considerable areas owing to the large dimensions of arboreal plants; in order to obtain statistically reliable results it is necessary to process a larger area in a forest than on farmlands with herbaceous vegetation, for the same number of plant samples. Furthermore, taking into consideration that the tree layer is a perennial formation, investigations into radionuclide migration in forest biogeocenoses must encompass a period of many years. Finally, one must take into account the difficulties involved in introducing radioactive substances into forest stands (including simulation of aerial influx of radioactive substances) from the standpoint of radiation safety for the experimenter.

Methodologically speaking, radioecological investigations into radionuclide migration in forests can be subdivided into three groups: 1) small-scale experiments with artificial contamination of small areas (a few dozen m^2) under arboreal vegetation (mainly young stands) with measured amounts of radionuclides; 2) large-scale experiments with introduction of relatively high concentrations of radioactive substances into forest cenoses, conducted on fairly large areas up to 1 hectare and more; 3) studies of the migration of radionuclides deposited on forest stands with global fallout.

Each such experiment usually aims at solving a particular problem. Small-scale experiments (such as those carried out under Timofeev and

Resovskii /9—11/ involving the introduction of various radionuclides into
the soil under the crowns of individual trees) offer the opportunity of detailed
studies into the buildup of a range of radionuclides in different arboreal
species and changes in their migration under the influence of environmental
factors controlled by the experimenter. Large-scale experiments may be
exemplified by research conducted at the Oak Ridge National Laboratory
involving artificial introduction of comparatively large amounts of ^{137}Cs
(about 1 mcurie/m^2) into adult forest stands over areas of several hundred
square meters /24/. The information elicited by these experiments enables
one to describe radionuclide migration in stands as a whole, which is primarily of practical interest. From the standpoint of the theory, these
experiments may yield information revealing laws governing the migration
of radionuclides in forest stands over a multiannual period.

Finally, studies of the distribution and redistribution in forest cenoses of
radionuclides deposited on forest vegetation with global fallout yield information on the behavior of different radioactive fission products in forest
biogeocenoses of different types.

Information concerning the migration of radionuclides in forest stands
with aerial methods of introducing radioactive substances into the cenoses
is of great practical interest, since the influx of radioactive substances into
the biosphere in this way is the most significant potential source of radioactive contamination of forest products and radiation damage to forest stands.

However, studies into the distribution, redistribution and circulation of
radionuclides deposited with global fallout encountered difficulties when
interpreting the migration paths of radionuclides in forest biogeocenoses
due to specific features of global fallout (the fallout occurs over many years
with large fluctuations not only from one year to another but also seasonal
fluctuations within the same year). At the same time, the gathering of
empirical data on the distribution of radioactive fission products deposited
with fallout due to nuclear weapon tests is of undoubted practical significance from the standpoint of radiation hygiene.

Comprehensive studies into the distinctive features of the migration of
radioactive fission products in various types of forest biogeocenoses, by
means of small-scale and large-scale experiments with artificial introduction of radionuclides into forest stands, and also by investigating the
migration of fission products deposited with global fallout, are a prerequisite for obtaining the fullest notions on the behavior of radioactive
substances in forests.

Thus, we must formulate the basic task of studies into radionuclide
migration in natural biogeocenoses, including forest ones, as a radioecological problem. This task consists in constructing a space-time
pattern of the migration of radionuclides introduced into a biogeocenosis
within the latter's boundaries over a period of time required by their almost
complete decay. The processes of radionuclide migration in forest stands
will now be examined in more detail.

Primary interception of radionuclides falling out of the
atmosphere in forest stands and migration of radioactive
substances in forest biogeocenoses

Fallout of radioactive particles from the atmosphere is one of the
principal ways for the influx of radioactive substances onto the plant cover.
Typical examples of aerial influx of radionuclides onto the ground are
deposition of radioactive aerosols formed in the course of nuclear weapon
tests (local and global fallout of radioactive fission products) and radionuclides discharged from the chimney stacks of nuclear enterprises. Such
discharges are especially dangerous when nuclear reactors are operating
in an emergency. Fallout of radioactive particles onto a terrain may be
one-time (for instance, in emergency situations) or prolonged (global fallout
following nuclear tests).

The fate of radioactive particles deposited from air onto the soil-plant
cover depends on many factors. In the case of horizontal transport of
radioactive substances with the air stream the deposition rate of radionuclides depends on the particle size, topographical and meteorological
conditions, and the type of plant cover. The presence along the horizontal
air streams of large forest massifs and even isolated trees causes
"filtration" of transported radioactive admixtures and their deposition on the
boles, branches, leaves and needles of trees and other surfaces. Thus, in
studies of the distribution of global radioactive fallout in forest stands, a
high concentration of fission products was observed in windward portions
of tree crowns as well as in the organs of trees that are directly exposed
to fallout /28/. However, on the whole the biogeophysical laws involved in
the interaction of horizontally moving radioactive particles transported by
air streams with forest stands still remain practically uninvestigated.

Under conditions of vertical transport of radioactive impurities onto the
ground as a result of turbulent diffusion or deposition of radioactive
aerosols under the force of gravity, the descending particles likewise
encounter a "filter," namely, aerial organs of arboreal vegetation possessing
a heavily ramified surface which sorbs the radionuclides. This process of
capturing radioactive particles falling out of the atmosphere onto the plant
cover is known as "primary interception" of radioactive substances by
vegetation. Quantitatively, it is described by the primary interception
coefficient $K_{p.i}$:

$$K_{p.i} = \frac{A_1}{A_0},$$

where A_0 is the quantity of radionuclides falling out per unit surface
(curie/km^2), while A_1 is the quantity of radionuclides primarily intercepted
by the plant cover (curie/km^2).

The extent of primary interception by plant cover of radioactive particles
falling out of the atmosphere depends on the dispersity of particles,
meteorological conditions and plant cover characteristics (density, closure,
morphological characteristics of plants, such as pubescent leaves, roughness
of the bole surface, etc.), and so on. The quantitative aspects of aerial
contamination of plant cover have been inadequately studied, in particular
the behavior of radionuclides in a cenosis in the initial period following

their fallout from the atmosphere, while information on the primary interception of radioactive particles in biogeocenoses of different types is very
scant. The majority of studies into extraradical contamination of vegetation
were carried out for herbaceous (cultured and natural) communities, whereas
aerial contamination of forest cenoses have hardly been dealt with in the
literature.

   The primary interception coefficient of radionuclides falling out onto the
plant cover as components of particles depends on the latter's dispersity.
It was experimentally demonstrated on herbaceous vegetation that particles
coarser than $40-45\mu$ are poorly intercepted by vegetation and are rapidly
deposited on the soil surface /12/. With respect to arboreal vegetation, it
may be assumed that also the presence of large unevenesses of the tree
boles and the roughness of branches may result in interception of even
coarser particles. Studies of interception for radioactive fission products
deposited as constituents of global fallout have revealed that, on the average,
25% of the total amount of falling-out radioactive substances is intercepted
by meadow vegetation (natural and cultivated meadows) and agricultural
crop stands /12/. Molchanov et al. /15/ studied the interception by forest
vegetation of various fission products within global fallout and found that the
primary interception coefficient for ^{90}Sr, ^{106}Ru, ^{137}Cs and ^{144}Ce could be
above 40% in the aerial portion of a pine-birch forest.

   In experiments simulating moist radioactive fallout onto advance growth
of pine, Tikhomirov et al. /19/ demonstrated that as much as 90% of the
falling-out radioactive substances are liable to be intercepted by the aerial
portion of arboreal plants.   These data suggest that forest stands possess a
much higher capacity for intercepting falling-out radioactive particles than
herbaceous communities.   In certain cases (very close coniferous stands)
all radioactive substances settling from the atmosphere are almost completely intercepted by the crowns of arboreal plants.

   Radionuclides that are constituents of radioactive particles primarily
intercepted on aerial organs of plants may afterward migrate to internal
tissues of plants (extraradical assimilation) or are removed from plants
(by rain, wind, with litter, etc.).   Extraradical assimilation of radionuclides
deposited on aerial parts of plants depends on the properties of particles
which include the radionuclides, the latter's physicochemical characteristics
and the properties of plants.   Experiments with herbaceous plants revealed
^{137}Cs to be the most mobile radionuclide for assimilation through foliage,
while less mobile radionuclides include ^{90}Sr, ^{95}Zr, ^{95}Nb, ^{106}Ru, ^{144}Ce, etc. /20/.
In the initial period the radioactive particles deposited on the plant cover are
only loosely bound with the plants and can be washed off by rain or blown
off by wind.

   In the case of deciduous species, the leaf fall in autumn plays an important
role in the migration of radionuclides primarily intercepted on aerial
portions of arboreal plants; in this period a significant part of radioactive
particles deposited in tree crowns may migrate to the forest floor.   Radionuclide migration related to biological shedding processes is slower in
coniferous stands, since the life-span of their needles is 3—4 years (sometimes as long as 7 years).

   The high sorption capacity of evergreen coniferous trees with respect
to radioactive particles deposited from the atmosphere, and the prolonged

retention of radionuclides on the perennial needles make it possible to use coniferous trees as a biological indicator of the radioactive pollution of a territory /33/. Collection of needles allows one to rapidly estimate the radiological situation.

The intensity of the migration processes of radionuclides in the first period of their deposition on the plant cover can be suitably described by resorting to the concept of "scavenging half-period," i.e., the time in which one-half of the entire amount of a given radionuclide is removed from vegetation by various processes (excluding radioactive decay). It has been demonstrated that the removal of different radionuclides from herbaceous vegetation approximately obeys an exponential law, the scavenging half-period being about 12—30 days for a fairly broad range of fission products. Simulating a one-time wet fallout for ^{89}Sr on young pine stands (age, 6—10 years), Tikhomirov et al. /19/ found that the scavenging half-period calculated for the aerial portion of a pine in its entirety was 2 weeks. This represents the lower boundary of scavenging half-periods for young forest stands. In very closed coniferous stands (spruce and fir) the scavenging half-period of arboreal plants may reach considerable values. However, as yet it is impossible to provide any quantitative estimates characterizing the contribution of different processes to the overall migration of radionuclides in forest stands, in the absence of certain starting data. The migration of radioactive particles in the first period after deposition of radioactive fallout on the crowns of arboreal plants has been studied very inadequately.

Nonetheless, from a radioecological standpoint, knowledge about radionuclide migration in the initial period is of special interest, since during that time a considerable fraction of deposited radioactive particles is concentrated in the crowns. Radionuclides intercepted by aerial portions of arboreal plants become sources of irradiation of their vitally important organs, and under these conditions the radiation effect manifests itself much more markedly than when radionuclides penetrate through the crowns and immediately proceed to the soil surface. Primary interception of a large quantity of radioactive particles on aerial organs of arboreal plants and slow scavenging of tree crowns play the decisive role in the dose buildup and development of radiation injuries to forests when the contamination source is represented by long-lived beta-emitters (such as ^{90}Sr), the radiation of which possesses lower penetrating capacity than gamma-radiation. It may be assumed that the radiation injury to plants, including arboreal ones, in the case of aerial type of influx of radioactive substances into biogeocenoses is mainly due to radiation of radioactive particles intercepted by aerial organs of plants. This is especially true in the case of coniferous stands distinguished by high interceptive capacity with respect to radioactive fallout and low rate of scavenging radioactive substances.

An additional factor responsible for heavier damage to coniferous in comparison to deciduous forests by fallout of radioactive aerosols is the higher overall radiosensitivity of coniferous species. Deciduous species are more stable toward irradiation, their crowns intercept smaller amounts of radioactive substances in comparison to the coniferous trees, and a considerable part of radionuclides originally sorbed on leaves are removed from the crowns to the forest floor by the leaf fall in autumn, which naturally reduces the irradiation dose on set buds and, on the whole, moderates the radiation injury to the tree layer. Moreover, deciduous trees remain defoliated for a considerable part of the year and consequently radioactive particles deposited from the atmosphere during that period settle directly on the forest floor. In the case of coniferous trees, the "filter of needles" may efficiently intercept radionuclides throughout the year.

Thus several factors (existence throughout the year of an efficient filter for radioactive particles, represented by the needles of gymnospermous coniferous species; prolonged retention of primarily intercepted radionuclides on their needles owing to their perennial nature; higher overall radiosensitivity of coniferous trees) make the coniferous forest the most radiation-sensitive type of forest cenoses.

In conclusion it should be stated that radionuclide migration in forest biogeocenoses in the initial period after deposition of radioactive particles from the atmosphere is of interest not only for forecasting possible radiation effects, but also for determining the radiological situation on a forested territory and also for regulating the uses to which contaminated forest products are put in such a period.

Absorption and accumulation of radionuclides by
forest vegetation from soil

Soil as a source of radioactive substances and the accumulation of radionuclides in forest vegetation through soil often play important roles in the subsequent fate of radioactive substances entering the forest biogeocenosis. In the case of aerial fallout of radioactive substances a fraction is deposited directly on the surface of the forest soil (more precisely, on the forest floor); moreover, a considerable proportion of radioactive aerosols intercepted by tree crowns and herbaceous vegetation subsequently find their way under the forest canopy and into the forest floor and soil, owing to the action of rain, wind, gravity and leaf fall in autumn. Having reached the forest floor and soil, radionuclides become available to the root systems of plants.

Thus, some time after the fallout of radioactive aerosols onto forest stands has ceased, during which time a significant part of the radioactive particles reach the forest floor and soil, radionuclide absorption by plants from the soil reservoir becomes an important, in some cases the main, source of the intake of radioactive substances by plants. Another typical situation in which migration from soil acquires importance in the intake of radionuclides by plants is exemplified by transport of radioactive substances with groundwater from "graveyards" of radioactive wastes /23/.

Studies of the accumulation of various radionuclides by forest vegetation from soil and examination of the concentration loci of radioactive substances taken up through the roots in a forest cenosis enable the internal irradiation doses to be found for different components of forest stands and the radiobiological and radioecological effect related to the intake of radionuclides from soil to be estimated. Furthermore, data on the accumulation of radionuclides by forest vegetation from soil are also of practical interest in preparing recommendations for the utilization of contaminated forest products in the period when migration through the root path plays a decisive role in the intake of radionuclides by plants.

Modern forest soil science has accumulated a large amount of empirical data describing the penetration into forest vegetation of a broad range of stable nuclides many of which belong among important nutrients, while the physiology of mineral nutrition of arboreal plants has elaborated pertinent theoretical concepts shedding light on the assimilation of ash elements and nitrogen by arboreal vegetation from soil /16—18/. This knowledge can be utilized in forecasting the root intake of radionuclides by arboreal plants.

Account should be taken, however, of certain specific features of radio-nuclide assimilation from soil by plants in comparison with stable nuclides, especially those due to ecologically nonequilibrium conditions governing the intake of radionuclides by plants. Artificial radionuclides initially appear as foreign components in external environments, they are characterized by specific physicochemical properties and are mainly concentrated in the upper soil layers. In the course of time they "age," entering exchange reactions of various types with chemically analogous stable nuclides found in the environment. An example of the "aging" of radioactive substances is the gradual conversion of fresh radionuclides to forms that are non-exchangeable and only sparingly available to plants. Furthermore, prolonged processes of horizontal and vertical migration result in a more uniform distribution of artificial radionuclides over the soil profile, in the course of time modifying their availability to the root systems of plants. Still another distinctive feature of the behavior of many radionuclides is their presence in soil in micromicro amounts. As a result, they participate in specific radiochemical reactions to which no complete analogy can be found in the behavior of macro amounts of stable nuclides.

Experimental studies of the absorption by plants of different fission products from soil have established that the majority of radioactive fission products (^{144}Ce, ^{95}Zr, ^{95}Nb, etc.) as well as many nuclides with induced activity (^{60}Co, ^{65}Zn) are usually very strongly sorbed by the soil absorption complex and are therefore little available to plants, including arboreal ones /10, 20/. The long-lived radionuclide ^{90}Sr is the fission product most actively assimilated by plants. Another long-lived fission product, ^{137}Cs, is sorbed much more strongly in soils than ^{90}Sr, and therefore its intake by the plants through their roots is relatively smaller (other conditions being equal) /20/.

Since arboreal vegetation is perennial, interest is concentrated on the in-take of long-lived radionuclides from soil. We shall examine in more detail the accumulation by forest vegetation of the two main long-lived radioactive fission products, ^{90}Sr and ^{137}Cs.

Absorption of ^{90}Sr by arboreal plants varies markedly with the species, deciduous species accumulating larger amounts of ^{90}Sr than coniferous species, because they are more calciphilic than the latter. Among coni-ferous tree species, high ^{90}Sr accumulation has been reported in spruce /7, 13/. In small-scale experiments, three years after the introduction of ^{90}Sr onto a solid surface in a young birch-aspen stand, the aerial organs of pine accumulated smaller amounts (by factors of 2—25) of the radionuclide than larch, while the deciduous species (birch and aspen) accumulated 1.5—8 times more of the radionuclide than larch. The buildup factor of

^{90}Sr in the needles of pine and larch was 0.011 and 0.24, respectively, as against 2.0 and 0.6 in the leaves of birch and aspen /11/.

The distribution pattern of ^{90}Sr taken up from soil in arboreal plants can also be estimated from information concerning the concentration of stable Sr in various organs of trees, since the intakes of ^{90}Sr and stable Sr by trees from soil obey the same laws. The highest concentration of stable Sr occurs in the leaves of trees, followed by branches, bark and roots, while the minimum concentration of stable Sr (approximately $^{1}/_{10}$ that in the leaves) was discovered in wood /1, 3, 8, 22, 26/. Deciduous species being more calciphilic in comparison to coniferous species, accumulate 2.5—5 times more stable Sr /3/.

The migration of radioisotopes of cesium (^{134}Cs and ^{137}Cs) in forest stands was studied at the Oak Ridge National Laboratory in the USA in large-scale model experiments with artificial introduction of radionuclides in forest cenoses (the cesium radioisotopes were first inoculated into the boles of trees) /24, 25, 29, 36/. These studies generally revealed a high mobility of cesium radioisotopes in the aerial part of forest biogeocenosis, whereas the same radioisotopes are strongly sorbed in soil. Thus, according to data fixing the migration pattern of ^{134}Cs in the second year after its introduction, the leaves may accumulate up to 40% of the total amount of the radionuclide introduced into the boles. Cesium radioisotopes are fairly readily leached from leaves by rain, the quantity of ^{134}Cs leached by rain being 15% of its content in the leaves. High mobility also remains a characteristic feature of cesium in the autumn redistribution of chemical elements in plants /34/. Some idea of the migration rate of radiocesium in a forest biogeocenosis may be obtained from the distribution setup of ^{134}Cs among the forest biogeocenosis components in the second year after inoculating the radionuclide into the boles of L i r i o d e n d r o n t u l i p i - f e r a (Figure 6.1). Cesium radioisotopes are also rapidly leached from the forest floor, and 20 weeks after its tagging with ^{134}Cs the content of this radionuclide is halved /30/. Cesium-137 is taken up by arboreal plants from soils in smaller amounts than ^{90}Sr. Thus, when ^{90}Sr and ^{137}Cs were introduced into young stands, the buildup of ^{137}Cs in birch and aspen leaves was only $^{1}/_{2}$—$^{1}/_{10}$ that of ^{90}Sr /11/, the difference being due to the stronger sorption of ^{137}Cs in the soil absorption complex in comparison to ^{90}Sr.

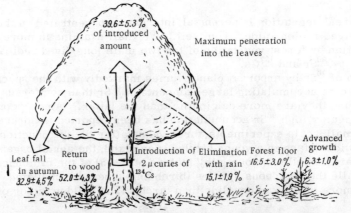

FIGURE 6.1. Distribution of ^{134}Cs among the forest biogeocenosis components in the second year after its introduction into the boles of trees /34/

A comparison of the migration capacity of ^{90}Sr and ^{137}Cs, the two most important long-lived artificial radionuclides, leads to the conclusion that ^{137}Cs is more mobile in migration through aerial organs of arboreal plants, but more ^{90}Sr is taken up by the plants from soil.

The multiannual cumulative buildup of radionuclides is characteristic of perennial orders of arboreal plants (wood, needles, branches, etc.), i. e., there is a gradual increase in the specific concentration of a radionuclide taken from soil through the root systems. From the standpoint of radiation hygiene it is of interest to forecast the possible maximum concentration of a radionuclide in perennial organs of a tree that can be obtained after a certain period of time. In such calculations, e. g., for ^{90}Sr, it is feasible to utilize the analogy in the behavior of stable Sr and its radioactive isotope. Assuming the assimilation and buildup of stable Sr by arboreal plants to proceed under equilibrium, and that of ^{90}Sr under nonequilibrium conditions, but with a trend toward attainment of an equilibrium state, the following equality may be regarded as valid:

$$\frac{^{90}Sr_{max} \text{ in a perennial organ of a tree}}{^{90}Sr \text{ in soil}} = \frac{Sr_{max} \text{ in a perennial organ of a tree}}{Sr_{stable} \text{ in soil}},$$

where ^{90}Sr$_{max}$ in a perennial organ of a tree signifies the possible maximum content of the radionuclide in this organ obtained after a certain period of time necessary for attainment of an equilibrium state between the radioactive and stable strontium isotopes. This ratio can be used for estimating the buildup of ^{90}Sr in wood /2/. Calculation data show satisfactory agreement with experimental data obtained in studying the buildup in the wood of various arboreal species of ^{90}Sr deposited on forest vegetation with global radio-active fallout. Thus, according to calculations, with 1 mcurie ^{90}Sr/km^2 the buildup of the radionuclide in wood is $2 \cdot 10^{-12}$ curie/kg, while the actual concentration of ^{90}Sr in wood turned out to be $1 \cdot 10^{-12}$ curie/kg.

In conclusion, it should be noted that the buildup of radioactive fission products by forest vegetation from soil has been far less thoroughly studied than the characteristic features of the assimilation of fission radionuclides by herbaceous (especially agricultural) vegetation. It is of interest to study in more detail such aspects as absorption from soil of the major radioactive fission products and nuclides with induced activity by different arboreal species in relation to the physicochemical properties of forest soils.

Distribution of radionuclides in forest stands and
circulation of radioactive substances in forest biogeocenoses

Compilation of a general scheme of radionuclide circulation in forest stands is one of the ultimate tasks of the division of radioecology concerned with migration of radioactive substances in biogeocenoses. The cycle of radionuclides in complex multicomponent systems, such as forest stands, can be regarded as multistage cyclic processes many times repeated. Methodologically speaking, studies of the circulation of radionuclides in forests call for large-scale experiments with artificial introduction of

sufficiently large quantities of radioactive substances over considerable
areas.   Such experiments are a prerequisite for tracing the space-time
pattern of the migration of radionuclides among the components of forest
cenosis taking into account the numerous food and biological chains.   Experiments of this kind have now been started, including [137]Cs /24, 25, 29, 36/
and [134]Cs /34, 35/.

Studies of the cycle of radionuclides in forest communities have yielded
extensive empirical material.   For radioecological purposes it is feasible to
represent the data describing the intensity of radionuclide circulation in
cenoses in such a form that would permit calculation of the content of a
given radionuclide in a given cenosis component at any time.   Such an
approach to stating the results of radioecological investigations into radionuclide migration in forest stands was used at the Oak Ridge National
Laboratory /29/.   By resorting to formulas of the indicated type one can
calculate the radiation doses on individual cenosis components in different
periods.

Besides large-scale experiments on the dynamics of the radionuclide
cycle, with artificial introduction of radioactive substances into forest
cenoses, information on the distribution of radionuclides among individual
components of forest cenoses can be obtained by studying global fallout.
Unlike model experiments, which are concerned with the kinetics and
dynamics of the processes involved in the cycle of nuclides, the majority of
papers dealing with the distribution of global fallout in forest stands provide
only a static distribution pattern of the fission products among the cenosis
components /4—6, 21, 27, 28, 31, 32/.   This is because investigations of this
kind as a rule did not include long-term observations on related studies into
the nature of global fallout and dynamics of fission product redistribution
in forest biogeocenoses.   These studies were often confined to short-term
(often one-time) measurements of the contents of fission products in
different parts of forest stands.   However, this information is also of value,
as it enables one in a first approximation to differentiate various cenosis
components with respect to their radionuclide contents, so permitting a rough
estimation of the irradiation dose.   Furthermore, data on the concentration
of radionuclides in different components of cenoses are useful for evaluating
the consequences of environmental pollution with radionuclides from the
standpoint of radiation hygiene.

As an example of radioecological studies into the behavior of fission
products deposited on forest vegetation with global fallout, one may cite
work done by Molchanov et al. /14/.   He studied the distribution pattern of
the major radioactive fission products in the main components of forest
vegetation and forest floor in forests of different natural zones in the
European USSR during 1964—65.   Molchanov noted that the highest concentration of nuclides was characteristic of the forest floor (^{90}Sr, [95]Zr+[95]Nb,
[106]Ru and ^{137}Cs averaged $n \cdot 10^{-8}$ curie/kg; [125]Sb, $n \cdot 10^{-9} - n \cdot 10^{-8}$ curie/kg),
while their minimum concentration was found in debarked bole wood
(^{90}Sr, [95]Zr+[95]Nb, [106]Ru and ^{137}Cs, $n \cdot 10^{-11}$ curie/kg; [144]Ce, $n \cdot 10^{-10}$ curie/kg;
[125]Sb, $n \cdot 10^{-12} - n \cdot 10^{-11}$ curie/kg).   An intermediate position with respect to
the concentration of fission products was occupied by needles, leaves,
branches and bark (^{90}Sr, [95]Zr+[95]Nb, [106]Ru and ^{137}Cs, $n \cdot 10^{-10} - n \cdot 10^{-9}$ curie/kg;
[144]Ce, $n \cdot 10^{-9} - n \cdot 10^{-8}$ curie/kg; [125]Sb, $n \cdot 10^{-10}$ curie/kg).

This distribution pattern of the major fission radionuclides among the fractions of forest vegetation attests to the important role played by extra-radical contamination of arboreal plants by fission products. It is natural to assume that, as the fallout intensity decreases, the relative contribution of soil to the intake of fission products by plants will increase in comparison to direct pollution from air.

In conclusion we examine certain common features of the migration and circulation of radionuclides in forest biogeocenoses and dwell briefly on the stages of migration processes in forests. In the space-time migration pattern of radioactive substances it is possible to distinguish several stages, their duration depending on such factors as type of penetration of radioactive nuclides into the external environment, especially into forest biogeocenoses. If the radioactive substances penetrate into the forest stands aerially, the fallout of radioactive aerosols being one-time, then the migration of radio-nuclides in forests can be differentiated, in a first approximation, into two major stages, nonbiological mechanisms of the migration of radioactive substances predominating in the first stage (mechanical interception of settling particles on the surface of forest vegetation, mechanical and physi-cochemical migration with rain, wind, etc., of radionuclides that have been only loosely sorbed on plants). After the bulk of radioactive substances deposited on the aerial parts of forest vegetation following a one-time fall-out has migrated under the forest canopy, i.e., to the forest floor and soil, the second stage of radionuclide migration in forest stands starts, in which the absorption of radioactive substances through the roots becomes the prevalent way of their penetration into vegetation.

The migration pattern of radionuclides is more complex in the case of long-term deposition of radioactive aerosols on forest stands. In this case temporal differentiation of migration into two stages becomes impossible, but the differentiation of migration processes into two large groups, those of physicochemical and mechanical migration on the one hand and biological transport on the other, is useful for interpreting the migration mechanisms and fate of radionuclides in cenoses and also for quantitative forecasts of the migration of radioactive substances. This is important from the radio-ecological standpoint, especially in an analysis of the possibility and scale of radiation injury when the radiation sources are represented by radio-nuclides migrating in the forest.

Forest radiation biogeocenology applied to radionuclide migration in forest stands

The current period in the development of the radioecology of the migration of radioactive substances in forests can be characterized as the stage of amassing empirical information on radionuclide transport pro-cesses in different components of forest communities. Over the last 10—15 years, an extensive body of data has been accumulated shedding light on various aspects of the general distribution of radionuclides in forests for different types of radioactive substance penetration into forest cenoses. It is important to note that at the current level of knowledge these aspects provide mainly a static picture, whereas information on the dynamics of radionuclide migration in forest cenoses is extremely sparse.

Processes of radionuclide transport in forests have been very little investigated, and then only with respect to some cenosis components; they have hardly been investigated at all with respect to the forest cenosis in its entirety. Therefore, a major task of forest radioecology is to study the circulation of radionuclides in space and time as applied to forest biogeo-cenosis as an independent lanscape unit. Such a development of forest radioecology conforms to the development logic of any science, from amassing of empirical data to theoretical generalizations, from investi-gation of isolated components of systems to studies of entire systems.

The interpretation of the fundamental laws governing the radionuclide cycle in forest biogeocenosis, treated as independent units in the biogeo-logical continuum of the biosphere, is one of the major tasks of forest radio-ecology. First, a general theory must be developed for the interaction of an atmospheric impurity with the plant cover (including forest stands), the atmosphere being an important source of influx of radioactive substances into forest communities. This task belongs to biogeophysics. Then, there is a need for interpreting the mechanisms and describing the quantitative behavior of a radioactive substance entering the aerial portion of stands. On the other hand, empirical information must be continually gathered because the available data is very sparse. The situation is more favorable in studies dealing with the buildup of radionuclides from soil, since in this case it is possible to resort to the results of numerous investigations into the cycle of stable nuclides in natural biogeocenoses. However, in this respect it must be remembered that assimilation of radioactive nuclides, in contrast to stable ones, takes place in an ecologically nonequilibrium state.

Achievements in the development of forest radioecology are linked to the degree of success in establishing contacts with other branches of science in solving fundamental problems. In considering the current tasks of radio-ecology as regards radionuclide migration in forest biogeocenoses, it is noteworthy that the problem of the cycle of chemical elements in the biosphere can now be formulated in its entirety and in separate biogeo-cenoses. The use of radioactive nuclides as tags in studies of the migration and cycle of chemical elements is a very promising method for investigating the transport of individual chemical elements along biological and food chains in biogeocenoses, interpreting the mechanisms of these processes, studying the biogeochemical migration cycles, examining the role played by different chemical elements from the biogeochemical standpoint, and so on. Such problems are thus closely related to problems in biogeochemistry, agrochemistry, physiology of mineral nutrition of plants, and soil science.

Application of computers is promising as regards research into the radionuclide cycle at the level of an individual biogeocenosis. Computers can be used to compute migration rates of radionuclides in complex multi-component systems (to which biogeocenoses belong) and to simulate the functioning of ecosystems themselves /29/.

From a methodological point of view, progress in studies of the radio-nuclide cycle in forests depends largely on the performance· of large-scale experiments with artificial introduction into forest cenoses of quantities of radionuclides representing sufficiently high activity. Experiments of this

kind can yield valuable information pertaining to individual processes
responsible for radionuclide migration, to the roles played by different
environmental factors in their transport, and so on.

References

1. Aleksakhin (Alexahin), R. M. and M. M. Ravikovich. — Pochvo-
    vedenie, No. 4:50. 1966.
2. Aleksakhin, R. M. and M. M. Ravikovich. — In: "V Nauchno-
    prakticheskaya konferentsiya po radiatsionnoi gigiene. Materialy
    k konferentsii 12 dekabrya 1967 g.," p. 6. Leningrad. 1967.
3. Aleksakhin, R. M. and M. M. Ravikovich. — Byulleten' MOIP.
    Otdel biologii, Vol, 73:118. 1968.
4. Alfimov, N. N. and P. N. Yagovoi. — Radiobiologiya, Vol. 6:601.
    1966.
5. Ananyan, V. L. — Doklady Akademii Nauk Armyanskoi SSR, Vol. 28,
    No. 5. 1959.
6. Gedeonov, L. I. and S. P. Rosyanov. — Pochvovedenie, No. 7:88.
    1968.
7. Karaban', R. T. and F. A. Tikhomirov. — Radiobiologiya, Vol. 7:275.
    1967.
8. Litovchenko, G. D. and S. A. Shipitsyn. — Zhurnal Obshchei Biologii,
    Vol. 21:297. 1960.
9. Makhonina, G. I. — Trudy Instituta Ekologii Rastenii i Zhivotnykh
    UF AN SSSR, No. 61:58. 1968.
10. Makhonina, G. I. et al. — DAN SSSR, Vol. 133:484. 1960.
11. Makhonina, G. I. et al. — DAN SSSR, Vol. 140:1209. 1961.
12. Milburn, G. M. and R. Taylor. — In: Voprosy radioekologii, p. 198.
    Moskva, Atomizdat. 1968.
13. Molchanov, A. A. et al.   Nekotorye zakonomernosti raspredeleniya
    vazhneishikh radioaktivnykh produktov deleniya, osedayushchikh
    iz atmosfery v sostave global'nykh vypadenii, v lesakh razlichnykh
    prirodnykh zon SSSR v 1964—1965 gg. (Some Laws Governing the
    Distribution of Major Radioactive Fission Products Deposited from
    the Atmosphere with Global Fallout in the Forests of Different
    Natural Zones of the USSR during 1964—1965). Moskva, Atomizdat.
    1968.
14. Molchanov, A. A. et al. — Lesovedenie, No. 6:18. 1968.
15. Molchanov, A. A. et al. — Lesovedenie, No. 3:13. 1970.
16. Sukachev, V. N. and N. V. Dylis (editors).   Osnovy lesnoi biogeotse-
    nologii (Principles of Forest Biogeocenology). Moskva, "Nauka."
    1964.
17. Remezov, N. P. et al.   Potreblenie i krugovorot azota i zol'nykh
    elementov v lesakh Evropeiskoi chasti SSSR (Consumption and
    Cycle of Nitrogen and Ash Elements in Forests of European USSR).
    Moskva, Izdatel'stvo MGU. 1959.

18. Rodin, L. E. and N. I. Bazilevich. Dinamika organicheskogo
        veshchestva i biologicheskii krugovorot zol'nykh elementov i
        azota v osnovnykh tipakh rastitel'nosti zemnogo shara (Dynamics
        of Organic Matter and Biological Cycle of Ash Elements and
        Nitrogen in the Major Type of Global Vegetation). Moskva-Lenin-
        grad, "Nauka." 1965.
19. Tikhomirov, F. A. et al.— In: Tezisy dokladov simpoziuma po
        migratsii radioaktivnykh elementov v nazemnykh biogeotsenozakh,
        p. 26. Moskva. 1968.
20. Yudintseva, E. V. and I. V. Gulyakin. Agrokhimiya izotopov
        strontsiya i tseziya (Agrochemistry of Isotopes of Strontium and
        Cesium). Moskva. Atomizdat. 1968.
21. Yagovoi, P. N.— Gigiena i Sanitariya, No. 2:32. 1967.
22. Alexahin, R. M. and M. M. Ravikovich.— In: Radioecological Con-
        centration Processes, p. 443.   Ed. by B. Aberg and F. P. Hungate.
        Pergamon Press. 1967.
23. Auerbach, S. I. and J. Olson.— In: "Radioecology," p. 509.   Ed. by
        V. Schultz and A. W. Klement. Reinhold. 1963.
24. Auerbach, S. I. et al.— Nature, Vol. 201:761. 1964.
25. Auerbach, S. I. et al.— In: "Radioecological Concentration Processes,"
        p. 467. Ed. by B. Aberg and F. P. Hungate. Pergamon Press. 1967.
26. Bowen, H. J. M. and T. A. Dymond.— Proc. Roy. Soc., Vol. 144B:355.
        1955.
27. Haussermann, W. and W. Morgenstern.— Atompraxis, Vol. 8:2.
        1962.
28. Ljunggren, P.— Nature, Vol. 184:4690. 1959.
29. Olson, J. S.— Health Phys., Vol. 11:1385. 1966.
30. Olson, J. S. and D. A. Crossley.— In: "Radioecology," p. 411.   Ed.
        by V. Schultz and A. W. Klement. Reinhold. 1963.
31. Rickard, W. H.— In: "Radioecological Concentration Processes,"
        p. 527.   Ed. by B. Aberg and F. P. Hungate. Pergamon Press. 1967.
32. Riekerk, H. and S. P. Gessel.— Health Phys., Vol. 11:1363. 1965.
33. Szepke, R.— Atompraxis, Vol. 14:391. 1965.
34. Witherspoon, J. P. jr.— Ecol. Monographs, Vol. 34:403. 1964.
35. Witherspoon, J. P. jr.— In: "Radioecology," p. 127.   Ed. by
        V. Schultz and A. W. Klement. Reinhold. 1963.
36. Witherspoon, J. P. jr. and G. N. Brown.— Botan. Gaz., Vol. 126:181.
        1965.

Chapter 7

ENVIRONMENTAL RADIOACTIVE POLLUTION AND MAN

Introduction

    Radiation hygiene is a science entrusted with protecting the population's health against effects of ionizing radiation. It has several tasks in common with radioecology, and in the first place comprehensive studies of the migration of radionuclides in the external environment along biological and food chains. Investigation of radionuclide penetration into the ultimate links of food chains (foodstuffs and human body) is of special interest to radiation hygiene. This chapter is concerned with the migration of certain, biologically most dangerous, radionuclides (mostly ^{90}Sr and ^{137}Cs) from the external environment into the human body in the light of tasks faced by radiation hygiene.

Radionuclide migration along food chains from the external environment into the human ration

    Radionuclide migration in external environments occurs via air, soil and water. This classification is arbitrary, since migration may often proceed along several links simultaneously. Nevertheless, their relative importance may differ according to several circumstances.

    A characteristic feature of pollution by air is the prevalent accumulation of radioactive substances on the surface of plants. Some of the radionuclides are removed from the plants by wind and rain, while others, in soluble state, penetrate into their tissues /4/; nevertheless, the bulk of radioactive substances is strongly sorbed on the surface.

    The public health significance of superficial contamination varies in the course of the growing season, increasing during the pasturing of dairy cattle and before harvest. The danger from direct contamination of plants decreases after harvest and in winter. However, radioactive particles deposited even in winter are largely sorbed on sod with the advent of spring, i. e., on the lower portions of the stalks of herbaceous vegetation, and become sources of contamination for young grass. Moreover, they are liable to penetrate into herbivores by direct licking, as is the case in early spring when the grass cover is scant, during droughts, etc.

    Long-lived radionuclides accumulate on the surface of perennial plants — a phenomenon that only recently received due attention. It is very distinct

in the case of coniferous plants (pine, fir, etc.), in which change of needles
occurs once in 3—4 years. Superficial contamination of their needles is
2—4 times heavier than that of the leaves of nearby deciduous trees /15/.

Public health significance of this phenomenon increases when the
corresponding parts of perennial plants are used for human food, such as tea.
In the period of intensive global fallout tea leaves contained several times
more ^{90}Sr than annual plants /11/.

An even more striking example of the effect of the cumulation of long-
lived nuclides deposited from air on the surface of perennial vegetation is
provided by the buildup of ^{137}Cs in lichens in the Far North, the life span of
these lichens being over 15 years. The migration of ^{137}Cs along the lichen-
reindeer-man food chain was responsible for heightened accumulation levels
of this nuclide in the organism not only of reindeer but also of reindeer
herders in the Far North /26, 41, 48/. Moreover, this group of the population was found to have a high content of ^{210}Pb, ^{210}Po and ^{90}Sr, likewise on
account of the specific features of the accumulation of these nuclides in
lichens and their migration along food chains. The slow growth of this
vegetation and its large specific surface due to the structure of lichens and
mosses is conducive to the accumulation of radioactive substances in the
lichens and mosses, irrespective of their latitudinal location. Thus, the time
factor controlling the life span of plants that are the starting link of food
chains along which radionuclides migrate from the environment into the
human body plays a very significant role in forming the human dose load.
The importance of this phenomenon can scarcely be overestimated, since
the human irradiation dose from incorporated natural radioactive substances
(^{210}Pb and its daughter nuclide ^{210}Po) under certain conditions is liable to be
significantly higher than the dose due to global fallout /35/. *

Unlike superficial contamination, the migration of radionuclides from
soils into plant tissues proceeds much more slowly and is controlled by the
intensity of metabolic processes in them.

In contrast to the superficial contamination of plants, which may ultimately
cause contamination of foodstuffs of plant origin by nearly all the nuclides
that are present in fallout, only some radionuclides are absorbed by plants
through their roots, on account of the soil separation. Depending upon the
nature of soils, the amount of ^{90}Sr of global origin migrating into agricultural crops in one year varies from 0.2 to 3% of its content in soil /7/.
Strontium-90 is more intensively absorbed by plants from light sandy soils
with low pH values, low in organic matter and calcium. Considerably
smaller amounts of ^{90}Sr are assimilated by plants from Ca-saturated soils
of heavy texture /4, 9, 32/. In the course of time ^{90}Sr in soils may be converted to forms that are only sparingly available to plants, and its intake by
the latter decreases /47/.

There exists an opinion that the intake of ^{137}Cs by plants from soils is
very slight and its migration into the fodder of dairy cattle is mainly due to
aerial contamination of grass /6, 31, 42/. It has been pointed out that the
intake of ^{137}Cs by plants varies with the type of soil /1, 4, 9/. However,
in some geochemical provinces the migration of these radionuclides from
soils plays the dominant role in contamination of locally produced foodstuffs
and is the source of ^{137}Cs incorporated in human tissues. An example is

---

* For a more detailed treatment of the migration of radionuclides in landscapes of the Far North the reader
is referred to Chapter 13. — Editor.

provided by extensive regions with predominance of sod-podzolic sandy soils formed on alluvial deposits /16, 24/.

Comparison of data characterizing the degree of migration of ^{137}Cs and ^{90}Sr from different types of soil into grass and milk reveals that the migration of ^{137}Cs into grass and milk from sod-podzolic and peaty sandy soils is considerably more intensive than the migration of ^{90}Sr (Table 7.1).

TABLE 7.1. Migration of ^{137}Cs and ^{90}Sr from different soils into grass and milk /16/

Soil type	^{137}Cs		^{90}Sr		^{137}Cs/^{90}Sr ratio in milk
	grass / soil	milk / soil	grass / soil	milk / soil	
Loamy sod-podzolic soils; chernozems; peat-bog soils	1.62	0.36	0.72	0.18	3.5
Sandy-loamy sod-podzolic soils; sod-peat soils; silty-boggy soils	7.95	5.45	2.34	0.32	25.2
Sandy sod-podzolic soils; peat-sandy soils	23.6	19.1	0.50	0.19	127.8

The increase in the intensity of ^{137}Cs migration from soils along food chains results in increased incorporation of this radionuclide in the human body. In certain cases, the content of ^{137}Cs in the body of adult persons reaches 100 ncuries, i. e., significantly above the mean values for the country, although the absolute values do not exceed the maximum permissible concentrations. Hence underestimation of the soil migration path of ^{137}Cs is unjustified for certain biogeochemical provinces. Under certain conditions, the geochemical factor plays a significant part in forming the dose load on the population due to incorporated ^{137}Cs.

The path traveled by radionuclides from a water body to man is very complicated. The simplest and at times the main connection between the water body and man depends on the use of water for drinking. Another path is represented by the water-fish—man chain. The existence of still longer food chains is likewise not precluded (water-microflora-zooplankton-fish-man; water-phytoplankton-fish-man; water-bottom sediments-benthos-fish-man, etc.), but these are comparatively unimportant.

All hydrobionts are capable of accumulating radioactive substances in their tissues by sorption processes, the buildup levels being largely dependent on the saline composition of water /12—14/. In particular, maximum ^{90}Sr concentrations occur in freshwater hydrobionts and minimum concentrations in marine hydrobionts. Equilibrium concentration of ^{90}Sr in soft tissues of freshwater fishes does not usually exceed 0.1 of its content in the water, while the discrimination factor of ^{90}Sr with respect to Ca varies from 0.35 to 0.75, averaging 0.4 /7/. The buildup factors of ^{90}Sr are 200—300 in bone tissue, about 10 in muscle tissue. The buildup of ^{137}Cs in the muscle tissue of freshwater fishes is considerably more intensive than that of ^{90}Sr, its buildup factor varying from $n \cdot 10^2$ to $n \cdot 10^3$ /31, 33, 44/.

Marine products used for human food present a greater variety. Besides, fishes, crustaceans (crabs, shrimps), mollusks (oysters, scallops), seaweed, etc., are of economic importance. However, the buildup levels of ^{90}Sr and ^{137}Cs in their tissues are considerably lower than in freshwater hydrobionts /25, 38/. At the same time, they contain higher concentrations of nuclides with induced activity (^{65}Zn, ^{60}Co, ^{59}Fe, etc.).

In addition to direct human intake of radioactive substances with products of water bodies, there are also more complex migration paths, for instance, when contaminated water and silt find their way from the water body to farmland (irrigation, inundation of floodplain meadows, river floods, etc.), resulting in contamination of agricultural produce. These aspects may be illustrated by data concerning ^{65}Zn concentration in the produce of farms using water from the Columbia River polluted with effluents from industrial nuclear reactors (Table 7.2).

TABLE 7.2. ^{65}Zn concentration in water, grass in pastures and agricultural produce samples in 1957 /39/

Samples	^{65}Zn concentration, μcurie/g	Buildup factor of ^{65}Zn, μcuries/g wet matter / μcuries/ml water
Irrigation water (unfiltered)	$1.88 \cdot 10^{-7}$	—
Grass in pastures	$8.29 \cdot 10^{-5}$	440
Beef	$5.23 \cdot 10^{-6}$	28
Milk	$4.88 \cdot 10^{-6}$	26
Green peas	$5.49 \cdot 10^{-7}$	2.9
Tomatoes	$4.58 \cdot 10^{-7}$	2.4

In the course of migration from the external environment, in every link of the biological or food chain natural agents cause some delay in the migration of radioactive substances, so that only part of the radionuclides reach the organism. The last link in the food-man chain offers no exception in this respect, but in this case the delay in the migration tempo is due to different agents.

Ordinary technological processing of raw foodstuffs and culinary processing of foodstuffs remove part of their radionuclide content. For instance, when grain of new harvest is cleaned from ordinary dust, part of the radioactive particles adhering to the grains is removed along with the dust. An even greater effect is achieved by husking the grain, the husks being the principal sorption locus of radioactive particles, the bulk of radionuclides (up to 80%) being removed with bran /30/. As a result, the majority of groats and the higher grades of flour (consequently also bread) have a considerably lower content of ^{90}Sr than the lower grades of flour which contain some bran. The usual culinary processing of vegetables (washing, paring) reduces their radionuclide content by 20—60% /3, 30/. Processing of milk to butter practically eliminates the ^{90}Sr.

Thus, in the last link of the food chain, in the processing of foodstuffs and food, there occurs further reduction in the amount of radionuclides.

Human intake of radioactive substances

The content of radioactive substances in the population's food ration is controlled by the contamination levels of its components and the structure of the diet, i. e., the assortment and quantity of consumed foodstuffs and drinking water.

The reader is reminded that the contamination level in foodstuffs is principally dependent on the nature and intensity of environmental pollution and the natural characteristics controlling the efficiency of the migration of radioactive substances along food chains. The structure of diet is a derivative of social-economic conditions and depends on the place of residence, nature of working activities, national features, etc. Thus, for the same levels of contamination of the external environment and foodstuffs with ^{90}Sr and ^{137}Cs of global origin the averaged indices characterizing the contents of ^{90}Sr and ^{137}Cs in the ration of inhabitants of countries in the northern hemisphere with different structures of diet differed by factors of 2—3 /31/. Somewhat less pronounced differences in the contamination levels of food rations with ^{90}Sr and ^{137}Cs (factors of 1.5—2) exist between different republics of the Soviet Union /11/, due mainly to national features in the diets of the rural population.

A still smaller variation in the mean levels of the contamination of rations with ^{90}Sr and ^{137}Cs has been determined for large cities of the USSR. This is due not only to the absence of any marked differences in the structure of diets of urban populations, but also to the peculiarities of food supplied to cities, arriving from different regions of the USSR and largely obliterating the differences due to different contamination levels in local products.

The parts played by individual ration components as "suppliers" of radionuclides are likewise controlled by the structure of diets. This is illustrated by data characterizing the ingestion of ^{90}Sr with different foodstuffs in the ration of populations of some countries listed in Table 7.3.

TABLE 7.3. Ingestion of ^{90}Sr with foodstuffs in the rations of inhabitants of different countries in 1963—1965 with respect to their total content in the ration, % /7/

Products	USSR* (urban and rural population)	USA (New York)	United Kingdom	Japan
Milk and dairy products	20	50	61	7
Grains (bread, bakery products, groats)	63	25	15	13
Vegetables and fruits	6	14	7	45
Root crops	6	5	5	20
Others (meat, fish)	5	6	12	15

* According to /11/.

It is seen from Table 7.3 that the importance of different products as suppliers of ^{90}Sr in the ration varies markedly, depending on the type of diet.

Thus, milk is the principal ^{90}Sr supplier in the ration in countries with a "western" diet, vegetables and root crops in the "rice diet," while bread and bakery products play the foremost part in the Soviet Union. This is due to national features of diet and to a somewhat higher consumption of bread in general, and dark bread in particular, in which the ^{90}Sr content is higher. However, even within the USSR these relationships vary widely, mainly owing to the national diets of different republics and in different natural-climatic zones of the USSR. For instance, venison is the predominant supplier of ^{137}Cs and ^{90}Sr in reindeer herders of the Far North. Although the diet structure of the rural population of Belorussia and the Ukraine is the same, the principal supplier of ^{137}Cs (up to 90%) in the ration of the inhabitants of certain areas in Polesie is milk, because of the properties pertaining to the migration of this nuclide from soils in that region.

Even under identical environmental pollution conditions, the importance of different foodstuffs as radionuclide suppliers and likewise the total content of ^{90}Sr and ^{137}Cs in the ration may vary widely depending on the structure of diets.

Different environmental pollution conditions may also significantly affect the importance of individual ration components as suppliers of radioactive substances. This is illustrated by the situation in a certain region prior to the beginning of global fallout caused by accidental contamination of a water body /21/.

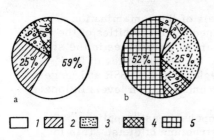

FIGURE 7.1. The roles played by individual ration components as ^{90}Sr suppliers for different contamination paths with respect to the total content:

a — average for the shore area; b — average for the USSR; 1 — water; 2 — fish; 3 — animal husbandry products; 4 — vegetables and root crops; 5 — bread and bakery products.

Since ^{90}Sr concentration in the water was rather low the water body was used for a variety of purposes (water supply, fishing, farming, etc.). In this case, the influx of ^{90}Sr into the local population's ration with its different individual components was found to be significantly different from the influx of the same radionuclide in the ration due to global fallout. As is seen from Figure 7.1, the predominant ^{90}Sr suppliers were drinking water (59% of ^{90}Sr) and fish (25%). Milk, meat and vegetables were only minor ^{90}Sr suppliers, while bakery products were altogether free from contamination, since the fields of grain crops were located outside the range over which radionuclides from the water body spread.

A more thorough analysis revealed that the significance of indices obtained for different population groups differed, depending on numerous circumstances. Thus, fish was the principal ^{90}Sr supplier in the ration of the 10—20% of the adult population engaged in fishing (Figure 7.2). In this case, the important factor was not only the degree of buildup of ^{90}Sr in the edible tissues of fish, but principally the frequency and quantity of fish consumption. However, in the case under consideration drinking water was the principal ^{90}Sr supplier in the ration of the majority of the local population, even taking into account the age distribution (Figure 7.3).

The part played by water as a supplier of radionuclides to the human body merits somewhat more detailed consideration. In the case of environ-

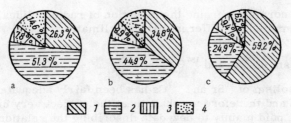

FIGURE 7.2. Water and foodstuffs as ^{90}Sr suppliers in the ration
(with respect to the total content):

a — ration of 1–5% of adult population in the locality; b — ration
of 10–20% of adult population in the locality; c — mean values
for the ration of the entire adult population of the locality; 1 —
water; 2 — fish; 3 — animal husbandry products; 4 — vegetables
and root crops.

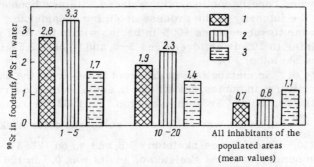

FIGURE 7.3. Ratios between ^{90}Sr contents in foodstuffs and water in the
daily rations of different age groups in the population:

1 — adults; 2 — children aged 1–4; 3 — children up to 1 year old.

ment polluted by radioactive fallout of global origin the contribution of
drinking water from open water bodies as a ^{90}Sr supplier to the human
ration is very minor, averaging about 5% of the content of this nuclide in
the ration for the entire USSR. The use of water from surface water bodies
subjected to local pollution may increase this contribution to 50%.

In calculations of the ingestion of radioactive substances by human
consumers with water, the mean-annual daily consumption is generally
assumed to 2.2 liters, including about 0.6 liter water from foodstuffs and
approximately 1.6 liters water used directly for drinking. This figure is
fairly well substantiated for temperate latitudes, but it does not fit the
natural-climatic conditions of the southern regions of the USSR, including
the Soviet Central Asian republics. According to available information
/23, 27, 28/, the consumption of drinking water by the adult rural population
in Uzbek SSR reaches 7 liters/day, taking into account the energy expenditures during the hot season of the year, averaging about 3 liters daily over
the year. The consumption of drinking water by urban inhabitants is
approximately only two-thirds of that, yet they likewise exceed the conventional standards. Hence there is the need for a differentiated approach

to the setting of norms governing the ingestion of radionuclides with water
and water consumption for different natural-climatic zones.

Human metabolism of ^{90}Sr and ^{137}Cs

Human metabolism of ^{90}Sr and ^{137}Cs has been fairly adequately described
in the literature and therefore we shall treat the subject very briefly.
Attention will be paid mainly to new data describing the relationship between
the environmental contents of radioactive substances and their contents in
the human body, elicited by recent field observations and investigations in
the USSR and elsewhere.

Strontium-90. The buildup of ^{90}Sr in the skeleton is the result of two
processes, its incorporation in new bones in the course of their formation
and continuous metabolism. The intensity of these processes varies with
age. In addition to the growth of new tissue, occurring only in children,
the already-formed bone is continuously replaced with new bones throughout
the human life, the intensity of the process declining considerably with age.
Renovation of bone tissue reaches 100% in babies within the first year of
their life, declining to 10% in children aged 3—8, and to only 2.5% in adults
(average for the skeleton) /7/.

The intensity of ^{90}Sr metabolism in different bones from the skeleton is
approximately uniform in babies, but differs in adults, with a corresponding
effect on the distribution of ^{90}Sr in the skeleton. In 1963, the ratio
$\dfrac{^{90}\text{Sr/Ca in individual bones}}{^{90}\text{Sr/Ca in entire skeleton}}$ in adults was as follows: vertebral column/skeleton = 1.7;
rib/skeleton = 1.0; cranial bones/skeleton = 0.6, and so on /22, 37, 45/. The
ratio ^{90}Sr/Ca in human teeth and skeleton of adults was 0.3 in 1963—1966 /19/.

Besides their theoretical significance, these figures are of practical
interest. For instance, they were widely used in the Soviet Union for monitoring the ^{90}Sr content in the human skeleton from analysis of individual bones
or extracted teeth. The contents of ^{90}Sr differ in different age groups in the
population, owing to the age-conditioned peculiarities of its metabolism in the
bone tissue. Strontium-90 concentration in the skeleton of a fetus is relatively
low and is directly related to its content in the mother's ration. Table 7.4
provides information on the magnitude of the observed ratio in the migration
of ^{90}Sr and Ca from the ration into the human skeleton.

TABLE 7.4. Magnitude of observed ratio in migration of ^{90}Sr and Ca from the ration into the human bone
tissue (mean values for 1963—1966)

Age group	Migration link	USSR (Moscow)	USA (New York)	United Kingdom	Denmark	Japan
Fetus	Bone tissue — mother's ration	0.05	0.17	0.13	0.18	0.05
Adults	Bone tissue — ration	0.02	0.05	—	0.04	0.03

Note. This table is based on /7/ for countries outside the USSR and on data provided by our laboratory
/17, 43/ for the USSR.

Table 7.4 shows that the "observed ratios" in Japan were close to those in the USSR, but were 2.5—3.5 times more in countries with a "western" diet. The differences are apparently due to the national features in the structure of diet in the Soviet Union. One such feature is the relatively large contents of bread, including that baked from dark grades of flour, in the ration, and therefore the ingested amount of ^{90}Sr is somewhat larger than in other countries. However, a considerably smaller amount migrates from the ration into the human skeleton, probably on account of the chemical nature of the compounds of ^{90}Sr in bread; phytin compounds probably play an important role in this respect. Consequently, the contents of ^{90}Sr in the skeleton of newborn babies as well as adult groups of the population in different countries of the northern hemisphere are fairly similar.

The dynamics of the ^{90}Sr content in the bone tissue of babies (in their first year) possesses its own peculiarities, due to the special nourishment of such babies (mother's milk, baby foods). Differences in the ingested amounts of ^{90}Sr between breastfed and other babies manifest themselves in a significant increase of ^{90}Sr concentration in the skeleton. In the case of Moscow babies this "jump" occurs at the age of 8 months, coinciding with their weaning and transfer to ordinary feeding (Figure 7.4). The considerable scatter of points is apparently due to differences in the consumption of mother's, fresh, and dry cow milk.

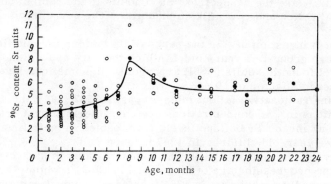

FIGURE 7.4. Dynamics of ^{90}Sr buildup in Moscow babies in relation to age:

O — data from individual samples; ● — mean data.

TABLE 7.5. ^{90}Sr content in the bone tissue of babies (in their first year) in some countries, Sr units

Year	USSR		United Kingdom	Denmark	USA (New York)	Norway
	mean for the country	Moscow				
1963	5.0	5.0	5.2	4.2	6.8	6.9
1964	5.4	5.9	8.6	6.5	7.9	8.9
1965	4.6	5.0	7.2	6.1	5.0	15.0

Note. The table is based on /7/ for countries outside the USSR and on data provided by our laboratory /17/ for the USSR.

In the case of rural babies the concentration peak of ^{90}Sr may be expected to appear later, because of the usually more prolonged suckling period.   In London babies the maximum ^{90}Sr concentration occurs at a considerably earlier age than in Moscow babies /40/, due to earlier weaning.   This phenomenon is due to social factors and is apparently the cause of the higher ^{90}Sr content in babies of certain countries (Table 7.5).

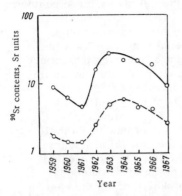

At the same time, even a relatively brief consumption of cow milk by babies in the first year of their life results in an appreciable buildup of ^{90}Sr in their bones, so that there is a satisfactory correlation between the contents of the nuclide in milk and babies' bones (the mean annual varies) (Figure 7.5).

FIGURE 7.5. ^{90}Sr content in milk (———) and bones (– – –) in Moscow babies in their first year

The mathematical expression describing the relationship between ^{90}Sr content in a baby's bones, its mother's ration and cow milk has the form

$$X = 0.06\,(X_1 + X_2),$$

where X, X_1 and X_2 are the respective ^{90}Sr contents in the babies' bone tissue, their mothers' ration and cow milk, in Sr units.   The correlation coefficient is 0.88 /29/.

With uniform ingestion of ^{90}Sr the peaks of its buildup in bone tissue occur in the age group 1–2 years old, in which they are approximately four times the level found in adults (USSR averages).   In some individuals differences reach a factor of 10 /46/.

With varying radioactive fallout intensity, as in 1962–1967, the age-contingent variations in ^{90}Sr concentrations are characterized by a gradual shift of the peak mean ^{90}Sr buildup levels from babies in their first year, in 1963, to five-year-old children, in 1967 (Figure 7.6).

The observed time-shifts of peak ^{90}Sr concentrations in Moscow children were due to the progressing age of those born in 1962, in whom ^{90}Sr concentration reached its maximum in 1964 (period of heaviest contamination of foodstuffs).   Although the ^{90}Sr content in the bone tissue of this group diminishes with every passing year due to natural metabolism, high ^{90}Sr concentrations are traced over the following five years.

It should be noted that in this age group of the population (1–4 years old) the intensity of ^{90}Sr migration from the ration into bone tissue is considerably higher than in other groups; this is also indicated by the relatively high value of the observed ratio, the mean value of which in 1963–1966 was 0.16, i.e., 8 times that for adults.

In considering the relationship between the migration of ^{90}Sr from the environment into the human body one must emphasize the importance of the time factor.   In this respect, in addition to the biological properties controlling the intensity of ^{90}Sr migration from the environment, the rate of bone tissue metabolism, etc., there are also factors affecting the consumption dates of contaminated products (storage, processing, transportation, etc.).   In the case of ^{90}Sr migration along the grass-cow-milk-man chain, an appreciable increase of its content in the bone tissue of children is observed only several months after the starting date of contamination of the grass. The "delay" effect in the buildup of ^{90}Sr in the skeleton is still more pronounced in adult population groups. One must bear in mind that in the period prior to 1967—1968 ^{90}Sr and ^{137}Cs were ingested by inhabitants of the USSR mainly with bakery products. Since grain crops are harvested once a year, while foodstuffs prepared from them are consumed from autumn and all through the following year and sometimes even longer, the cause for the significant delay and time-spread of the rise of ^{90}Sr concentration in the bone tissue of adults becomes clear. Comparison of the dates of the establishment of maximum ^{90}Sr concentrations in the human ration and bone tissue indicates that this "delay" amounts to one year as the average for adult population groups for the entire USSR, but increases to two years in the case of the inhabitants of Moscow.

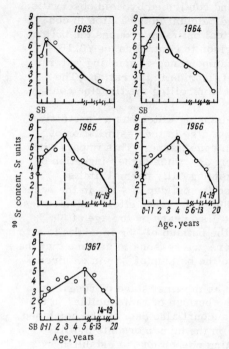

FIGURE 7.6. Variation of ^{90}Sr concentration in the bone tissue of different age groups in the Moscow population during 1966—1967 (SB denotes stillborn babies)

According to observations, ^{90}Sr contents in the bone tissue of similar groups in the population of different geographical regions of the USSR differ significantly. The maximum ^{90}Sr concentration in the human skeleton occurs in the northern regions of European USSR (where it approaches the indices for Norway), and somewhat in Belorussia, while minimum concentrations occur in the Ukrainian steppe and plains of Soviet Central Asia and Kazakhstan. Within each of these zones there are in turn regions with different ^{90}Sr concentrations in identical age groups of the population. At present, it would be a rather complicated task to analyze in detail the causes of these phenomena, in view of the variety of geographical and natural factors as well as the population's social-economic conditions.

Comparison of data describing ^{90}Sr content in the food ration and human bones in these zones does not yet yield any correlation for all cases. This circumstance only emphasizes the dependence of ^{90}Sr buildup in the human body on a variety of factors, the relative importance of which is liable to vary under different specific conditions.

As already mentioned, the specific significance of foodstuffs making the principal contributions to ^{90}Sr in the ration differs in various regions of the USSR. Consequently it may be assumed that the degree of resorption of ^{90}Sr in the gastrointestinal tract as well as its buildup in the bone tissue depend to a certain extent on the nature of chemical compounds of the nuclide in different foodstuffs. Moreover, certain mineral components in the ration have a "protective" effect on the buildup of ^{90}Sr in bone tissue. It has been experimentally demonstrated on animals and confirmed by field observations on humans that the presence of fluorine and calcium in the ration in con- centrations corresponding to the physiological optimum reduces the buildup of ^{90}Sr in bone tissue by 20–30% in comparison to control cases /8,10,43/. In this connection one must emphasize the importance of drinking water, which is usually the principal "supplier" of fluorine in the ration. The importance of local geochemical conditions controlling the fluorine con- centration in water thus becomes clear.

It is of interest to note that calcium in drinking water has a "protective" effect not only against the buildup of ^{90}Sr, but also in the assimilation of ^{226}Ra, limiting their incorporation in human bone tissue. This impact with respect to ^{226}Ra is most pronounced in early age, when higher ingestion of calcium with drinking water within the physiologically normal limits (17% more than in the control area) reduced the buildup of ^{226}Ra in the bones of younger children by approximately 50–80% /18/. The effect becomes less pronounced with increasing age and almost disappears by the age of 30.

Since the presence of calcium reduces the buildup of ^{90}Sr in adults to a greater extent than the buildup of ^{226}Ra, there are reasons to assume that the protective effect of calcium with respect to the buildup of ^{90}Sr in children should be even stronger.

The above data should be regarded only as the first stage of research into the effect of geochemical factors on the buildup of radionuclides in the human body. Further studies, taking into account also data on the combined effect of stable micro- and macroelements on the human organism, will make it possible to reveal equally interesting phenomena, to establish pertinent laws, and to utilize the data to forecast the consequences of the effect of ionizing radiation and prevent such an effect. Knowledge of the distribution of radioactive substances in the environment and human popu- lation is important when evaluating the radiation situation under conditions of radioactive environmental pollution.

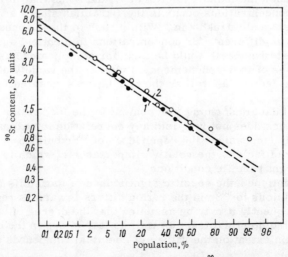

FIGURE 7.7. Occurrence frequency of different ^{90}Sr concentrations in the bone tissue of the adult Moscow population in 1963 (1) and of the USSR population in 1964 (2)

The occurrence frequency of different ^{90}Sr concentrations in the bones of any one of identical age groups of the population on the scale of a country, a region or a city obeys the normal distribution law /5, 20/ (Figure 7.7). This equally applies to combined populations of several countries /7/. It follows that the mean concentrations are exceeded in individual cases by factors of 4—7 with probability 10^{-3}—10^{-5}. The risk due to the penetration of ^{90}Sr into the organism can then be estimated.

Cesium-137. This substance mainly enters the human organism perorally with foodstuffs and is concentrated, similar to potassium, mainly in the muscle tissue. According to the latest data, ^{137}Cs concentration in muscles approaches its concentration in blood /49/. Its concentration in bones and marrow is approximately $\frac{1}{20}$ of the latter. The biological half-life of ^{137}Cs with respect to the human organism is comparatively short, 70—140 days in adults /7, 26/, and is approximately halved during pregnancy (down to 30—70) /7/. The biological half-life of ^{137}Cs reaches its minimum in babies one month old, being 7—10 to 21—25 days according to different authors /31, 34/. The lower levels of the ingestion of ^{137}Cs in fetuses and babies and also its relatively rapid metabolism ensure minimum ^{137}Cs buildup levels for this age group. The concentration of ^{137}Cs increases with age and reaches its maximum in adults, the increase being directly proportional to age for young age groups (up to 18 years) /2/. However, the highest ^{137}Cs concentration occurs in children, cow milk being the principal source of this nuclide.

The predominant localization of ^{137}Cs in the muscle tissue, which is better developed in men than in women, explains the 1.4—1.6 times higher content of ^{137}Cs in the body of men compared with women.

The high resorption rate of ^{137}Cs in the gastrointestinal tract /7/ ensures its rapid migration from the ration into the organism. There is a direct relationship between the influx of ^{137}Cs and its content in the human body. Consequently, the observed differences in ^{137}Cs content in the organisms of identical age groups in the populations of different geographical zones are due to the different contamination levels of the rations. The degree of stability of ^{137}Cs concentration in the human population, and the nature and dynamics of seasonal variations in the content of this radionuclide are mainly controlled by the contamination of foodstuffs, which are the principal suppliers of ^{137}Cs in the ration. The comparatively small seasonal variations of ^{137}Cs concentration in the majority of the adult population of the Soviet Union with a marked trend toward its gradual obliteration are due to the fact that the principal suppliers of ^{137}Cs to the human organism are bakery products. The variation of ^{137}Cs concentrations in these products, usually occurring once yearly, is related to the arrival of grain of the new crop.

Populations with a different structure of diet, including that ingesting ^{137}Cs mostly with milk (West European countries and the USA), display fairly pronounced seasonal variations of ^{137}Cs content in the organism, following the ^{137}Cs concentration dynamics in milk, on account of the variations in the concentration of this radionuclide in the fodder of dairy cattle /36/.

The direct relationship between ^{137}Cs content in the human rations and organism seems to be the principal controlling factor in the overall distribution pattern of ^{137}Cs in similar population groups.

In the Soviet Union, where ^{137}Cs is principally ingested with bakery products (as an average for the entire country), the similar distributions of ^{137}Cs contamination levels in the majority of similar population groups are readily explained by the lognormal distribution of the occurrence frequencies of different ^{137}Cs contamination levels in bakery products.

According to /7/, the probability of a ^{137}Cs buildup in some individuals that is triple the mean value for the population group of the same type in a given region is $1.5 \cdot 10^{-3}$.

References

1. B a k u n o v, N. A. Vliyanie svoistv pochv i pochvoobrazuyushchikh mineralov na postuplenie tseziya-137 v rasteniya (Effect of the Properties of Soils and Soil-Forming Minerals on the Intake of Cesium-137 by Plants). Thesis. Moskva, Moskovskaya Sel'skokho-zyaistvennaya Akademiya imeni Timiryazeva (TSKhA). 1967.

2. B e l l e, Yu. S. et al. Soderzhanie tseziya-137 i kaliya u naseleniya SSSR v 1962—1966 gg. (Contents of Cesium-137 and Potassium in the USSR Population in 1962—1966). Moskva, Atomizdat. 1967.

3. B e l o v a, O. M. et al.— Gigiena i Sanitariya, No. 3:40. 1967.

4. G u l y a k i n, I. V. and E. V. Y u d i n t s e v a. Radioaktivnye produkty deleniya v pochve i rasteniyakh (Radioactive Fission Products in Soil and Plants). Moskva, Gosatomizdat. 1962.

5. D i b o b e s, I. K. et al. Global'nye vypadeniya strontsiya-90 na terri-torii Urala v period 1961—1966 gg. (Total Fallout of Strontium-90 Over the Urals in 1961—1966). Moskva, Atomizdat. 1967.

6. Report of the UN Scientific Committee on Effects of Atomic Radiation, XVII Session, A/5216, Vol. 3, Supplement F. New York. 1962.

7. Report of the UN Scientific Committee on Effects of Atomic Radiation, A/6314. New York. 15 July 1966.

8. K a r m a e v a, A. N. et al.— In Sbornik: "Raspredelenie, biologicheskoe deistvie i uskorenie vyvedeniya radioaktivnykh izotopov," p. 167. Moskva, "Meditsina." 1966.

9. K l e c h k o v s k i i, V. M. (ed.). O povedenii radioaktivnykh produktov deleniya v pochvakh, ikh postuplenii v rasteniya i nakoplenii v urozhae (Behavior of Radioactive Fission Products in Soils, Their Intake by Plants and Buildup in Crops). Moskva, Izdatel'stvo Akademii Nauk SSSR. 1956.

10. K n i z h n i k o v, V. A.— Gigiena i Sanitariya, No. 7:39. 1967.

11. K n i z h n i k o v, V. A. et al.— Gigiena i Sanitariya, No. 1:11. 1968.

12. L e b e d e v a, G. D.— In Sbornik: "Radioaktivnye izotopy v gidrobiologii i metody sanitarnoi gidrobiologii," p. 52. Moskva-Leningrad, "Nauka." 1964.

13. L e b e d e v a, G. D.— Radiobiologiya, Vol. 6:556. 1966.

14. M a r e i, A. N. et al. Radiatsionnaya gigiena (Radiation Hygiene), Vol. 2. Moskva, Medgiz. 1962.

15. M a r e i, A. N.— Sbornik Referatov po Radiatsionnoi Meditsine, Vol. 4:223. Moskva, Medgiz. 1961.

16. Marei, A. N. et al.— Gigiena i Sanitariya, No. 1:61. 1970.
17. Marei, A. N. et al. Strontsii-90 v kostnoi tkani naseleniya Sovetskogo Soyuza (1957—1967 gg.) (Strontium-90 in the Bone Tissue of the USSR Population in 1957—1967). Moskva, Atomizdat. 1968.
18. Marei, A. N. and A. N. Karmaeva.— Gigiena i Sanitariya, No. 8:44. 1966.
19. Marei, A. N. et al. Issledovanie ekstragirovannykh zubov kak metod massovogo kontrolya za soderzhaniem strontsiya-90 v organizme lyudei (Investigations of Extracted Teeth as a Routine Monitoring Technique for Strontium-90 in the Population). Moskva, Atomizdat. 1965.
20. Marei, A. N. et al. Soderzhanie strontsiya-90 v kostnoi tkani nase-leniya SSSR (materialy 1964 g.) (Strontium-90 in the Bone Tissue of the USSR Population for 1964). Moskva, Atomizdat. 1965.
21. Marei, A. N. and M. M. Saurov.— In: Report of Symposium on Re-moval of Radioactive Debris in Seas, Oceans, and Surface Waters. IAEA, Vienna. 1966.
22. Marei, A. N. et al. K voprosu raspredeleniya strontsiya-90 v skelete vzroslogo cheloveka (Distribution of Strontium-90 in an Adult's Skeleton). Moskva, Atomizdat. 1965.
23. Molchanova, O. P. et al.— Voprosy Pitaniya, Vol. 5:53. 1936.
24. Moiseev, A. A. et al. Osobennosti migratsii global'nogo tseziya-137 iz dernovo-podzolistykh peschanykh pochv po pishchevym tsepoch-kam v organizm cheloveka (Features of the Migration of Global Cesium-137 from Sandy Sod-Podzolic Soils Along Food Chains Into the Human Body). Moskva, Atomizdat. 1967.
25. Polikarpov, G. G. Radioekologiya morskikh organizmov (Radio-ecology of Marine Organisms). Moskva, Atomizdat. 1964.
26. Tamzaev, P. V. et al. Osnovnye itogi radiatsionno-gigienicheskikh issledovanii migratsii global'nykh vypadenii v priarkticheskikh raionakh SSSR v 1959—1966 gg. (Principal Results of Radiation Health Studies into the Migration of Global Fallout in Near-Arctic Regions of the USSR in 1959—1966). Moskva, Atomizdat. 1967.
27. Rozanova, E. F. Materialy k razrabotke gigienicheskikh normativov pit'evogo rezhima rabochikh v usloviyakh zharkikh oblastei Sovetskogo Soyuza (Data Pertaining to the Development of Hygienic Standards for the Drinking Regimen of Laborers in Hot Regions of the USSR). Thesis. Moskva. 1954.
28. Sadykov, A. S. Vliyanie chaya na organizm cheloveka i zhivotnykh v usloviyakh zharkogo klimata (Effect of Tea on Humans and Animals in Hot Climates). Thesis. Tashkent. 1939.
29. Stepanov, Yu. S. and R. M. Barkhudarov. Nekotorye zakonomer-nosti zagryazneniya ob"ektov vneshnei sredy strontsiem-90 v period stratosfernykh vypadenii (Some Laws Governing Environmen-tal Pollution with Strontium-90 During Stratospheric Fallout). Moskva, Atomizdat. 1969.
30. Todd, F. A. Protection of Food and Water Against Contamination by Radiation.— Proc. IAEA and WHO Seminar, 20 November 1963. Geneva, WHO. 1966.
31. Eisenbud, M. Environmental Radioactivity. New York, McGraw-Hill. 1963.

32. Yudintseva, E. V. and I. V. Gulyakin.— Informatsionnyi Byulleten'
     Radiobiologiya, No. 9:26. 1966.
33. Beninson, D. et al.— In: Report of Symposium, 16—20 May 1966,
     Report No. 72/19.   Vienna, IAEA. 1966.
34. Bengtson, L. G. et al.   Maternal and Infantile Metabolism of
     Caesium.— In: "Assessment of Radioactivity in Man," Vol. II,
     p. 21. IAEA, Vienna. 1964.
35. Blanchard, R. L.— Nature, Vol. 211:995. 1966.
36. Boni, A. L.— Health Phys., Vol. 12:501. 1966.
37. Bryant, F. J. and J. F. Loutit.— In: Atomic Energy Research
     Establishment, Report AERE-R-3718. Harwell. 1961.
38. Bryan, G. W. et al.— In: Report of Symposium, 16—20 May 1966, Report
     No. 72/39. Vienna, IAEA. 1966.
39. Davis, J. J. et al.— In: "Proc. 2nd Intern. Conf. Peaceful Uses Atomic
     Energy," p. 393. Geneva. 1958.
40. Fletcher, W. et al.— Brit. Med. J., Vol. 2:1225. 1966.
41. Hanson, W. C.— Amer, J. Veterin. Res., Vol. 27(116):359. 1966.
42. Johnson, J. E. et al.— Soil Sci. Soc. Amer. Proc., Vol. 30:416. 1966.
43. Knizhnikov, V. A. and A. N. Marei.— In: Strontium Metabolism,
     edited by Lenihan, J. M. A. et al. New York, Academic Press.
     1967.
44. Kolehmainen, S. et al.— Health Phys., Vol. 12:917. 1966.
45. Kulp, J. L. et al.— Science, Vol. 132, No. 448. 1960.
46. Kulp, J. L. and A. K. Schulert.— Science, Vol. 136:619. 1962.
47. Roberts, H. jr. and R. G. Menzel.— J. Agric. and Food Chem.,
     Vol. 9:95. 1961.
48. Svensson, G. K. and K. Liden.— Health Phys., Vol. 11:1393. 1965.
49. Yamagata, N. et al.— Bull. Inst. Publ. Health, Vol. 15(4):173. 1966.

Chapter 8

EFFECT OF IONIZING RADIATION ON POPULATIONS
(RADIATION-GENETIC ASPECTS)

Introduction

The expansion of human economic activity on an unprecedented scale
results in very significant alterations in the biosphere. Vernadskii /9/
observes that mankind as a whole is becoming a powerful geological force;
it, its thought and actions are faced with the problem of a biosphere which is
being modified as a result of free-thinking humanity following its own
interest. The effect of chemical and radiation factors on living organisms
and their communities is currently assuming special importance. The main
feature in the effect of these factors (mostly ionizing radiation) on living
organisms is their ability to cause profound changes in the genetic material,
interfering with its normal organization. These changes, when transmitted
to subsequent generations, are liable to significantly alter the entire course
of species evolution. Therefore, comprehensive research must be conducted into the specific genetic-evolutionary patterns involved in the effect
of new environmental factors on the biosphere. These problems pertain
to the field of evolutionary genetics, a new trend in science based on the
theory of evolution and population genetics.

The historical beginnings of modern population genetics date back to
1908 when Hardy, and later Weinberg, provided the purely mathematical
substantiation of the principal according to which the genetic structure of
any population obeys Mendel's laws of heredity applicable not to an individual crossing but to a group whose heredity has become interrelated by
uncontrolled crossing. The distribution of genotypes discovered in populations and known as the Hardy-Weinberg Law is fundamental to analysis
of the genetic structure of populations.

However, the origin of population genetics as a new branch of science
is related to the publication of Chetverikov's paper /42/, which not only
revealed the significance of the Hardy-Weinberg Law to the problem of
population genetics but also formulated the basic principles of population
genetics proper and indicated their significance to analysis of evolutionary
problems. A few years later several works appeared which laid the
cornerstones of the theory of modern population genetics /12, 17, 66, 80, 186/.
Foundations were laid for experimental population analysis from the
genetic standpoint.

In the course of the following thirty years, and especially in the last
decade, the development of population genetics was extremely rapid. The
investigators discovered factors ensuring the flourishing of species and
their communities.

Natural populations are the real groups of organisms in which the evolution begins and develops. Individual variability of organisms in populations serves merely as an elementary source for the start of the evolutionary process in a population. Data on quantitative regularities in the appearance of mutations, inheritance mechanisms in populations and in selection, rearranging the intrinsic content of the hereditary basis of the whole population, are extremely important in evolutionary theory. This is why the mathematical and general genetic examination of the theory of population genetics is most important. At the same time the theoretical work on this problem must clearly be based on experimental research into the evolution of populations in nature.

The purpose of this investigation is to throw light on the fundamental influence of long-term irradiation on populations of different organisms. Before examining the effect of radiation on populations it is necessary to dwell on radiosensitivity of cells and organisms, since alterations occurring in these life organization levels provide the starting point for genetic transformation of populations. Experimental studies into the effect of short-term irradiation on populations yield valuable information in this respect. Studies of the reaction of populations to short-term irradiation make it possible to estimate the duration of preservation in the population of genetic modifications produced by irradiation and to trace the effect of these modifications on the adaptive population genotype. Therefore presentation of data on the effect of long-term irradiation on populations will be preceded by examination of the results of short-term irradiation experiments.

The accumulated information concerning the effects of long-term irradiation on populations provides an approach to a mathematical description of the dynamics of mutation processes in irradiated populations. We attempted to describe the dynamics governing the accumulation of the genetic load in populations for the case of populations of unicellular green algae. In conclusion we shall examine data on the adaptation of populations to long-term irradiation.

Factors affecting radiosensitivity of organisms

Radiation damage to various biological systems (cells, organism, population) is primarily related to damage to genetic material. On the cellular level the leading role of damage to genetic material in radiation damage to the cell becomes evident if the radiosensitivities of the nucleus and cytoplasm are compared. When a cell is subjected to irradiation, the radiation energy is liberated within the cell more or less uniformly. The biological consequences of irradiation depend on the degree to which this energy uniformly liberated in the cell injures the latter's various organoids. A large amount of information indicates that injuries to the cell nucleus play a decisive role in radiation damage /3—5, 124, 172, 187/. For instance, a H a b r o b r a c o n cell is killed when its nucleus is hit by a single alpha particle but $21 \cdot 10^6$ alpha particles are required to produce the same effect by irradiation of cytoplasm /175/. Between 15 and 20 protons hitting a small area are sufficient to produce an appreciable effect on a chromosome.

No such effect could be detected when a similar area on the tip of a nuclear spindle or in the cytoplasm was hit with thousands and hundreds of thousands of protons /188/. In experiments with nuclear transplantation in Amoeba proteus it was demonstrated that irradiation of the cytoplasm was of minor significance, whereas an injury to its nucleus killed the cell /121/.

Comparison of radioresistance of organisms of varying degrees of complexity reveals a relationship between the radioresistance of organisms and the amount of genetic material in the cells and its organization.

Radiosensitivity is largely determined by such parameters as the molecular weight of nucleic acids /60, 68, 78, 135/, composition of bases in DNA /86, 84/, single- or double-stranded nature of nucleic acids /85, 112, 166—168/, and the total amount of nucleic acids in chromosomes /85, 88, 166, 167/. In the higher organisms a relation can be traced between the radiosensitivities of different species and the size of their chromosomes /44, 119, 139, 160/.

The number of chromosomes plays an important part in radiosensitivity. Plants with the same mean nuclear volume but different numbers of chromosomes differ in their radiosensitivities, the species with the larger number of chromosomes being the more resistant. According to Evans and Sparrow /64/ this difference is due to variation in the mean loss of genetic material per one chromosome split. Among the radiosensitivity-controlling characteristics, these authors also mentioned the position of centromere in the chromosome, the type of centromere and the amount of heterochromatin.

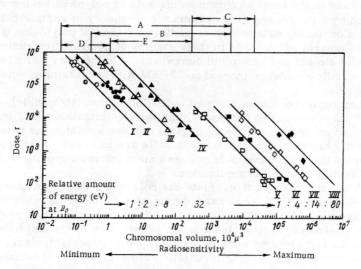

FIGURE 8.1. Radiosensitivity as a function of the chromosomal volume for prokaryotes (A), bacteria (B), eukaryotes (C), viruses (D) and yeasts (E). The points lie on eight regression lines with a slope of -1. The figure shows the mean relative amount of energy (eV) absorbed by a single chromosome at D_0 /158/:

I — predominantly RNA-viruses; II and III — mostly DNA-viruses; IV — viruses, bacteria, yeasts; V — yeasts, bacteria, mammals; VI — bacteria, algae, amphibians; VII — bacteria, plants; VIII — algae, ferns.

Sparrow et al. /156, 158/ demonstrated the radiosensitivity D_0 of cells to be inversely proportional to the volume of chromosomes in the interphase. This was demonstrated for a large group of organisms ranging from protozoans to the most complex forms. A graphical expression of the relationship between D_0 and the mean volume of a single chromosome in the interphase in these organisms on a logarithmic scale is provided by a series of eight regression lines (radiotaxa) with slope -1 shown in Figure 8.1.

The radiosensitivity of living organisms is largely dependent on their ploidy, the diploid organisms being as a rule considerably more stable than haploid ones. The relationship is exemplified by the comparative radioresistances of diploid and haploid yeasts /22, 25, 91, 92, 111/. Polyploid series display an optimum ploidy level with respect to radioresistance, differing in different species. For instance, among the yeasts, for which strains ranging from haploid to hexaploid were obtained, triploids proved the most radioresistant among the homozygous forms /91/, and diploids among hybrid forms /111/. Tetraploid cells proved resistant to irradiation among heteroploid forms /24/. Further increase of ploidy in these cases resulted in successive decline of radiostability.

Tul'tseva and Astaurov /40/ demonstrated the radioresistance of silkworm embryos (Bombyx mori L.) to increase with increasing ploidy. Their radiostability increases upon transition from the diploid to triploid levels. Further, a less marked increase in radiostability occurs as the ploidy increases from triploid to tetraploid embryos.

Stadler /161/ first demonstrated on polyploid series of oats and wheat that an increase in the level of chromosome sets in polyploid series was accompanied by a decrease in the occurrence frequency of radiational mutations. Convincing data on higher radioresistance of polyploids were obtained by Sakharov et al. /36/ in their studies of comparative radioresistances of diploid and tetraploid buckwheats. In the Chrysanthemum genus radioresistance increases 22-fold with increasing species ploidy /159/.

The higher radiostability of polyploids (including autotetraploids) depends not only on genomic defense which impedes mutations /67/, but also on the peculiar phenomenon of physiological defense by which the very appearance of chromosomic mutations in cells are impeded. Sakharov et al. /36/ suggested that autotetraploids possess mechanisms impeding realization of potential into actual replication.

As yet, it is still difficult to evaluate the role of aneuploidy in changing the radioresistance. On the one hand, duplication of individual chromosomes must increase the radiostability, but on the other hand, aneuploids possess an unbalanced genomic system, leading to degeneration in the majority of cases. The second circumstance is probably more important, since available information concerning irradiation of certain lines of tumor cells from mice containing different numbers of chromosomes show lines possessing a balanced number of chromosomes to be more radiostable /123/. For instance, a series of irradiations of an unstable line of cells in a culture containing an unbalanced set of 63 chromosomes produced stable lines of 44 chromosomes /83/.

Duplications of loci in a chromosome that are not accompanied by an increase in the number of chromosomes, i. e., those affecting the unbalancing of a genome to a lesser degree than in the case of heteroploidy, probably play a somewhat different role in modifying the radioresistance than heteroploidy. It may be assumed that duplications of genes controlling processes in the cell should result in a relative increase of its radioresistance.

Consider now a modification of the radioresistance of organisms in relation to their position in the phylogenetic system.

Preobrazhenskaya and Timofeev-Resovskii /33, 34/ studied the radiosensitivity of seeds of 120 species of plants. They reported a relationship between the degree of radiostability of species and their position in the phylogenetic system. All the investigated coniferous species were radiosensitive, over half of the investigated monocotyledon plants were radiosensitive and only one species was radioresistant, whereas about one-half of the investigated dicotyledon plants were radiostable and only few species were radiosensitive.

Terzi /166—168/, Kaplan and Moses /85/ and also Kondo /88/ analyzed radiosensitivity data for cells of different organisms beginning with viruses and ending with mammal and human cells. It was demonstrated that the cells of living organisms fall into four groups with respect to their stability to radiation, closely related to the quantity and qualitative status of DNA in them /85/. The first group comprises viruses with single-stranded DNA, the second group viruses with double-stranded DNA, the third group haploid microorganisms, and the fourth group comprises diploid organisms. Within every group there is a distinct dependence of resistance on the quantity of nucleotides per nucleus; the slopes of the corresponding curves (on a logarithmic scale) relate variations in radiosensitivy to those in the amount of nucleotides and are the same for all the groups.

Underbrink et al. /173/ compared the results reported by Sparrow et al. /158/ with those reported by Kaplan and Moses /85/ and Terzi /168/, and investigated the relationship between the content of nucleotides in the nucleus and the chromosomal volume. It was demonstrated that in the majority of prokaryotes, i. e., organisms possessing a nucleus resembling the material nucleus in organization, there was a linear correlation between radiosensitivity and the chromosomal volume (with slope -1). A similar regression line was obtained also for eukaryotes, but in their case the amount of nucleotides per unit chromosomal volume is smaller. Comparison of the radioresistance of cells as a function of the total content of nucleotides in the chromosomes allows all the organisms to be subdivided into groups, the radiosensitivities of which are described by four parallel regression lines with slope -1 (Figure 8.2). Thus, the same organisms form four or eight groups depending on the parameters taken into account (chromosomal volume or the number of nucleotides in the chromosomes). At present it is not clear what structural or cytogenetic features are involved in such regrouping of organisms into different numbers of radiotaxa. The author pointed out that the grouping of organisms based on the volume of interphase chromosome was probably better than their grouping by the content of nucleotides in the chromosome or nucleus (data reported by Kaplan and Moses), because the intact interphase chromosome is characterized by the effective volume of nucleic acid. Moreover, the volume of chromosomes is more easily determined for the majority of organisms than their contents of nucleotides.

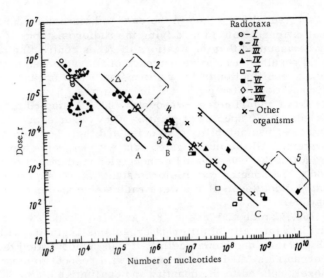

FIGURE 8.2. Radiosensitivity as a function of the number of nucleotides per chromosome. The slope of the regression lines is -1 /173/:

A — mostly RNA-viruses; B — 2-S-DNA-viruses, bacteria, yeasts; C — yeasts ($\geqslant 2x$), bacteria, lower plants, animals; D — bacteria, algae, plants ($\geqslant 2x$); 1 — microviruses (RNA); 2 — 2-S-DNA-viruses; 3, 4 — yeasts ($\geqslant 2x$); 5 — higher plants; I–VIII — see Figure 8.1.

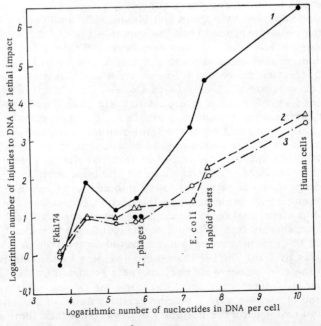

FIGURE 8.3. Stability of DNA to damage induced by UV radiation (1), X-rays and gamma-radiation (2) and ^{32}P (3) as a function of the content of nucleotides in DNA per cell /88/

Kondo /88/ demonstrated an increased resistance of DNA to lethal injuries with increasing complexity of genome (Figure 8.3), testifying to the improvement of reparative systems in the cell with increasing complexity of the genome in the course of evolution. The author assumed that besides the reparative aspects it is necessary to consider the tolerance mechanisms with respect to DNA damage. The tolerance in this case is defined as the ability of the cell to maintain its integral capacity for normal functioning, i. e., its capacity for fission, DNA replication, etc., in spite of sustained damage. In Kondo's opinion the tolerance of organisms increases with their progressing evolutionary complications.

Heretofore we have been concerned with cytogenetic factors influencing the radiosensitivity of organisms, meaning mostly short-term irradiation. In long-term irradiation of populations and cenoses, besides the above enumerated factors controlling the radiosensitivity of organisms there appear additional factors that are liable to modify the effect of a given dose on the organism. Sparrow and Woodwell /159/ divided the ecological factors affecting organism responses under the influence of radiation into the following four groups: 1) those modifying the dose; 2) those modifying the dose rate or fractionation; 3) those modifying the contribution to the dose from given kind of ionizing radiation under mixed irradiation; 4) those affecting the recuperation from radiation damage.

In considering the effects of long-term irradiation on populations and cenoses one should remember that the sexual reproduction process is one of the first to sustain injury /79, 155/, due to certain features of meiosis. The volume of the nucleus and size of chromosomes in meiosis are larger than in somatic cells, rendering the nucleus in meiosis more sensitive to radiation /154, 160/. The duration of the meiosis stages (especially the prophase) in plants is usually considerably longer than the mitotic fission time, likewise increasing the probability of damage to genetic material /155/. The radiation damage is also enhanced by the prolonged ripening of the seeds of plants, in the course of which there occur only a very small number of cell fissions, and the state of rest of the seeds /159/.

At present, the results of experiments on determining the comparative radioresistances of organisms of different biological complexities allow forecasts of the effect of short-term and long-term irradiation on various microorganisms, plants and animals /85, 88, 166—168, 173/.

Radioresistance norm

The principal property of radioresistant forms, attracting the attention of a majority of researchers, is their resistance to many chemical and physical agents, such as hydrogen peroxide, antibiotics and sulfamide preparations, chemical mutagens, UV radiation /31, 50, 81, 82, 89, 91/. This is not contradicted by the finer discrimination of the stability of series of radioresistant strains to such inactivating agents /73—76/. Mutations improving an organism's radioresistance represent a rearrangement of the metabolism which results in a lesser vulnerability of the entire organism and better reparative capacities.

As already noted, the stability of cells and organisms to radiation varies with varying sensitivity of genetic material in the cellular and ontogenetic cycles, and also with variation in metabolic activity of cells and various tissues of the organism. Most frequently the physiological aspects of radio-stability are secondary, controlled by genotype. The sensitivity of cells and organisms to the effect of ionizing radiation also varies with variations in the intensity of various modifying environmental factors /37, 39/. Ulti-mately, every organism or population possesses a characteristic regular radioresistance range (radioresistance norms). The radioresistance norm, as one of the concrete manifestations of the norm of genotype reactions, is an integral part of the possible radioresistance levels of the given genotype depending on the environmental conditions and cellular, ontogenetic and other stages. Reliable data on differences in the radioresistance norms of different organisms cannot be obtained unless the organisms with different genotypes are placed under identical conditions with respect to both external and internal factors. Under such conditions the radioresistance of orga-nisms is related, on the one hand, to the qualitative composition of the geno-type, i. e., to the degree of perfection with which the cell metabolism con-trolled by the genotype is capable of restoring or replacing damaged struc-tures and the related cell functions; on the other hand, it is also related to the amount of radiation-sensitive targets.

Effect of short-term irradiation on populations

In considering the effect of radiation on living organisms under natural conditions one must take into account that the population, not an individual organism, is the subject of the evolutionary process. The genetic pro-perties of a population are incomparably in excess of the genetic potentials of individual organisms. A Mendelian population represents a definite level of organic integration which is transformed under the influence of natural selection based on its intrinsic regularities.

Consider the data on the effect of short-term irradiation of populations. Early work on the genetics of irradiated populations was performed by Wallace and King /176—180/. Their experiments were conducted on an experimental population of D. melanogaster, made up of individuals free from lethals and semilethals in the second chromosome. The experi-ment showed that under the influence of ionizing radiation the irradiated populations developed numerous lethals, semilethals and other limitations impairing the viability. After a certain number of generations, populations cope with the damage even in the case of long-term irradiation, and they get rid of harmful mutations after the irradiation has ceased. In some cases radiational mutations may be advantageous to the population.

Similar results were also obtained on experimental populations of fruit fly in subsequent investigations following Wallace's work /6, 7, 54, 133, 134, 170/.

Stone and Wilson /163/ studied the consequences of nuclear tests carried out on Marshall Islands in 1954 and 1956 on fruit fly populations. The population on Bikini Island underwent direct irradiation and radiation from

fallout, those on other islands only radiation from fallout. The populations
were collected in 1955, 1956 and 1957. On the whole, there was considerable
damage to genetic structures in all the populations. The population on
Rongelap Island recovered its genetic characteristics after 26 generations.
The population on Bikini Island sustained the gravest damage and displayed
traces of genetic injury even in 1957.

Winge et al. /181/ and also Marques et al. /104/ investigated the effect
of irradiation on natural populations of D. w i l l i s t o n i inhabiting small
forest "islands" in Brazilian llanos. After the captured flies reproduced
in the laboratory they were irradiated and then introduced into the populations of forest "islands." Population counts under natural conditions
showed that they were predominantly composed of individuals from the
laboratory. The experiments revealed that natural selection eliminated
many lethal and semilethal mutations produced by irradiation. Moreover,
harmful radiational mutations were incorporated in the genotype of natural
populations owing to selection for heterozygosity. The consequences of
irradiation persisted for a very long time, as was shown by low viability
of the population for 35 generations.

This aftereffect of radiation is related to the protracted elimination of
genetic load due to the effect of ionizing radiation on a Mendelian population.
Genetic load as defined by Crow /57/ is the fraction of population for which
adaptation to its environment has been reduced in comparison to the viability
of individuals with the optimum genotype. Studies of natural populations
revealed genetic load as an inevitable feature of an evolving population
/13, 14, 58, 59/. Its sources lie in the inevitable process of natural mutation.
The latter's rate is in itself subject to natural selection /13, 43, 164/ and
serves the purpose of adaptations ensuring the population's adequate reactions to its habitat.

Kimura /87/ distinguishes the following three types of genetic load:
1) mutational, reducing the environmental adaptation for a part of the population by elimination of recurrent harmful mutations through these individuals; 2) splitting-off load, reducing the adaption of part of the population
by splitting-off homozygotes in reproduction of harmful mutations in
heterozygotes possessing heterosis; 3) substitutional, reducing that part of
the population which possesses an allel displaced by another, superior allel
in the course of natural selection.

The aftereffect of radiation on the fruit fly population, manifesting itself
in reduced fertility, is related to the splitting-off genetic load.

Dobzhansky /59/ subdivides adaptive diversity of individuals and the
corresponding diversity of genotypes into the genetic elite, genetic norm,
and genetic load. These categories and the corresponding genotypes
usually change place when the environmental conditions vary. The genetic
elite is extremely specialized and is well-adapted only to specific conditions.

One must bear in mind that in addition to mutations, which make up the
genetic load of a population and which are transmitted to the following
generations, an important feature in the case of multicellular organisms
is the disintegrating effect of radiation via somatic mutations dislocating
the complex system of regulatory mechanisms controlling the normal
morphogenesis.

Let us consider the dynamics governing the elimination from a population of a genetic load arising from the population's exposure to ionizing radiation. For this purpose we shall resort to data on populations of vegetatively reproducing unicellular algae Chlorella vulgaris, strain LARG-1, and Chlamydomonas reinhardi, strain pf-19 (+) /47, 48/, as well as on fruit fly populations. The unicellular algae proved to be a convenient object for studying some aspects of population genetics. On mineral culture media it is possible to reveal a broad spectrum of visible mutations of unicellular algae characteristic of natural populations. The rapid succession of generations in vegetatively reproducing unicellular algae makes it possible to obtain 30—40 generations (duplication cycles for the number of cells in the population) in the course of one or two months in a periodically trans- ferred culture. In this manner one can trace the mutation process dynamics following a one-time irradiation, or, in the case of a long-term irradiation, from the time of exposure until the population stabilization period. The performance of such research on fruit fly populations usually takes two or three years, only part of the resultant mutations (recessive visible and lethal mutations in individual chromosomes) being most commonly analyzed in such experiments.

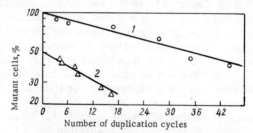

FIGURE 8.4. Displacement of mutant cells of the clones U-1 (1), and U-4 (2) by normal chlorella cells in model populations. For U-1, the initial ratio of the number of mutant to normal cells is 49:1, and for U-4, 1:1.

To begin with, we consider the simplest case, namely, the dynamics governing the elimination from a population of a single mutant clone. Figure 8.4 presents data on elimination of two mutant chlorella clones U-1 and U-4 produced by means of UV irradiation. These clones, differing from the starting strain LARG-1 in their light green color, do not revert, as was checked in prolonged cultivation of the mutant clones (some 60 dup- lication cycles). The fraction of mutant clone in its mixture with the strain LARG-1 is seen to decrease exponentially. By our calculations, the same regularity is obeyed when normal fruit flies displace the Bar and ebony fruit fly in model fruit fly populations /19, 93, 94/. Figure 8.5 presents (on a semilogarithmic scale) data on displacement of ebony fruit flies by normal fruit flies in quantitatively stable model populations /19/. Approxi- mately 250 days after the beginning of the experiment the number of ebony flies decreased to 10% and less of the total number of flies in the box, then became stabilized at the 8—9% level. The decrease in the relative presence of ebony flies in the course of time is exponential, and the period of half-elimination of the mutant flies from the population is 80—90 days.

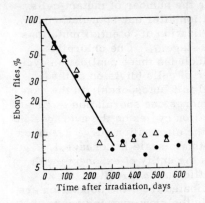

FIGURE 8.5. Displacement of ebony flies by normal flies in numerically stable model populations of D. melanogaster /19/:

• — experiment No.1; △ — experiment No.2.

The above data imply that every mutant clone, artificially added to a population or arising in it as a result of exposure to mutagenic agents and which undergoes negative selection, is eliminated from the continuously reproducing population in the course of time (in successive generations) by exponential law.

In natural populations a researcher deals with an enormous number of mutations continuously arising and undergoing selection. Every mutation possesses its own peculiar adaptive value, its own parameters of growth rate and elimination. We shall conduct a general examination of the fate of mutations induced by radiation in the instance of populations of unicellular algae.

The size of populations usually exceeds hundreds of millions of individuals and consequently approaches infinity. Under the influence of ionizing radiation the population develops n mutant clones the number of which is likewise fairly large ($n \to \infty$). Let a serial number $i = 1, 2, \ldots, n$ be assigned to every mutant clone, the i-th clone being assigned the number M_i. The entity of these mutant clones, under definite environmental conditions inferior to the normal population with respect to growth rate, constitutes the population's genetic load, according to Crow's definition /57/. In accordance with the data shown in Figure 8.5, every mutant clone will be eliminated from the continuously reproducing population according to the law

$$M_i = M_{i0} e^{-k_i x} , \tag{8.1}$$

where M_{i0} is the starting fraction of mutant clone in the population in question; k_i is the selection rigidity coefficient for the given mutant clone; x approximates the number of generations or the number of duplications of the number of individuals in the population for algae dividing into 4—8 daughter autospores (the number of duplication cycles for the number of cells in the population will be used by us as the population reproduction index).

The total number M of visible mutant cells in the population in a given duplication cycle for the number of cells following irradiation is equal to all the mutant clones M_i that have remained in the population at the time of this duplication cycle. Strictly speaking, M is equal to the sum total of a finite number of exponents, i.e.,

$$M = \sum M_i = \sum_{i=1}^{n} M_{i0} \, e^{-k_i x} = M_1 e^{-k_1 x} + M_2 e^{-k_2 x} + \cdots$$

$$\cdots + M_n e^{-k_n x}. \tag{8.2}$$

Figure 8.6 presents data on the elimination from a chlamydomonas population of all visible mutant clones produced by short-term X-ray irradiation of the population with dose 10 kr. The starting number of mutant cells must be the number of mutant cells in the population after one sporulation, i. e., after the realization of potential mutations arisen in the stage G_1. One chlorella sporulation includes three duplication cycles of the number of cells (division of the maternal cell in 8 autospores); in the case of chlamydomonas one sporulation is equal to 2.5 duplication cycles on the average. It is seen that the elimination curve drawn on a semilogarithmic scale is concave, i. e., it represents the entity of exponents with different coefficients k. At first, the mutants eliminated from the population are predominantly the slow-growing ones (dwarf and certain pigment mutations), while subsequent duplication cycles eliminate mutants with reproduction rates that are somewhat different from normal.

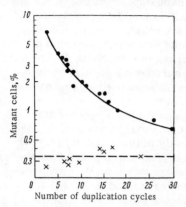

FIGURE 8.6. Elimination of mutant cells from a population of Chlamydomonas reinhardi, strain pf-19 (+), following an X-ray irradiation with dose 10 kr:

● — experimental population; × — control population.

The number of mutant clones in a fairly large population is high. Therefore equation (8.2) will contain a very large number of terms, so that it is practically impossible to determine the integral mutational process in the population with respect to time by summing the exponents. We present the elimination pattern for mutant cells on a logarithmic scale (Figures 8.7 and 8.8). It follows that there is a linear relationship between the logarithm of the fraction of mutant cells remaining in the population and the logarithm of the number of duplications of the number of cells in the population occurring since the beginning of the experiment. The slope of regression lines is the same for chlorella and chlamydomonas, and hence it may be concluded that the spectra of induced mutations are the same for both species with respect to the characteristic "relative reproduction rate." The slope angle of regression lines is 45°, indicating that the elimination of all mutant clones from chlorella and chlamydomonas populations obeys the simple law $M = M_0 x^{-1}$. Thus the elimination of genetic load in a fairly large population proceeds exponentially.

The question arises as to the condition under which the exponential elimination curves for individual mutant clones produce an exponential function. Consider the elimination process for an infinitely large number of mutant clones $n \to \infty$, i. e.,

$$M = \lim_{n \to \infty} \sum_{i=0}^{n} M_{i0} \, e^{-k_i x}. \tag{8.3}$$

Here it is convenient to introduce a transformation to an integral with respect to i in line with the analogy between improper integrals and functional series /41/.

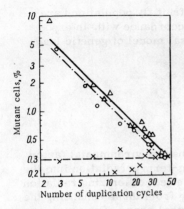

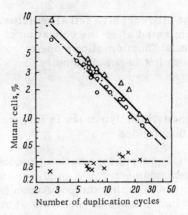

FIGURE 8.7. Elimination of mutant
cells from populations of Chlorella
vulgaris, strain LARG-1, following
X-ray irradiation with doses of 10 kr (O)
and 20 kr (Δ):

x — control population.

FIGURE 8.8. Elimination of mutant cells
from populations of Chlamydomonas
reinhardi, strain pf-19 (+1), following
X-ray irradiation with doses of 10 kr (O)
and 20 kr (Δ):

x — control population.

For comparison with experimental data, it is possible to make certain simplifying assumptions with respect to M_{i0} and k_i. Assuming that $k_i \sim i$, it follows that

$M = \int_0^\infty M_{i0}e^{-ix}di$. Let M_{i0} be integrable as a function of i. It is bounded, while

function e^{-ix} does not change its sign in the interval $[0, \infty]$. Therefore it is
possible to make use of the first mean value theorem for improper integrals
/41/. If function M_{i0} is continuous, then

$$M = \int_0^\infty M_{i0}e^{-ix}\,di = M_{i0av}\int_0^\infty e^{-ix}\,di =$$

$$= M_{i0av}\left(-\frac{1}{x}e^{-ix}\right)_0^\infty = M_{i0av}\,x^{-1}. \tag{8.4}$$

Agreement of experimental data with theoretical results on the basis of
our assumption $(k_i \sim i)$ indicates the existence in real populations of an exponential relationship between the growth rate coefficient of mutant clones
produced in the population by its irradiation and the probability of appearance
of these clones.

The question now arises as to the extrapolation of the results obtained
concerning populations of unicellular algae to those of other organisms.
Consider Wallace's data on the elimination of recessive lethal and semilethal mutations from irradiated fruit fly populations /178/. In this case
the relative elimination rate is considerably lower than in the chlorella
and chamydomonas populations, probably because the elimination of mutations proceeds mainly via individuals that are homozygotic with respect to
the recessive mutation, constituting about $\frac{1}{4}$ of the population, i.e., the elimination rate in this case is considerably lower than in the algal populations. In general, the elimination of all recessive mutations for fruit
fly populations can be described by the formula

$$M = M_0 x^{-p}, \tag{8.5}$$

where p is the coefficient describing the slope of the elimination curve for
mutant individuals when log M is plotted as a function of log x. The fact

that all recessive lethal mutations in irradiated fruit fly populations are eliminated after the cessation of irradiation in accordance with an exponential function allows one to construct a universal model of genetic load dynamics in populations.

Genetic load dynamics in long-term irradiation
of populations

Long-term irradiation produces a new permanent environmental factor. The fundamental new condition introduced by long-term irradiation into the factors affecting the population is the mutagenic effect of ionizing radiation producing a broad spectrum of mutations, ranging from those enhancing the viability and radioresistance to lethal and semilethal ones. The appearance of these mutations may cause profound changes in the irradiated populations.

The primary results of long-term irradiation of populations is an increased concentration in them of lethals, semilethals and other mutations detracting from viability. With a constant irradiation intensity the concentration of mutations constituting the population's genetic load becomes stabilized after several generations at a certain level which is proportional to the irradiation dose, as was demonstrated in Wallace's experiments on D. melanogaster populations /176—178/. The results of analyzing the incidence of lethals in five irradiated populations are shown in Figure 8.9. The incidence of lethals increased most significantly in populations Nos. 5 and 6, which were subjected to long-term irradiation with a dose rate of 5.1 r/hr in every population. Population No. 7, subjected to long-term irradiation with dose rate 0.9 r/hr, produced an appreciably smaller incidence of lethals. Population No. 1, in which the males and females were subjected to one-time doses of 700 r and 1,000 r, respectively, exhibited a higher incidence of lethals than the controls in the first five generations. Control population No. 3 was at first free from lethals, but subsequently under the pressure of the natural mutational process it acquired a stable 25—35% concentration of lethals. It is seen that populations Nos. 5 and 6 reached a stable concentration of lethals in the 60th—70th generation after the beginning of the experiment.

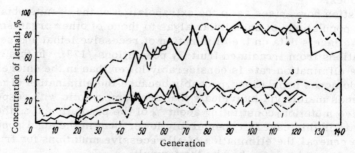

FIGURE 8.9. Concentration of lethal mutations in irradiated experimental populations of D. melanogaster over 140 generations /19/:

1—3 — populations Nos. 1, 3 and 7, respectively; 4 and 5 — populations Nos. 5 and 6, respectively.

Similar data were obtained from studies into the effect of long-term irradiation on chlorella populations (Figure 8.10), with an equilibrium mixture of radionuclides ^{90}Sr—^{90}Y added to the nutrient medium as the irradiation source. The dose rates of irradiation of chlorella cells with beta particles was 56 and 560 rad/day, depending on the employed radionuclide concentrations.

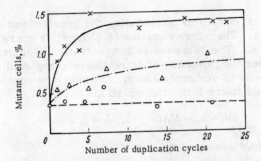

FIGURE 8.10. Dynamics of the buildup of mutant cells in chlorella populations in long-term exposure to beta radiation from ^{90}Sr—^{90}Y added to the nutrient medium:

^{90}Sr—^{90}Y concentrations in the medium, 2 μcuries/ml (Δ) and 20 μcuries/ml (x); dose rate, 56 rad/day (Δ) and 560 rad/day (x); o — controls.

Stabilization of the number of mutant cells during irradiation occurs at the level 0.6—0.8% for irradiation with dose rate 56 rad/day and 1.3—1.5% with dose rate 560 rad/day. Attention is drawn to the fact that concentration of visible mutations in the chlorella population subjected to long-term irradiation reaches a stable level already by the 10th duplication cycle for the number of cells in the population, whereas the stabilization in the fruit fly populations occurs in the 60th—70th generations. This is probably due to the larger mutational "capacity" of the diploid genome of the fruit fly in comparison to the haploid one in the chlorella.

The experimental data on chlorella populations makes possible a general consideration of the building-up of mutant individuals in populations of unicellular algae exposed to long-term irradiation. Similar to the derivation of equation (8.4), we shall first trace the fate of the i-th clone in the population. The level of the natural mutational process in this clone is b_i. In long-term exposure to radiation in every duplication cycle for the number of cells in the population there appears a certain additional quantity of induced mutant individuals of the clone.

Suppose the clone is not subject to the pressure of selection. In this case the amount of mutant cells of the clone will asymptotically approach c_i. The rate of approach to c_i will be controlled by the coefficient l_i, which is proportional to the irradiation dose. The differential equation describing the relationship between the amount of mutant individuals accumulated in a population and the number of duplication cycles for the number of cells in it has the form

$$dM_i = (c_i - M_i)\, l_i dx. \tag{8.6}$$

The initial condition, in accordance with the above notation, is

$$M_i|_0 = b_i.$$

The solution of this equation is

$$M_i = c_i - (c_i - b_i) e^{-l_i x}. \qquad (8.7)$$

In real populations, mutations constituting the genetic load are subjected to negative selection. Therefore equation (8.6) must be corrected for the pressure of selection. The decrease in concentration of mutant individuals of a given clone under the influence of selection in every duplication cycle for the number of cells is designated as k_i. Concurrent consideration of these processes leads to the following equation:

$$dM_i = (c_i - M_i) l_i dx - k_i (M_i - b_i) dx \qquad (8.8)$$

with initial condition $M|_0 = b_i$.
Hence equation (8.8) yields

$$M_i = \frac{c_i l_i + b_i k_i}{l_i + k_i} - (c_i - b_i) \frac{l_i}{l_i + k_i} e^{-(l_i + k_i) x};$$

$$M_i|_\infty = \frac{c_i l_i + b_i k_i}{l_i + k_i}. \qquad (8.9)$$

We now consider some of the simplest cases.
1. If $l_i = k_i$, then

$$M_i = \frac{c_i + b_i}{2} - \frac{c_i - b_i}{2} e^{-2 l_i x},$$

i.e., the incidence of mutant cells of the given clone in the population is stabilized at the level $\frac{c_i + b_i}{2}$.

2. If $l_i \gg k_i$, then

$$M_i = \frac{l_i}{l_i} \left(\frac{c_i + b_i \frac{k_i}{l_i}}{1 + \frac{k_i}{l_i}} \right) - (c_i - b_i) \times$$

$$\times \frac{l_i}{l_i \left(1 + \frac{k_i}{l_i}\right)} e^{-l_i x} = c_i - (c_i - b_i) e^{-l_i x}.$$

In this case the level of mutant cells of the given clone in the population is described by equation (8.7).
3. If $k_i \gg l_i$, then $M_i = b_i$, i.e., the level of mutant cells in the clone is stabilized in the populations within the level of the natural mutational process of the clone.

We have considered the buildup dynamics of mutant cells of a single mutant clone under exposure to long-term irradiation. In actual continuously irradiated populations one deals with a fairly large number of different mutant clones. Every one of these clones, possessing different viabilities, has its own selection rigidity coefficient k_i.

Let us introduce the following notation: for the i-th clone,

$$\frac{c_i l_i + b_i k_i}{l_i + k_i} = A_i;$$

$$\frac{(c_i - b_i) l_i}{l_i + k_i} = B_i;$$

$$l_i + k_i = C_i.$$

Then equation (8.8) yields

$$M_i = A_i - B_i e^{-C_i x}. \tag{8.10}$$

For n clones,
$$M = \sum_{i=1}^{n} (A_i - B_i e^{-C_i x}).$$

When $n \to \infty$,

$$M = \lim_{n \to \infty} \sum_{i=0}^{n} (A_i - B_i e^{-C_i x}) = \int_0^\infty (A_i - B_i e^{-C_i x})\, di =$$

$$= \int_0^\infty A_i di - \int_0^\infty B_i e^{-C_i x}\, di. \tag{8.11}$$

It is noteworthy that agreement with experimental data is achieved under the following assumptions:
if

$$l_i, k_i \sim i; \quad c_i, b_i \sim e^{-i};$$
$$k_i = ki; \quad l_i = li; \quad b_i = be^{-i}; \quad c_i = ce^{-i},$$

then

$$A_i \sim e^{-i}; \quad C_i \sim i; \quad B_i \sim e^{-i}; \quad M_i \sim e^{-i} - e^{-i} e^{-ix}.$$

In deriving equation (8.4), it was noted that the assumption $k_i \sim i$ fully fits real populations.   Analysis shows that the assumptions with respect to l_i, c_i and b_i likewise fit reality when treating the integral mutational process in populations.

Substituting the above expressions in equation (8.11) and integrating gives

$$\int_0^\infty A_i di = \int_0^\infty \frac{c_{i0} l_{i0} + b_{i0} k_{i0}}{l_{i0} + k_{i0}} e^{-i} di = \frac{cl + bk}{l + k};$$

$$\int_0^\infty B_i e^{-C_i x} di = \frac{(c-b) l}{l + k} \int_0^\infty e^{-i} e^{-(l_{i0}+k_{i0})ix} di = \frac{(c-b) l}{l + k} \times$$

$$\times \int_0^\infty e^{-(l_{i0}x + k_{i0}x + 1) i} di = \frac{l(c-b)}{l + k} \cdot \frac{1}{x(l + k) + 1}.$$

Thus, the genetic load dynamics in algal populations exposed to long-term irradiation under the above assumptions is described by the equation

$$M = \frac{cl + bk}{l + k} - \frac{l(c - b)}{l + k} \cdot \frac{1}{x(l + k) + 1}. \tag{8.12}$$

In deriving equation (8.12), we made use of the case $p=1$, when the slope of the regression line describing the elimination of mutant cells from the population is 45° (see equation (8.4)). As already pointed out, the slope of the elimination regression line for fruit fly populations is $p<1$ (see equation (8.5)). Therefore, by introducing index p (corresponding to the fruit fly population) in equation (8.9), it is possible to obtain an equation describing the genetic load dynamics in crossing populations. It is of interest to describe the variation in the concentration of mutant individuals in populations subject to irradiation of varying intensity, for instance, in cases of exponentially decreasing or increasing doses. Cases of this kind may occur in areas polluted with artificial radionuclides and likewise in prolonged space flights.

A description of the genetic load dynamics in populations with allowance for parameters p (slope of the regression line describing elimination of mutant individuals from a population) and l (in buildup intensity of mutant individuals in the population, proportional to the irradiation dose) forms the purpose of our further investigations.

Effect of ionizing radiation on populations
of laboratory mammals

At this time, the attention of many researchers is being attracted by radiation genetics of populations of laboratory mammals. The principal target of these investigations is a scientific prognosis of the genetic consequences of the effects of ionizing radiation on humans /169/. However, before discussing aspects related to radiation genetics of mammal populations we must briefly examine the possibility of estimating the human genetic effects of ionizing radiation.

A scientifically validated estimate of genetic danger calls for information on the dose dependence of the yield of mutations, the role of the time factor in the mutational process and the relative genetic effectiveness of different types of radiation. This information should apply to different types of mutations. Scientific information of this kind regarding humans is very limited.

Currently available information concerning the direct effect of radiation on the genome of human sexual cells is scant. Shifts in the numerical ratio of sexes have been known to occur in the progeny of persons who were exposed to radiation for therapeutic purposes, in reactor breakdowns, nuclear explosions, or who came in contact with radiation in the course of their work. However, the available information is insufficient for a quantitative estimate of the genetic effect of radiation. Information on the incidence of chromosomic diseases, teratisms, stillbirths, spontaneous abortions, etc., in the progeny of irradiated persons is likewise scant and contradictory.

In view of the insufficiency of correct information concerning the effect of ionizing radiation on the genome of human sexual cells, the data obtained on mammals are extremely important. Recent achievements pertaining to the radiation genetics of mammals have revealed the basic regularities governing the effect of ionizing radiation on the genome of sexual cells of mice and made possible a preliminary extrapolation to man of experimental data obtained on mice.

The maximum danger for future humanity lies in mutations passing through the sieve of meiosis and capable of being transmitted from one generation to another, i. e., the gene mutations, and mutual translocations among chromosomal aberrations. Results that are currently most suitable for extrapolation are those elicited by studies of the induced frequency of visible mutations in several loci in mice /132/. As yet, extrapolation of data on chromosomal aberrations and karyotype mutations is practically impossible because of inadequate information on the subject.

In determining the degree of human genetic danger of ionizing radiation one must usually resort to certain assumptions which include, for instance, linear dependence of the yield of mutations on the dose and absence of an influence of irradiation intensity, while calculations are based on data obtained for radiation of low specific ionization density acting on a single generation. At the same time, the results of recent investigations have shown that reducing the dose rate reduces the incidence of recessive mutations in seven loci in mice /127/. The effect was discovered only in irradiation of spermatogonia and oocytes, but was absent in irradiation of postmeiotic stages.

In estimating the human genetic effect of ionizing radiation it is especially important to study the mutations in spermatogonia and oocytes, i. e., in the most immature cells serving as the replenishment source of mature cells. The genetic radiosensitivity of sexual cells in different stages of gametogenesis varies considerably. In an overwhelming majority of cases the sexual cells of males form the following series of increasing genetic radiosensitivity: spermatogonia, spermatocytes, spermatozoa, spermatids. The most radiosensitive stage in females is that preceding ovulation — diakinesis-metaphase I, then oocytes in the stage of anaphase I and metaphase II of meiosis; finally, the genetic radiosensitivity of early stages of oocytes turned out to be the least. In this connection it should be borne in mind that the genetic radiosensitivity of sexual cells is unrelated to the radiosensitivity determined from their survival /125/.

The most extensive investigations into radiation genetics of mammals were carried out by a group of Swedish scientists under Lüning /95—100, 141, 142/. The work was begun in 1959 /95/ and the results of analysis for 13 generations have now been published. The investigation was conducted on mice of the line CBA, the starting population consisting of 125 pairs. Two males and two females were reserved from every pair for propagating the population, the crossing being random. At the age of 60 days the males' testicular area was subjected to X-ray irradiation with a dose of 276 r. The males were crossed with nonirradiated females 90 days after the irradiation, i. e., the fecundation was performed by spermatozoids that had been irradiated in the stage of spermatogonia. An analogous control population was maintained in parallel.

The effect of radiation was determined from the numerical size of litters, the number of litters over 100—170 days, the mortality of baby mice from birth to weaning, the occurrence frequency of dominant and recessive lethals and radioresistance. The frequency of dominant lethals was determined from postimplantational embryonal mortality observed in uterotomy on pregnant females. The frequency of recessive lethals was observed in crossings of the type brother X sister. In this case, in addition to the postimplantational mortality a count was also taken of the death of embryos prior to implantation, determined from the numerical ratio of the corpora lutea in the ovary and the implantation loci in the uterus.

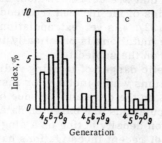

FIGURE 8.11. Difference between experimental and control populations of mice with respect to numerical size of litters (a), number of baby mice per female (b), and mortality (c)

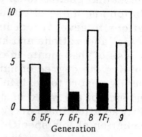

FIGURE 8.12. Decrease in numerical size of litters in mice with respect to controls in the 5th—9th generations of the irradiated population (the difference between experimental and control populations, %, is marked on the y-axis):

□ — direct irradiation of males;
■ — offspring from irradiated males.

The results of studies of the first nine generations revealed a slight decrease in the numerical size of litters, a decrease in the total number of offspring, and an increase in mortality of baby mice /97, 99/. As seen from Figure 8.11, the difference with respect to the controls did not exceed 5—10% and, most important, did not increase from one generation to another. In order to isolate the results of irradiation of males directly in a given generation and the built-up effect due to the influence on preceding generations, the data on offspring of irradiated males were compared with those for offspring of males that remained nonirradiated in the given generation. From Figure 8.12 it is seen that the decrease in the numerical size of litters was mainly due to the effect of radiation directly on the male of the given generation, whereas the effect from exposure of preceding generations was very slight and did not increase from one generation to another. Analysis of postimplantational mortality of embryos did not reveal any differences between the experimental and control populations.

Studies of embryonal mortality in crossings of the type brother × sister establish the presence of radiation-induced recessive lethals in the experimental population /98/. Taking into account the spontaneous lethals, the decrease in the survival of embryos was only 1.14% on the average for 4—7 generations. The total difference with respect to the control is significant, but the separate differences for every generation are statistically nonsignificant. Scaled to 1 r per genome the frequency of recessive lethals amounted to $8-20 \cdot 10^{-5}$, based on a mean accumulated dose of 1,300 r. It is of interest to note a trend toward accumulation of the number of recessive lethals in the generations. However, the differences between generations are statistically nonsignificant and were discovered only in determinations of the survival of embryos with reference to the number of corpora lutea, not by analysis of postimplantational mortality alone.

Special experiments were conducted in order to find out whether recessive lethals could manifest themselves in the heterozygote /100/, as in the case of fruit flies /16, 65, 113, 114/. For this purpose, families with embryonal mortalities >15% and <10% were isolated from the brother × sister crossings. It was assumed that the families were heterozygotic with respect to recessive lethals in the first but not in the second case. The females and males from these families were crossed with unrelated animals. Higher embryonal mortality was not discovered either before or after implantation in the groups with suspected presence of recessive lethals in the heterozygote. Consequently, the authors concluded that there is no manifestation of recessive lethals in the heterozygote in mice.

Studies were likewise made of the effect of repeated irradiation in several generations on such an integral index as radioresistance. Determination of LD_{50} for males from the 13th and 14th generations in the experimental and control groups did not reveal any differences between them. Likewise no difference was detected in the lifespans of experimental and control animals in the same generations following fractionated irradiation with dose 1,400 r /141/. No differences were detected in studies of the effect of irradiation with doses of 65 and 100 r on the productivity of females (determined from the numerical size of litters and total number of offspring) between animals of the 13th and 14th generations in the experimental and control populations /142/, neither were there any shifts in the ratio of sexes in the irradiated populations /96/.

Thus, the principal conclusion to be drawn from the experiments is absence of buildup of mutations through several generations of repeatedly irradiated populations; alternatively, such a buildup must be extremely insignificant.

At approximately the same time similar investigations were conducted in the USA by a group of scientists under Spalding /143—153/ on mice of the line PFM. Every generation of males was subjected to whole-body irradiation with a dose of 200 r at the age of 26 days; 25—34 days later they were crossed with sisters and with unrelated females. After 10 generations a subline was isolated in order to determine the elimination rate of genetic damage; the subline was no longer irradiated. Studies were made of productivity (by several indices), lifespan, and radioresistance. By now results have been obtained from more than 50 generations.

Fecundity studies revealed some decrease with respect to the age of females at the first litter, weight at birth, and incidence of cannibalism. At the same time, females from the irradiated population produced a larger number of litters and enjoyed a more prolonged reproductive period. Studies of the numerical size of litters, early postnatal mortality and lifespan did not reveal any differences between animals of the experimental and control populations.

The radioresistance was estimated from the lifespan of mice placed in a gamma-field (dose rate, 4 rad/hr), which was lower in the irradiated population (for both inbreeding and cross breeding), being 715 hr as against 906 hr in the control population. Lower radioresistance, persisting over

25 generations, was also found in the sublines of animals that had been irradiated in only 10 generations.

Examination of the weight of different organs and of the entire body, deposition of fat, activity of cell enzymes and blood picture in animals of the 25th—37th generations revealed lesser fat deposits, smaller body weight and somewhat larger weight of kidneys in mice 6—8 months old in the experimental population. No differences were detected in the weight of the other organs. The only shifts in the blood picture in the experimental population of mice were decrease of WBC due to a decrease in lymphocytes and an increase in the mean volume of erythrocytes.

Thus, the results of these investigations revealed differences only with respect to the level of radioresistance in the population of mice exposed to ionizing radiation.

In experiments conducted by a group of researchers in the Jackson Laboratory /70—72,136/ the gonads of male mice of the line C_{57} or hybrids bred with the participation of several inbred lines were subjected to irradiation with doses of 50 or 100 r per generation. Seven weeks after the irradiation the males were crossed with unrelated females and observed for 11 generations. Fecundity was investigated by means of such tests as the number of effective crossings, the numerical size of litters, the total number of offsprings, the reproductive period duration, the mortality of baby mice and embryonal mortality, which in this case might be a consequence of dominant lethals. The offspring of the irradiated population exhibited a reduced duration of reproductive period and a reduced total number of offspring. However, no differences were detected with respect to the numerical size of litters and embryonal mortality. In another series of experiments the same lines of mice were subjected to local irradiation of testicles with doses of 200 r (C_{57}) and 900 r (hybrids) for 7 generations. Studies of the same indices of fecundity and shifts in the numerical ratio of sexes did not reveal any differences in comparison to the nonirradiated population. Investigation of the influence of all the above indicated doses on lifespan after 7—11 generations revealed no reduction. The only change was a buildup of mutations leading to some reduction in the animals' reproductive capacity.

Searle et al. /137,138/ subjected male and female mice of the line C_3H/HeH to long-term gamma-irradiation with ^{60}Ca, dose rate 1 r/day, throughout their lifespan. The mean irradiation dose was about 80 r per generation. The indices of the effect of gamma irradiation were fecundity, productivity from three litters, numerical size of litter, postnatal mortality, and embryonal mortality for different types of matings and skeletal anomalies. Analysis of 24 generations showed the numerical size of the first and second litters to be almost unchanged but it decreased beginning with the third litter, apparently due to the effects of ionizing radiation directly on the oocytes of early stages in the females. The embryonal survival on the 14th day of pregnancy was considerably reduced in comparison to the controls for both inbreeding and crossbreeding. Calculated incidence of recessive lethals was $2.94 \cdot 10^{-2}$ per gamete. It is noteworthy that postnatal mortality during the period from birth to weaning was almost halved in the irradiated population, apparently due to heterosis along newly

arising mutations. Studies of skeletal anomalies did not establish any
reliable effect of irradiation, on account of extensive variability of these
characters in the nonirradiated animals.

The experimental results thus revealed some shifts in populations under
the influence of repeated irradiation, but their magnitude does not lend itself readily to comparison with changes reported by other researchers
from analogous experiments and also from experiments with short-term
irradiation, since in this case all the stages of sexual cells, not spermatogonia alone, were exposed to irradiation.

Gowen and Stadler /69/ irradiated mice of the five lines Bab, BaB, E, S, N
for 10 generations with gamma-quanta of ^{60}Co at dose rate 2.6 r/day,
followed by crossbreeding. They detected some reduction of lifespan, but
did not detect any changes in many other characters.

Touchberry and Verley /171/ subjected mice of the line C_{57} at age
60—70 days in every generation to X-ray irradiation with doses of 50, 100,
150 and 200 r (males) and 20, 30, and 40 r (females). Ten to fourteen days
after irradiation the mice were mated in all possible variants (crossbreeding). Some decrease was observed in the numerical size of litters,
becoming more pronounced from one generation to another, but it was
difficult to determine to what extent it was caused directly by the irradiation
of the animals themselves and to what extent it reflected the effect accumulated through generations. Postmeiotic cells were exposed in these experiments.

Japanese researchers /116, 165/ subjected male and female mice to
long-term gamma-irradiation for three generations. Comparison of the
results of short-term and long-term irradiation within a single generation
showed a lower incidence of recessive lethals in the case of long-term irradiation. However, one must bear in mind the difficulties involved in making
such a comparison, in view of the different stages that were actually exposed to irradiation in each generation.

Chapman et al. /55/ irradiated male (in the stage of spermatogonia) or
female rats aged 10—14 weeks to fractionated irradiation with doses 100,
150 and 200 r. The males were mated with nonirradiated unrelated females
for 12 weeks, whereas the females were mated immediately after irradiation.
Studies of the numerical size of litters, postnatal survival and shift in the
numerical ratio of sexes did not reveal any cumulative effect of irradiation
in the first four generations.

Brown /52/ subjected 11 generations of rats of the line Holtzman to daily
gamma-irradiation with dose rate 2 r/day. In this case all stages of the
male and female sexual cells were subjected to the irradiation. Analysis
of 10—13 generations revealed a decrease in the numerical size of litter,
which persisted also in crossing with nonirradiated animals. There was no
change in the ratio of sexes. Investigation of the animals with respect to
various physiological tests led the investigators to conclude that there was
only a small shift with respect to the drowning time.

Newcomb and McGregor /117/ subjected the gonads of male rats aged
90 days to a one-time irradiation with dose 600 r for 12 generations. They
were mated directly after irradiation, i. e., fertilization was achieved by
spermatozoids that had been irradiated in their postmeiotic stages. The
test was provided by the learning time in a maze. The animals' orientation
capacity deteriorated with increased dose accumulated by the gonads.

Studies of the body weight in the offspring of the 13th generation of this population /118/ showed that at the age of 90 days the animals from this population were heavier than the controls, and the difference persisted in comparisons of the weight of animals from litters of the same size.

Although the experimental results obtained on mice and rats are some-what contradictory, their analysis permits the preliminary conclusion that the buildup of genetic damage proceeds much more slowly than could be expected from the results of experiments with irradiation of animals in a single generation. It is important that the buildup of genetic damage was mainly observed when irradiation had affected postmeiotic cells, and it was absent or very slight when irradiation had affected spermatogonia. The experimenters were practically unable to detect any distinct increase in the buildup of genetic damage in generations with increasing accumulated dose on spermatogonia.

It must be emphasized that no buildup effect could even be expected in respect to the majority of tests. Thus, a decrease in the numerical size of litters and embryonal mortality are related to major chromosomal aberrations and confined to the progeny of directly irradiated animals, but do not occur in the generation following the irradiated one /97, 101/. In these tests, the absence of cumulative effects of damage during irradiation of spermatogonia was primarily due to elimination of injuries in different gametogenesis stages, and possibly also in the course of ontogenesis and even postnatal development.

It is now well known that populations of spermatogonia are extremely heterogenous with respect to radiosensitivity /120/. Thus, spermatogonia of some types are killed by doses of the order of only a few roentgens, while others survive doses of several hundred roentgens (man) or even several thousand roentgens (mouse). The production of these radiostable cells is responsible for repopulation of the germinal epithelium after irradiation. There probably exists a positive correlation between the radiosensitivity of spermatogonia, determined from their survival, and mutability. As a result, only radiostable cells with relatively small number of mutations survive irradiation with large doses. This is responsible for the peculiar nature of the dose-effect relationship when spermatogonia are exposed to radiation /126, 127/. For irradiation with doses of the order of several hundred roentgens the yield of mutations is a linear function of the dose, after which the yield of mutations declines as the dose increases. Thus, the process of selective destruction of spermatogonia after irradiation is an evolutionary defense mechanism shielding the population against an excessive number of mutations. Furthermore, in the case of small doses the quantity of mutations may decrease due to the recovery of premutational injuries /125—131/. Consequently, the maximum number of mutations in a popu-lation should be expected in repeated fractionated irradiations, when the survival rate of spermatogonia has not yet been excessively reduced but the dose is still sufficiently large to suppress the activity of the rehabilitative system. This is borne out by data elicited by experiments involving irra-diation of a single generation of animals /127/. Mutations preserved in the surviving spermatogonia must then pass through several mitotic fissions and also through meiosis. Obviously, only point mutations and very few chromosomal aberrations (translocations, inversions) are preserved in this case.

It is of interest that in irradiated spermatogonia the number of mutations is the same throughout the reproductive period /140/, whereas in the case of females the number of mutations dwindles markedly, beginning with the third litter after irradiation /129/. It is likewise possible that the mutations passing through the "sieve" of mitosis and meiosis may be eliminated also in a later period. A discrepancy has recently been discovered between the amount of mutual translocations (caused by irradiation of spermatogonia) detected cytologically (in the diakinesis stage-metaphase of the first meiotic split) and genetically (from the number of semisterile individuals in the progeny) /138/. In the first case the number of detected translocations is larger, suggesting selection in postmeiotic stages of spermatogonesis or else selection of mature spermatozoids with translocation on account of their reduced fertilizing capacity. Thus, there possibly exists also a second, postmeiotic barrier (or even barriers) interfering with mutations, or else there is a selection in later stages. Some investigators /101, 137/ have thus reported high mortality of baby mice in the progeny of irradiated males in the early postnatal period. However, these data were mostly obtained in experiments with irradiation of a single generation of animals.

All the above information indicates that the effect of selection should be expected already in the first generation of mammals with respect to the majority of investigated indices. Nevertheless, it does not preclude the possibility that studies of point mutations will enable one to detect buildup of mutations in generations and manifestation of newly arising mutations in the heterozygote similar to the case with the fruit flies.

Studies of this kind call for better techniques of quantitative reckoning of mutations. The currently available techniques for detecting recessive lethals in mammals are extremely labor-consuming and imperfect. For instance, there are doubts concerning the possibility of calculating the yield of induced lethals in the case of repeated irradiation in generations based on cumulative doses. Such calculation presumes a linear dose-effect relationship, yet, as already pointed out, the dose dependence in irradiation of spermatogonia is nonlinear, being of an unusual nature.

The above facts point to the need for further developing various aspects of the radiation of population genetics for mammals. Available data do not yet make it possible to look into the need for revising the role played by the natural radiation background in spontaneous mutagenesis, and to assess the human genetic danger from small doses of irradiation in relation to the presumed slow buildup of induced mutations in mammal populations.

Genetic adaptation of populations to long-term irradiation

In a natural environment, ionizing radiation is among the ecological factors affecting populations in a variety of directions. In order to estimate the part played by ionizing radiation, it is necessary to isolate from the entirety of ecological factors the specific action mechanisms of radiations on the populational and cenotic levels. Taking into consideration the genetic load, alteration of coadaptive ties among individuals as well as other changes

in irradiated populations, it is important to describe all these changes, which
may be termed radiation damage to the population. In considering the
radiation effect on the cenotic level, Aleksakhin /1, 2/ introduced the concept
of "radiation damage to cenosis." The magnitude of radiation damage to a
population depends, on the one hand, upon the irradiation intensity and, on the
other, upon the radioresistance norm of the population, including the possible
radioresistance levels of the population's genetic system under given environmental conditions /46/. The increase in genetic load, or (more broadly)
radiation damage to the population is the negative aspect of the effect of
long-term irradiation on populations. In itself, an increase of genetic load,
manifesting itself in the appearance of a broad spectrum of forms less
adapted to environmental conditions in comparison to the normal form,
intensifies the selection. Under these conditions, the selection will adopt
any mutation reducing the percentage of eliminated individuals, i. e., reducing
the population's genetic load. Such mutations, ensuring the populations'
adaptation to conditions of long-term irradiation, may be termed radioadaptational. Such mutations do not necessarily improve the radioresistance
of individuals in the population. The formation of radioadaptation is facilitated by the broad spectrum of mutations, including those improving coadaptive ties in populations, producing new systems of balanced polymorphism, etc. At the same time, it is clear that the principal role in the
formation of radioadaptation is played by mutational transformation of the
genotype of individuals ultimately increasing the population's radioresistance.
As regards radioresistance mutations, selection produces a new system of
coadaptive ties in the population in line with the latter's modified genetic
content. Thus, in considering the effect of long-term irradiation on the
populations one must bear in mind that the radioresistance mutations only
represent a part of positive genetic modifications that become involved in
the complex process leading to the radioadaptational genotype.

One fundamental theoretical problem in radiation population genetics is
the examination of the role played by the natural radiation background in
the evolution of populations and species. Several authors working in the
1930s demonstrated that the natural radiation background was capable of
producing mutations, but their incidence is very small and hardly exceeds
0.001 of the incidence of all spontaneous mutations /49, 115/. Further
investigations demonstrated the importance of the life-cycle duration of
living organisms and their radiostability in estimating the effect of a higher
radiation background. Accordingly the radiation background was assigned
an important place in the natural mutational process /15, 16/.

Investigations in areas with high natural radiation background are of
primary interest when solving such an important problem as the part
played by the natural radiation background in life evolution on Earth, first
formulated in the general theoretical aspect by Vernadskii /8/. On these
territories, characterized by high concentration of natural isotopes of uranium,
thorium, radium and other elements, living organisms have been exposed to
higher levels of long-term irradiation than in other regions of the globe,
for prolonged geological time. Comprehensive study of populations of
different living organisms inhabiting such areas and comparison of the
results with those for control populations shed light on the consequences
of prolonged exposure of living organisms to small doses of irradiation.

Grünberg /77/ investigated the effect of a high radiation background on black rats (Rattus rattus L.) on a small island in Kerala State (India). The high radiation background on that island (dose rate of external irradiation about 1,300 mrad/year, approximately 7.5 times above the control background) is due to the presence of monazite sands containing thorium phosphate. Nine hundred rats were investigated in eight areas on the island in order to investigate the effect of the high radiation background. The rats had been unwittingly brought there by man 300 years ago and since then 500 generations of rats had received the total dose of about 500 r, as against 67 r in the control. Fifteen measurements of rat skeletons and 6 measurements of their teeth served as the principal indices of the effect of long-term irradiation. No significant differences were established between the experimental and control rat populations. In view of the low dose rate on the island, it was suggested that mutations produced by the ionizing radiation were either eliminated by natural selection or the increase in the genotypic variability was compensated by the decrease in paratypic variability. Comparative studies of the teeth measurements in the rats of investigated populations revealed that the selection with respect to these characters had probably proceeded differently in the different populations /174/; the variability with respect to these characters turned out to be higher in the rats exposed to the higher radiation background.

Verkhovskaya et al. /10/ discovered higher incidence of sterility, smaller weight of testicles, heavier damage to spermatozoids and sperm ductules in striped field mice in areas with high radiation background (up to 4—8 mr/hr).

Raushenbakh and Monastyrskii /35/ and Monastyrskii /28/ observed sousliks and striped field mice living in areas isolated from one another and differing only in a higher (6—7-fold) natural radiation background in some of the areas due to thorium-bearing minerals forming monazite placers. Differences were detected between the populations with respect to the numerical ratio of sexes. In the case of animals from the experimental areas, the females comprised 63% of the sousliks and 64% of the striped field mice, compared to the 1:1 ratio of the sexes in the same species in control areas. Fractionated irradiation of animals from the experimental and control areas in vivarium revealed the former's higher radioresistance. Thus, 33 and 46%, respectively, of the sousliks from the experimental area and only 11 and 15%, respectively, from the control area survived irradiation doses of 2,850 and 1,800 r (at the end of the experiment). In addition 20 and 40%, respectively, of the striped field mice from the experimental area and only 0 and 20%, respectively, from the control area survived irradiation with the same doses (at the end of the experiment). Animals from the experimental areas were distinguished by a higher redox potential; moreover, they were found to comprise two mutations: polydactylia and color mutation, which, in the authors' opinion, indicates a higher intensity of the mutational process in the experimental than in the control areas.

Higher radiostability of Andropogon filifolius growing on uranium-enriched soils was reported by Mewissen et al. /110/.

Merzari /109/ discovered a series of anomalies in several plants collected on areas with high uranium concentrations. Osborne /122/ studied P e n s t e m o n v i r e u s growing in a thorium-bearing area with high natural radiation background (0.05—0.4 mr/hr) and found that the plants in this area were distinguished by a variety of teratisms and abnormalities, presumably related to the effect of ionizing radiation. In the same area, experiments were conducted toward understanding the effect of different radiation levels on mutational processes in a spiderwort /107, 108/. There were significant differences in the incidence of mutations of the flower when the plants were grown under long-term irradiation with dose rates of 0.1 and 0.25 mr/hr.

On Niue Island in the Pacific, where the natural radiation background is rather high owing to high radioconcentrations in its soil, the local inhabitants (45,000 families) are distinguished by lower fecundity in comparison to inhabitants of neighboring islands /105, 106/.

Of considerable interest are data dealing with the effect of radioactive wastes from nuclear industry on microorganisms, plants and animals. For instance, an investigation of the qualitative and quantitative composition of microflora and microfauna populations in freshwater bodies at Oak Ridge contaminated with radioactive wastes detected no morphological, physiological or genetic changes /90/.

On the other hand, Lozina-Lozinskii and Aleksandrov /27/, Lozina-Lozinskii /26/ and Kiselev et al. /21/ observed high radioresistance in microorganisms isolated from radioactive springs.

Blaylock /51/ and Nelson and Blaylock /29/ reported data concerning the effect of environmental pollution with radioactive wastes on populations of C h i r o n o m u s t e n t a n s. Large quantities of its aquatic larvae inhabit the bottom sediments of White Oak Lake, White Oak Spring and Clinch River used since 1943 for the disposal of radioactive wastes from the Oak Ridge National Laboratory. Approximately 130 generations of the midge over a period of 22 years were exposed to irradiation with dose rate 230 rad/yr. In 1960—1964 a determination was performed of the incidence of aberrations in the chromosomes of salivary glands of these larvae, collected by seasons in the contaminated and control habitats. In all the populations there were six endemic heterozygotes by inversions, three of which were relatively frequent (0.09—0.22), but the frequency did not vary with the seasons and did not undergo any considerable changes in the irradiated populations over a period of five years. The difference in the incidence of endemic inversions between the irradiated and control populations was insignificant. The conclusion was that ionizing radiation increased the frequency of chromosomal aberrations in the investigated populations but the aberrations were eliminated by selection.

A comparative investigation was made of the populations of mammals (cotton rats and P e r o m y s c u s l e u c o p u s mice) in areas contaminated with radioactive wastes and on a control territory /61/. Differences were detected between the populations with respect to the ratio of sexes, the duration of reproductive period and certain other indices, but final conclusions concerning the effect of radioactive contamination were made difficult owing to the complex interaction of other ecological factors affecting the populations.

Thus, in connection with the above, it may be presumed that evolutionary transformation of the genotype of a population inhabiting an area with a high radiation background takes place only in the presence of sufficiently high doses of long-term irradiation producing a definite selection pressure. The effect of ionizing radiation as a new environmental factor will be significant for evolution of populations when its effect on living organisms is commensurable with the effect of other factors all of which determine the manifestation of natural selection.

Increasing radioresistance of populations in long-term irradiation

The earliest works which provided the start for extensive studies into the genetic control of radioresistance were those by Witkin /184, 185/, who produced by means of UV-irradiation a stable strain E. coli B/r, which provided the material for investigations of the radioresistance phenomenon. Following Witkin, experimental increase of radiostability was observed also by other researchers in different organisms. The isolation and investigation of radioresistant strains by repeated irradiation was carried out on yeasts /102, 103/, coliform bacteria /30/, the blue-green alga Anacystis nidulans /89/, as well as other microorganisms /11, 20, 62, 63/. Zhestyanikov /18/ studied the variation of radioresistance in a continuous culture of E. coli strains B, B/r and K-12 subjected to continuous gamma-irradiation for 22 months. After 2 months the daily transferred cultures exhibited an increased radioresistance which proceeded to increase. An extensive body of factual data on the selection of radioresistance forms in microorganisms was gathered together by Sokurova /38/.

Platonova and Sakharov /31/ carried out experimental selection for radiostability of diploid and autotetraploid forms of buckwheat. Increased radioresistance was exhibited by diploid buckwheat after only a single selection, and by tetraploid buckwheat after two or three selections, the increase in stability being related to selection of forms possessing chromosomes that are stable to irradiation.

Radiation-stable colonies of cells from mice of strain L were isolated from surviving cells after irradiation with dose 1,000 r /182/. The radiostable strain R_1 and its derivatives R_2 and R_3 had survival curves with smaller slope than the original strain L. Cytogenetic investigation of strains L and R_3 showed them to possess characteristic, chromosomal markers. Nonirradiated cultures of strain L contain some 3% of cells with markers that are typical of the radioresistant strain R_3, in which these cells rise to 71.4% /183/.

Thus, increase in radioresistance has been observed in experiments on organisms situated on different levels of phylogenetic organization. In all the experiments, radioresistant forms were isolated after a single or repeated irradiation, acting as a selection factor for radiostable variants, arising spontaneously or under the influence of ionizing radiation.

In natural selection, the processes in irradiated populations do not relate to separate mutations but to the entire integrated adaptive genotype. For instance, evolution of populations under the influence of ionizing radiation is in many respects related to the appearance of mutations in polygenic systems. Buzzati-Traverso and Scossiroli /53/ demonstrated on hybrid lines of D. melanogaster that mutations in a polygenic system arising under the influence of ionizing radiation are capable of ensuring progress of populations under the influence of natural selection. These results are of fundamental significance, demonstrating the possibility of progressive evolution based on radiational mutations.

In this connection it is of interest to trace the establishment of an adaptive genotype of populations under natural environmental conditions and in exposure to long-term irradiation of fairly high intensity, capable of producing a directed selection toward the population's increased radio-resistance.

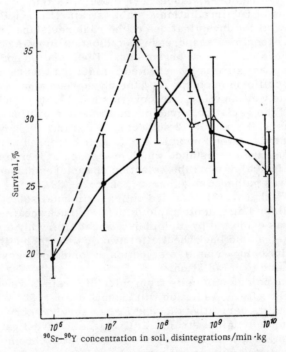

FIGURE 8.13. Survival of chlorella as a function of the $^{90}Sr-^{90}Y$ concentration in soil, with additional provocative radiation of 30 kr in 1962 (●) and 1963 (Δ)

Consider the radioresistance data for chlorella populations (Chlorella vulgaris Beijer) existing in long-term exposure to high radiation background. The chlorella forms were isolated from samples taken from soils to which ^{90}Sr in equilibrium with ^{90}Y had been added, and from control plots.

A total of 192 chlorella strains were isolated. Many of the strains isolated from areas with high ^{90}Sr content were mutant (pigment mutations). The high level of mutant clones in the populations indicated strong mutational pressure due to the mutogenic effect of beta-radiation from ^{90}Sr—^{90}Y.

During six years following the addition of ^{90}Sr to the soil the chlorella populations went through some 200 generations. The isolated chlorella strains were subjected to additional provocative X-ray irradiation with 30 kr. Data on the radioresistance of the chlorella populations are shown in Figure 8.13. The strains are grouped according to different ^{90}Sr concentrations in the soil. Attention is drawn to the increased radioresistance at all points in the chlorella forms isolated from soils with high ^{90}Sr concentrations. One must point out the characteristic radioresistance peak and medium ^{90}Sr—^{90}Y concentration in the soil in 1962 of $5 \cdot 10^8$ disintegrations/min·kg. The calculations show that the dose rate of external irradiation of chlorella cells at this concentration of ^{90}Sr—^{90}Y amounts to about 10 rad/day. In 1963 this peak shifted toward a lower ^{90}Sr—^{90}Y concentration in the soil ($5 \cdot 10^7$ disintegrations/min·kg). A gradual decline in the radioresistance of chlorella is typical for the highest ^{90}Sr—^{90}Y concentrations in the soil.

Hence, there exists an optimum ^{90}Sr—^{90}Y concentration in soil for selection of chlorella for radioresistance ($5 \cdot 10^7$—$5 \cdot 10^8$ disintegrations/min·kg). At ^{90}Sr—^{90}Y concentrations exceeding the optimum concentration for the breeding of chlorella for radioresistance, the genetic load on the population probably increases considerably owing to accumulation of a large number of mutations impairing its overall viability. Stability-impairing mutations arise in forms with radioresistance considerably above that of the controls. Identical data on the radioresistance peak for two years in soils with high ^{90}Sr—^{90}Y concentration testify that the population reached a radiostability level at which stabilization takes place between radioresistance mutations and continuously arising mutational load.

According to calculations, the chlorella population existing in the presence of a high radiation background is approximately 1.5—2 times more radiostable than the controls. The mean LD_{50} of the control strains is 12.7 kr, whereas for strains isolated from areas to which ^{90}Sr has been added the mean LD_{50} varied between 15 and 25 kr, depending on the ^{90}Sr concentration. The LD_{50} for the 10 most radioresistant strains averaged 42.8 kr.

The upper temperature of the growth boundary of strains was determined in the control and experimental populations. Figure 8.14 shows data summarizing the findings for all the investigated strains. The experiments yielded the same regularity as the radioresistance studies; there was a concentration of ^{90}Sr in the soil that was optimum for selection for thermal stability, coinciding with the optimum concentration of this nuclide for selection for radioresistance. In this case, the high radiation background is also present, but the provocative background facilitating the preeminent selection of thermostable chlorella forms is absent. It may therefore be concluded that an increase in the radioresistance of chlorella forms is accompanied by an attendant increase in the stability of these forms toward high temperature. The thermal stability of pigment mutants,

usually considerably smaller than in the controls, likewise increases with
increasing ^{90}Sr concentration in the soil, exceeding the thermal stability of
control strains.   This testifies that the pigment mutants in soils with
different ^{90}Sr concentrations originate from those radioresistant (and thermostable) forms which had already "conquered" areas with such ^{90}Sr concentrations in the soil.

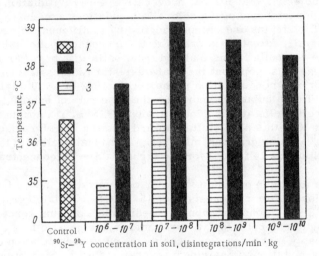

FIGURE 8.14. Thermal stability of experimental and control chlorella strains:

1 — control population; 2 — phenotypically normal forms of experimental population; 3 — pigment mutants of experimental populations.

The control and experimental strains were tested for productivity by two
criteria: the growth rate coefficient $\alpha = \dfrac{\log N_t - \log N_0}{t}$ (where N_t is the number
of cells in the culture after time t, days) and chlorella productivity me^{α}
(where m is the dry weight of 10^8 cells of the strain).   The mean growth
rate of experimental strains (in the exponential phase of culture growth)
was 1.99 ± 0.02, as against 1.71 ± 0.02 for control strains ($P < 0.01$).   The
mean productivity of experimental strains was 13.5 ± 0.82 mg/day, as
against 10.6 ± 1.1 mg/day in the controls ($P = 0.05$).   These data point to a
higher productivity of chlorella strains inhabiting soils with higher ^{90}Sr
concentrations.

Thus, the radioresistant chlorella strains prove likewise stable to high
temperature and possess higher productivity.   Therefore it may be concluded
that in the presence of a higher radiation background there occurs a mutational rearrangement of the genotype, enhancing the overall viability of
organisms.

The increase in the radioresistance of chlorella populations is probably due to gene mutations. Polyploid forms of chlorella were not detected in either the control or experimental populations. Investigation of a series of E. coli mutants showed their radiosensitivity to be probably controlled by several loci /73—76/. This conclusion was based on data concerning the differential sensitivity of these mutants to different radiations, radiomimetics and antibiotics. An examination of the stabilities of individual chlorella strains to X-rays and beta-radiation of ^{90}Sr—^{90}Y reveals that some strains that are resistant to ^{90}Sr—^{90}Y prove unstable toward X-rays. Moreover, radiostable strains differ in their productivities and thermal stabilities. This circumstance indicates the existence of several genes or gene systems (polygenes) the mutation of which reduces the damaging effect of ionizing radiation. In the case of mutation of certain genes, some characters such as thermostability and productivity are affected more, and in the case of mutation of other genes they are less affected. Evolution toward increased stability to high radiation background will evidently proceed by improvement of these gene systems in generations. The above data testify that radioresistant strains possess different genotypes. We deal with a process of broad polymorphism, not only with respect to visible mutations but also to mutations all of which control the population's radioresistance.

Analysis of the data led us to conclude that in considering the effect of long-term irradiation of organisms and populations under natural conditions it is feasible to introduce the concept of the norm of radioresistance of organisms and populations. The radioresistance norm is the concrete expression of the fundamental constant "genotype reactions norm" and reflects the change in the sensitivity of the genetic content in ontogenetic and populational cycles and during a variation in environmental conditions. From the existing concept of ecological radiosensitivity, taking into account the change in radiosensitivity under the influence of a broad spectrum of environmental conditions /2/, the concept of radioresistance norm differs by emphasizing the internal genetic factors modifying the radiosensitivity.

The discovered differences in radioresistance between the control and experimental chlorella populations represent differences in their radioresistance norms. They reflect the profound mutational transformations of the population's genotype occurring during exposure to a high background of ionizing radiation. Since the sequence of these transformations is related to a gradual rise of radioresistance norms of individual lines and entire populations, this process may be termed progressive evolution of the population's radioresistance norm.

Conclusion

The problem of the radiation of population genetics in our days is very important for understanding the nature of biological effects of ionizing radiation on populations of living organisms. There are two major aspects in this field of research. First, it is necessary to establish the scale of genetic damage to populations under the influence of different irradiation

doses, in the presence of the entire complexity of ecological and biocenotic
relationships, involving analysis of the appearance and fate of genetic load
in populations. Study of the radiation of population genetics of mammals
brings us nearer to an assessment of the human genetic danger of radiation.

The other major problem is analysis of adaptive evolution of populations
in response to a high ionizing radiation background. The question arises as
to the adaptation of populations to the effects of ionizing radiation. Experimental studies of chlorella populations in habitats with high radiation background revealed real shifts in the population's radiosensitivity. The
general theory of appearance and selection of radiostability mutations shows
such an adaptive evolution in response to ionizing radiation to be fully
possible. Its forms must be very different in organisms of different species.

At present, much interesting data have already been obtained in the field
of radiation of population genetics. We are now entering a new phase in
the development of this problem. The time has come to implement a broad
front of research into problems of radiational genetic load and radioadaptational evolution, influencing several agricultural aspects which must be
reckoned with in the agricultural development of areas possessing high concentrations of radioactive substances.

The influence of ionizing radiation on the heredity and evolution of
organisms should be investigated under natural conditions, by tracing the
transformations of irradiated populations in all their biocenotic relationships. In conjunction with model experiments, such a research program
will ensure successful progress in the problem of radiation genetics of animals, plants and microorganisms, this being one of the major theoretical
and practical trends in modern biology.

References

1. Aleksakhin, R. M. — In Sbornik: "Problemy sozdaniya zamknutykh
   ekologicheskikh sistem," p. 203. Moskva, "Nauka." 1967.
2. Aleksakhin, R. M. — In Sbornik: "Problemy radioekologii," p. 8.
   Moskva, Atomizdat. 1968.
3. Astaurov, B. L. — Zhurnal Obshchei Biologii, Vol. 8:421. 1947.
4. Astaurov, B. L. — In Sbornik: "Pervichnye mekhanizmy biologi-
   cheskogo deistviya izluchenii," p. 140. Moskva, "Nauka." 1963.
5. Astaurov, B. L. — Byulleten' MOIP, Otdel biologicheskii, Vol. 63:35.
   1958.
6. Bileva, D. S. — In Sbornik: "Vliyanie ioniziruyushchikh izluchenii na
   nasledstvennost'," p. 177. Moskva, "Nauka." 1966.
7. Bileva, D. S. — Genetika, Vol. 4:66. 1967.
8. Vernadskii, V. I. Biosfera (The Biosphere), Vol. 1 and 2. Moskva-
   Leningrad, Nauchno-khimiko-tekhnicheskii otdel VSNKh. 1926.
9. Vernadskii, V. I. Khimicheskoe stroenie biosfery Zemli i ee okru-
   zheniya (Chemical Structure of the Earth's Biosphere and its
   Surroundings). Moskva, "Nauka." 1965.

10. Verkhovskaya, I. N. et al.— Radiobiologiya, Vol. 5:5. 1965.
11. Graevskii, E. Ya. and E. G. Zinov'eva.— DAN SSSR, Vol. 121:837.
    1958.
12. Dubinin, N. P.— Zhurnal Eksperimental'noi Biologii, Vol. 7:5. 1931.
13. Dubinin, N. P.— Biologicheskii Zhurnal, Vol. 5:939. 1936.
14. Dubinin, N. P.— Zhurnal Obshchei Biologii, Vol. 9:203. 1948.
15. Dubinin, N. P.— Botanicheskii Zhurnal, Vol. 43:1093. 1958.
16. Dubinin, N. P. Evolyutsiya populyatsii i radiatsiya (Evolution of
    Populations and Radiation). Moskva, Atomizdat. 1966.
17. Dubinin, N. P.— Biologicheskii Zhurnal, Vol. 5:5. 1932.
18. Zhestyanikov, V. D.— Radiobiologiya, Vol. 3:847. 1963.
19. Zurabyan, A. S. and N. V. Timofeev-Resovskii.— Zhurnal
    Obshchei Biologii, No. 5:612. 1967.
20. Kashkin, K. P.— In Sbornik: "Voprosy radiobiologii," Vol. 4:313.
    Leningrad, TsNIRRI. 1961.
21. Kiselev, P. N. et al.— Mikrobiologiya, Vol. 30:207. 1961.
22. Korogodin, V. I.— Trudy MOIP, Vol. 7:181. 1963.
23. Korogodin, V. I. Problemy postradiatsionnogo vosstanovleniya
    (Problems of Postradiational Rehabilitation). Moskva, Atomizdat.
    1966.
24. Korogodin, V. I. et al.— Radiobiologiya, Vol. 3:39. 1963.
25. Korogodin, V. I. and Liu Ai-shen.— Tsitologiya, Vol. 1:379. 1959.
26. Lozina-Lozinskii, L. K. — Tsitologiya, Vol. 3:154. 1961.
27. Lozina-Lozinskii, L. K. and S. N. Aleksandrov.— Tsitologiya,
    Vol. 1:84. 1959.
28. Monastyrskii, O. A. Issledovaniya radiochuvstvitel'nosti zhivotnykh
    iz raionov s normal'nym i povyshennym estestvennym fonom
    ioniziruyushchei radiatsii (Investigations into the Radiosensitivity
    of Animals from Areas with Normal and High Natural Ionizing
    Radiation Background).— Thesis. Novosibirsk. 1968.
29. Nelson, D. J. and B. G. Blaylock.— In Sbornik: "Voprosy radio-
    ekologii," p. 162. Moskva, Atomizdat. 1968.
30. Pershina, Z. G.— In Sbornik: "Izmenchivost' mikroorganizmov,"
    Vol. 2:24. Moskva. 1957.
31. Platonova, R. N. and V. V. Sakharov.— Radiobiologiya, Vol. 2:595.
    1962.
32. Platonova, R. N. and V. V. Sakharov.— Genetika, Vol. 3:56. 1965.
33. Preobrazhenskaya, E. I.— "Byulleten' Nauchno-tekhnicheskoi in-
    formatsii po agronomicheskoi fizike," No. 10:53. 1962.
34. Preobrazhenskaya, E. I. and N. V. Timofeev-Resovskii.—
    DAN SSSR, Vol. 143:1219. 1962.
35. Raushenbakh, Yu. O. and O. A. Monastyrskii.— In Sbornik:
    "Vliyanie ioniziruyushchikh izluchenii na nasledstvennost',"
    p. 165. Moskva, "Nauka." 1966.
36. Sakharov, V. V. et al. Radiatsionnaya genetika (Radiation Genetics),
    p. 346. Moskva, Izdatel'stvo Akademii Nauk SSSR. 1960.
37. Dirovov, B. N. and V. V. Khvostova.— In Sbornik: "Ioniziruyushchie
    izlucheniya i nasledstvennost'," Vol. 3:176. Moskva, Izdatel'stvo
    Akademii Nauk SSSR. 1960.

38. Sokurova, E. N. — Uspekhi Sovremennoi Biologii, Vol. 53:69. 1962.
39. Tarasov, V. A. — Uspekhi Sovremennoi Genetiki, No. 1:106. 1967.
40. Tul'tseva, N. M. and B. L. Astaurov. — Biofizika, Vol. 3:197. 1958.
41. Fikhtengol'ts, G. M. Kurs differentsial'nogo i integral'nogo
        ischisleniya (A Course in Differential and Integral Calculus).
        Moskva, Fizmatgiz. 1962.
42. Chetverikov, S. S. — Zhurnal Eksperimental'noi Biologii, Vol. 2:4.
        1926.
43. Shapiro, N. I. — Zoologicheskii Zhurnal, Vol. 17:592. 1938.
44. Shapiro, N. I. et al. — Zhurnal Obshchei Biologii, Vol. 21:104. 1960.
45. Shevchenko, V. A. — In Sbornik: "Upravlyaemyi biosintez," p. 262.
        Moskva, "Nauka." 1966.
46. Shevchenko, V. A. — Informatsionnyi byulleten' radiobiologiya,
        No. 10:60. 1967.
47. Shevchenko, V. A. — Uspekhi Sovremennoi Genetiki, No. 1:246. 1967.
48. Shevchenko, V. A. et al. — Genetika, Vol. 5:61. 1969.
49. Efroimson, V. P. — Zhurnal Obshchei Biologii, Vol. 1:3. 1931.
50. Adler, H. J. — Genetics, Vol. 48:881. 1963.
51. Blaylock, B. G. Disposal of Radioactive Wastes into Seas, Oceans
        and Surface Waters. — Proc. of Symposium, Vienna, 16—20 May
        1966, p. 337. Vienna, IAEA. 1966.
52. Brown, S. O. — Genetics, Vol. 50:1101. 1964.
53. Buzzati-Traverso, A. A. and R. E. Scossiroli. — Adv. Genet.,
        Vol. 7:47. 1955.
54. Carson, H. L. — Genetics, Vol. 49:521. 1964 .
55. Chapman, A. B. et al. — Genetics, Vol. 50:1029. 1964.
56. Chaffee, R. R. et al. — Genetics, Vol. 53:875. 1966.
57. Crow, J. F. — Human Biol. Vol. 30:1. 1958.
58. Dobzhansky, T. — Genetics, Vol. 24:391. 1939.
59. Dobzhansky, T. — Am. Naturalist, Vol. 98:151. 1964.
60. Donini, R. and H. T. Epstein. — Virology, Vol. 26:359. 1965.
61. Dunaway, P. B. and S. Kaye. — In: Radioecology, New York, Reinhold
        Publ. Corp. Washington D. C. Amer. Inst. Biol. Sci. 1963.
62. Erdman, J. E. — Canad. J. Microbiol., Vol. 7:199. 1961.
63. Erdman, J. E. et al. — Canad. J. Microbiol., Vol. 7:207. 1961.
64. Evans, H. J. and A. H. Sparrow. — Brookhaven Symposia in Biology,
        Vol. 14:101. Brookhaven Natl. Lab. U. S. AEC report BNL-675
        (C-31). 1961.
65. Falk, R. — Mut. Res., Vol. 4:805. 1967.
66. Fisher, R. A. The Genetical Theory of Natural Selection. Oxford.
        1930.
67. Fröier, K. et al. — Hereditas, Vol. 28:165. 1942.
68. Ginoza, W. — Nature, Vol. 199:453. 1963.
69. Gowen, J. W. and J. Stadler. — Genetics, Vol. 50:1115. 1964.
70. Green, E. L. — Ibid., p. 423.
71. Green, E. L. and E. P. Les. — Ibid., p. 497.
72. Green, E. L. et al. — Ibid., p. 1053.

73. Greenberg, J.— Genetics, Vol. 49:771. 1964.
74. Greenberg, J.— Genetics, Vol. 50:639. 1964.
75. Greenberg, J. and K. P. Woody.— J. Gen. Microbiol., Vol. 33:283. 1963.
76. Greenberg, J. and K. P. Woody.— Radiation Res., Vol. 20:350. 1963.
77. Grüneberg, H.— Nature, Vol. 204:222. 1964.
78. Guild, W. R.— Radiation Res., Vol. 9:124. 1958.
79. Guneckel, J. E. and A. H. Sparrow.—Brookhaven Symposia in Biology, p. 252. Brookhaven Natl. Lab. U. S. AEC report BNL-258 (C-19). 1954.
80. Haldane, J. B. S.   The Causes of Evolution. London. 1932.
81. Haynes, R. H.—Biophys. Society Meeting, Chicago, 26 Feb. 1964.
82. Haynes, R. H. and W. R. Juch.— Proc. Nat. Acad. Sci. USA, Vol. 50:839. 1963.
83. Horikana, M. and J. D. Tsutomusugahara.— Radiation Res., Vol. 22:478. 1964.
84. Kaplan, H. S et al.— J. Cell. Comp. Phys., Vol. 64:69. 1964.
85. Kaplan, H. S. and L. E. Moses.— Science, Vol. 145:21. 1964.
86. Kaplan, H. S. and R. Zavarine.— Biochem. Biophys. Res. Commun., Vol. 8:432. 1962.
87. Kimura, M.— J. Genet., Vol. 57:21. 1960.
88. Kondo, S.— Japan. J. Genetics., Vol. 39:176. 1964.
89. Kumari, H. D.— Ann. Bot., 27, 723. 1963.
90. Lackey, J. B. and C. F. Bennet.—In: Radioecology, New York, Reinhold Publ. Corp., Washington, D. C., Amer. Inst. Biol. Sci. 1963.
91. Laskowski, W. Z.— Naturforsch., Vol. 156:495. 1960.
92. Latarjet, R. and B. Ephrussi.— Compt. rend. Acad. Sci., Vol. 229:306. 1949.
93. L'Héritier, Ph. and G. Teissier.— Compt. rend. Soc. Biol., Vol. 117:1049. 1934.
94. L'Héritier, Ph.— Compt. rend. Soc. Biol., Vol. 124:881. 1937.
95. Lüning, K. G.— Hereditas, Vol. 46:668. 1960.
96. Lüning, K. G.— U. N. Document, A/AC. 82. 1962.
97. Lüning, K. G.— Hereditas, Vol. 50:361. 1964.
98. Lüning, K. G.— Mut. Res., Vol. 1:86. 1964.
99. Lüning, K. G. and W. Sheridan.— Genetics, Vol. 50:1043. 1964.
100. Lüning, K. G. and W. Sheridan.— Mutation Res., Vol. 3:340. 1966.
101. Lyon, M. F. et al.— Genet. Res., Vol 5:448. 1964.
102. Maisin, I. et al.— Compt. rend. Soc. Biol., Vol. 149:227. 1955.
103. Maisin, I. et al.— Rev. Belg. Pathol. Med. Exptl., Vol. 25:392. 1956.
104. Marques, E. K. et al.— Proc. XVI Intern. Congr. Genet. The Hague, Vol. 1:76. 1963.
105. Marsden, E.— Nature, Vol. 187:192. 1960.
106. Marsden, E.— University of Chicago Press Chapt., Vol. 50:87. 1964.
107. Mericle, L. W. et al.— Radiation Res., Vol. 19:185. 1963.
108. Mericle, L. W. and R. P. Mericle.— Health Phys., Vol. 11:1607. 1965.

109. Merzari, A. H. — Cited in: Gopal-Ayengar, A. K.   Effects of
      Radiation on Human Heredity, p. 115.   Geneva, World Health Orga-
      nization. 1957.
110. Mewissen, D. J. et al. — Nature, Vol. 183:1449. 1959.
111. Mortimer, R. K. — Radiation Res., Vol. 9:312. 1958.
112. Levinson, W. and H. Rubin. — Virology, Vol. 28:533.   1966.
113. Müller, H. J. — Amer. J. Hum. Genet., Vol. 2:111. 1950.
114. Müller, H. J. — Amer. J. Publ. Health, Vol. 54:42. 1964.
115. Müller, H. J. and L. M. Mott-Smith. — Proc. Nat. Acad. Sci.
      USA, Vol. 16:277. 1930.
116. Muramatsu, S. et al. — Int. J. Rad. Biol., Vol. 6:45. 1963.
117. Newcombe, H. B. and J. F. McGregor. — Genetics, Vol. 50:1065.
      1964.
118. Newcombe, H. B. and J. F. McGregor. — Genetics, Vol. 52:851.
      1965.
119. Nubom, N. — Bot. Not., Vol. 109:1. 1956.
120. Oakberg, E. F. — Genetics, Vol. 52:463. 1965.
121. Ord, M. J. and J. F. Danielli. — Quart. J. Microscop. Sci.,
      Vol. 97:29. 1956.
122. Osburn, W. S. — Cited in: Klement, A. W.   Health Phys., Vol. 11:1265,
      1965.
123. Revesz, L. et al. — Nature, Vol. 198:260. 1963.
124. Rogers, R. W. and R. C. von Bortsel. — Radiation Res., Vol. 7:484.
      1957.
125. Russell, L. B. and C. I. Saylors. — In: Repair from Genetic
      Radiation Damage and Differential Radiosensitivity in Germ Cells,
      ed. by F. H. Sobels.   New York, Pergamon Press. 1963.
126. Russell, W. L. — Genetics, Vol. 41:658. 1956.
127. Russell, W. L. — In: Repair from Genetic Radiation Damage and
      Differential Radiosensitivity in Germ Cells, ed. by F. H. Sobels.
      New York, Pergamon Press, p. 205. 1963.
128. Russell, W. L. — Nucleonics, Vol. 23:53. 1965.
129. Russell, W. L. — Proc. Nat. Acad., Sci. USA, Vol. 54:1552. 1965.
130. Russell, W. L. — Genetics Today, Vol. 2. N. Y. Pergamon Press. 1965.
131. Russell, W. L. — Pediatrics, Vol. 41:223. 1968.
132. Russell, W. L. et al. — Science, Vol. 128:1546. 1958.
133. Sankaranarayanan, K. — Genetics, Vol. 50:131. 1964.
134. Sankaranarayanan, K. — Genetics, Vol. 54:121. 1966.
135. Schambra, P. E. and F. Hutchinson. — Radiation Res., Vol. 23:514.
      1964.
136. Schlager, G. et al. — Mut. Res., Vol. 3:230. 1966.
137. Searle, A. G. — Mut. Res., Vol. 1:99. 1964.
138. Searle, A. G. et al. — Ann. Human Genet., Vol. 29:111. 1965.
139. Sharma, A. K. and A. K. Chatterji. — Nucleus, Vol. 5:67. 1962.
140. Sheridan, W. — Mut. Res., Vol. 2:67. 1965.
141. Sheridan, W. — Mut. Res., Vol. 4:675. 1967.
142. Sheridan, W. and C. Rönmback. — Mut. Res., Vol. 4:683. 1967.
143. Spalding, J. F.   Effect of Ionizing Radiation on the Reproductive
      System, Ed. by W. D. Carlson and F. X. Gassner, p. 147. Pergamon
      Press. 1964.

144. Spalding, J. F. et al. — Health Phys., Vol. 10:273. 1964.
145. Spalding, J. F. et al. — Genetics, Vol. 50:1179. 1964.
146. Spalding, J. F. and M. R. Brooks. — Proc. Soc. Exptl. Biol. Med.,
        Vol. 119:922. 1965.
147. Spalding, J. F. et al. — Genetics, Vol. 54:756. 1966.
148. Spalding, J. F. et al. — Proceed. XII Int. Congr. Genetics, p. 101.
        Tokyo. 1968.
149. Spalding, J. F. — Radiation Res., Vol. 15:329. 1960.
150. Spalding, J. F. et al. — Genetics, Vol. 46:129. 1961.
151. Spalding, J. F. and V. G. Strang. — Radiation Res., Vol. 16:159.
        1961.
152. Spalding, J. F. et al. — Radiation Res., Vol. 18:479. 1963.
153. Spalding, J. F. et al. — Genetics, Vol. 48:341. 1963.
154. Sparrow, A. H. et al. — Radiation Botany, Vol. 1:10. 1961.
155. Sparrow, A. H. and V. Pond. —In: Conference on Radioactive Isotopes
        in Agriculture, p. 125. Michigan State University, 12—14 January
        1956, USAEC report TID-7512. 1956.
156. Sparrow, A. H. et al. — Radiation Botany, Vol. 8:149. 1968.
157. Sparrow, A. H. and L. Schairer. — In: 2nd Intern. Conf. Peaceful
        Uses Atomic Energy, 13 September 1958, Vol. 27:335. 1958.
158. Sparrow, A. H. et al. — Radiation Res., Vol. 32:915. 1967.
159. Sparrow, A. H. and G. M. Woodwell. Radioecology. New York,
        Reinhold Publ. Corp. Washington, D. C., Amer. Inst. Biol. Sci.
        1963.
160. Sparrow, A. H. —Brookhaven Symposia in Biology, Vol. 14:76. Brook-
        haven Natl. Lab., USAEC report BNL-675 (C-31). 1961.
161. Stadler, L. J. — Proc. Nat. Acad. Sci. USA, Vol. 15:876. 1929.
162. Stadler, L. J. and J. W. Gowen. — Genetics, Vol. 46:901. 1961.
163. Stone, W. and F. Wilson. — Proc. Nat. Acad. Sci. USA, Vol. 84:565.
        1958.
164. Sturtevant, A. H. — Biol. Bull.,. Vol. 73:542. 1937.
165. Sugahara, T. — Genetics, Vol. 50:1143. 1964.
166. Terzi, M. — Nature, Vol. 191:461. 1961.
167. Terzi, M. — J. Radiation Biol., Vol. 5:403. 1962.
168. Terzi, M. — J. Theor. Biol., Vol. 8:233. 1965.
169. The Evaluation of Risks from Radiation. Pergamon Press. 1966.
170. Torroja, E. — Genetics, Vol. 50:1289. 1964.
171. Touchberry, R. W. and T. A. Verley. — Genetics, Vol. 50:1187.
        1964.
172. Ulrich, H. — Radiation Res., Vol. 9:196. 1958.
173. Underbrink, A. G. et al. — Radiation Botany, Vol. 8:205. 1968.
174. Van Valen, L. and R. Weiss. — Genetical Res., Vol. 8:3. 1966.
175. Von Borstel, R. C. and R. W. Rogers. — Radiation Res., Vol. 8:248.
        1958.
176. Wallace, B. — Genetics, Vol. 36:612. 1951.
177. Wallace, B. — Science, Vol. 115:487. 1952.
178. Wallace, B. — J. Genet., Vol. 54:280. 1956.
179. Wallace, B. and J. C. King. — Amer. Naturalist, Vol. 85:209. 1951.
180. Wallace, B. and J. C. King. — Proc. Nat. Acad. Sci., USA, Vol. 38:
        706. 1952.

181. W i n g e , O. et al.— Experientia, Vol. 17:406. 1961.
182. W h i t f i e l d , J. F. and R. H. R i x o n.— Experimental Cell Res.,
        Vol. 9:531. 1960.
183. W h i t f i e l d , J. F. and R. H. R i x o n.— Experimental Cell Res.,
        Vol. 23:412. 1961.
184. W i t k i n , E. M.— Proc. Nat. Acad. Sci. USA, Vol. 32:59. 1946.
185. W i t k i n , E. M.— Genetics, Vol. 32:221. 1947.
186. W r i g h t , S.— Genetics, Vol. 16:97. 1931.
187. Z i r k l e , R. E.— J. Cell. Comp. Phys., Vol. 2:251. 1932.
188. Z i r k l e , R. E. et al.— Radiation Res., Vol. 9:206. 1958.

Chapter 9

EFFECT OF IONIZING RADIATION
ON FOREST BIOGEOCENOSES

Introduction

Forest occupies a special place in the general problem of the effect of ionizing radiation on the biosphere. Interest in the biogeophysical conse- quences of irradiation of forests is due to the forest as an ecological system being extremely sensitive to ionizing radiation and radioactive contami- nation, first owing to the high radiosensitivity of arboreal plants, and second owing to the pronounced interceptive capacity of crowns with respect to radioactive fallout.

The forest reacts in different ways to irradiation as a result of the complex structure of the forest biogeocenosis. The latter is a multilayer ecological system, its components being interrelated and also related with their environment by numerous direct ties and feedbacks /9, 10/. In some forest types it is possible to distinguish as many as five or six vegetation layers. The layer of arboreal plants is the principal layer of a forest biogeocenosis. Under its canopy there are specific environmental con- ditions, a microclimate and water regime of soil that are peculiar only to forests. Under the forest canopy there are the plants of the lower layers, including the underwood, advanced growth and herbaceous vegetation, mosses, lichens, algae and fungi. As a rule, they are represented by species for which the environmental conditions under the forest canopy represent the optimum. The animal kingdom (insects, birds, mammals) are likewise represented by species that are best adapted to forest conditions.

The components of a forest cenosis may be interrelated in a variety of ways, which include alimentary relationships between different groups of organisms, direct ties and feedback between the layer of arboreal plants and the microclimate under their canopy. An important role is played by interrelationships among animals and plants on the one hand, and the geo- physical environment on the other, manifesting themselves in the organisms' reactions being attuned to the photic, temperature and other signals received from the geophysical environment and signaling imminent changes in it /18/. There are various types of symbiosis and competition among different species of animals and plants, etc.

Under normal conditions these relationships, with all their diversity and complexity, are balanced in such a way that a forest biogeocenosis consti- tutes a relatively stable system functioning without irreversible disturbances for many decades. However, there may arise situations in which the state

of relative ecological equilibrium is disturbed to such an extent that the
forest cenosis is damaged or even destroyed. The history of forestry
knows many cases of this kind, caused by an attack on the forest of insect
pests under climatic and biotic conditions favorable to their mass proli-
feration. Occasionally such pests destroy forests over enormous territories. Other potent factors (forest fires, prolonged droughts, bogging of
soil, excessive felling, etc.) may likewise bring about destruction of forest
stands or a change in their structure.

Ionizing radiation likewise belongs among the harsh factors. Biogeophysical consequences of an irradiation of a forest may vary. In the first
place, irradiation may have a direct effect on individual components of the
forest cenosis, especially on organisms of high radiosensitivity. With
sufficiently large doses irradiation may prove lethal, so that the radiosensitive species become eliminated from the irradiated cenosis. Such
effects will be termed primary, since they are due to the direct effect of
ionizing radiation.

However, the consequences of forest irradiation are not confined to the
primary effects. Elimination of radiosensitive species from the community
inevitably unbalances the previously established interrelationships among
different groups of organisms, which in turn leads to further rearrangement
of the forest cenosis and a change in its structure. Changes of this kind
will be termed secondary; they arise as a consequence of primary disturbances in the cenosis. Therefore problems pertaining to the effect of
ionizing radiation on forests as a whole should be solved by resorting to the
concepts and methods of modern biogeocenology /10/. At the same time,
the achievements of other branches of ecology, and also of biogeochemistry,
geophysics and biogeography, should likewise be used.

Radiosensitivity of arboreal plants

A forecast of possible primary changes in an irradiated forest cenosis
requires in the first place a knowledge of the radiosensitivity of the principal forest-forming species.

The most complete and thorough investigation of the radiosensitivity of
arboreal plants was carried out by Sparrow et al. in the Brookhaven
National Laboratory for Nuclear Research (USA). They succeeded in
clarifying important features in the radiosensitivity of plants and in establishing the principal radiosensitivity-controlling factors /14, 49, 73—79/.

Among investigated arboreal species, coniferous species proved to be
the most highly sensitive to ionizing radiation in all stages of development.
Critical irradiation doses for seeds of various species of pine, cedar,
spruce, fir and larch, reducing the germinative capacity by 50% and causing
most of the sprouts to be unviable, are 600—6,000 r in short-term irradiation
(Table 9.1). Damaging doses for seedlings and adult coniferous trees are
as a rule smaller (Table 9.2). The seedlings on the majority of deciduous
species are killed after the seeds have been irradiated with doses of 10,000 r
and higher. Heavy damage to crowns of deciduous trees is caused by

TABLE 9.1. Dose of short-term gamma-irradiation reducing the germinative capacity of air-dry seeds of arboreal plants by 50%

Tree or shrub	Critical dose, kr	Source
Siberian larch (Larix sibirica)	>1; <5	/12/
	>0.7; <1.5	/4/
Pines:		
Siberian stone (Pinus sibirica)	>1; <5	/12/
Scotch (P. sylvestris)	>3; <5	/12/
	>0.7; <1.5	/4/
	1.0	/21/
	0.9–1.5	/38,71/
	0.9; 1.2	/28,29/
eastern white (P. strobus)	0.7	/29/
slash (P. elliottii)	2.0	/45/
	1.6	/72/
	13.0	/43/
shortleaf (P. echinata)	1.6	/72/
longleaf (P. palustris)	1.6	/72/
	2.0	/45/
pitch (P. rigida)	3.5–5.0	/50/
red (P. resinosa)	4.9	/28/
jack (P. banksiana)	6.3; 4.9	/28,29/
Spruces:		
Crimean (Picea pallasiana)	>3; <5	/16/
Norway (P. excelsa)	0.8–3.0	/38,71/
	>0.7; <1.5	/4/
Siberian (P. obovata)	>5; <10	/12/
white (P. glauca)	1.1	/29/
black (P. mariana)	3.7	/29/
Italian cyprus (Cupressus sempervirens)	5.0	/12/
Alder (Alnus)	>5; <10	/71/
European white birch (Betula verrucosa)	>5; <10	/12/
European black currant (Ribes nigrum)	5	/12/
Sundry apple trees	2–10	/12/
Siberian peashrub (Caragna arborescens)	>5; <10	/12/
Littleleaf linden (Tilia cordata)	>15	/12/
Common honeylocust (Gleditschia triacanthos)	>20	/12/
Maples:		
boxelder (Acer negundo)	<20	/12/
silver maple (A. saccharinum)	<10	/40/
red (A. rubrum)	>10; <100	/40/
sugar (A. sacharum)	>10; <100	/40/
Oaks:		
white (Quercus alba)	<10	/40/
swamp chestnut oak (Q. prinus)	<10	/40/
black (Q. velutina)	<10	/40/
Black locust (Robinia pseudoacacia)	<10	/40/
American elm (Ulmus americana)	<10	/40/
Slippery elm (U. fulva)	<10	/40/
White ash (Fraxinus americana)	>10; <100	/40/
Eastern black walnut (Juglans nigra)	>10; <100	/40/
American planetrees (Platanus occidentalis)..	>10; <100	/40/

TABLE 9.2. Damaging irradiation doses for arboreal plants

Tree or shrub	Development phase	Physiological status at the time of irradiation	Irradiation regime	Dose, r	Effect of irradiation	Source
Pines:						
eastern white (Pinus strobus)	Seedlings	Vegetative phase	Brief irradiation (< 5 days)	600	LD_{100}	/75/
slash (P. elliottii)	"	"	Same	380—630	LD_{50}	/26/
				900	Killing	/45/
longleaf (P. palustris)	"	"	"	730	LD_{50}	/26/
				720	Killing	/45/
Japanese black (P. thunbergi)	"	"	"	1,000	LD_{100}	/62/
Yew podocarpus (Podocarpus macrophylla)	"	"	"	575	LD_{50}	/26/
Zamia (Zamia floridana)	"	"	"	610	LD_{100}	/26/
Giant sequoia (Sequoia gigantea)	"	"	"	1,120	30% suppression of growth	/83/
American elder (Sambucus canadensis)	"	"	"	2,000	LD_{50}	/79/
Southern red oak (Quercus rubra)	"	"	"	8,000	LD_{50}	/79/
Yellow birch (Betula lutea)	"	"	"	8,000	LD_{50}	/79/
Maples:						
sugar (Acer saccharum)	"	"	"	8,000	LD_{50}	/79/
red (A. rubrum)	"	"	"	10,000	LD_{50}	/79/
White ash (Fraxinus americana)	"	"	"	10,000	LD_{50}	/79/
Pines:						
slash (P. elliottii)	"	Rest phase	"	560	LD_{50}	/26/
eastern white (P. strobus)	"	"	"	1,000	Killing	/74/
Scotch (P. sylvestris)	"	"	"	1,100	LD_{50}	/29/
red (P. resinosa)	"	"	"	1,300	LD_{50}	/29/
jack (P. banksiana)	"	"	"	1,300	LD_{50}	/29/
Japanese black (P. thunbergi)	"	"	"	1,600	LD_{50}	/29/
				715	Pollen not formed	/61/

Yew (Taxus media)	"	"	"		800	Killing	/74/
Yew podocarpus (Podocarpus macrophylla)	"	"	"		860	LD$_{50}$	/26/
White spruce (Picea glauca)	"	"	"		1,300	LD$_{50}$	/29/
Balsam fir (Abies balsamea)	"	"	"		1,020	Killing	/74/
Japanese larch (Larix leptolepis)	"	"	"		1,050	"	/74/
Eastern arborvitae (Thuja occi-dentalis)	"	"	"		1,250	"	/74/
Pines:							
pitch (P. rigida)	Adult trees	Vegetative phase	"		1,500	"	/74/
	"	"	Long-term irradiation with dose rate 3 r/day		7,400	LD$_{50}$	/78/
	"	"	Same		5,400	Yield of seeds reduced by 50%	/78/
eastern white (P. strobus)	Seedlings	Throughout the year	"	12	25,500	Killing	/75/
Aleppo (P. halepensis)	Adult trees	Vegetative phase	"	20	6,760	"	/55/
	Seedlings	"	"	20	9,100	"	/75/
	"	"	"	36	920	"	/34/
	"	"	"	20	2,240	"	/34/
	"	"	"	10	3,400	"	/34/
	"	"	"	20	6,800	"	/34/
Italian stone pine (P. pinea)	"	"	"	25–42	400–700	Growth reduced by 50%; death of terminal bud	/57/
longleaf (P. palustris)	"	"	"	540	9,100	LD$_{100}$	/57/
Alder (Alnus)	"	"	"	60–160	30,000–90,000	LD$_{100}$	/59/
Birch (Betula)	"	"	"	60–160	30,000–90,000	LD$_{100}$	/59/
Poplar (Populus)	"	"	"	60–80	95,000–60,000	LD$_{100}$	/59/

irradiation with doses of 2,000—10,000 r.   Comparison of these data shows
that the principal forest-forming coniferous species are 5—10 times more
sensitive to irradiation than deciduous species.

TABLE 9.3. Relationship between radiosensitivity of arboreal plants and size of chromosomes /79/

Species of arboreal plant	Volume of a single chromosome during the vegetation period, μ^3	LD_{50} for short-term irradiation, r	
		prognosis (from the size of chromosomes)	experimental
Eastern white pine (Pinus strobus)	61.1±3.9	570±250	473±7
Eastern larch (Larix laricina)	46.3±2.3	690±260	705±40
Balsam fir (Abies balsamea)	46.4±2.6	690±260	894±28
Pines:			
ponderosa (P. ponderosa)	45.6±2.1	700±260	818±32
red (P. resinosa)	45.1±2.5	700±260	784±25
Eastern hemlock (Tsuga canadensis) ...	44.9±2.8	720±270	690±32
Spruces:			
Norway (Picea abies)	42.6±2.5	730±270	1,100±22
white (P. glauca)	42.2±1.9	740±270	710±23
Colorado (P. pungens)	40.7±1.6	760±280	1,186±40
red (P. rubens)	36.9±1.5	820±300	1,027±32
Japanese larch (Larix leptolepis)	35.2±1.7	850±310	834±67
Yews:			
Canada yew (Taxus canadensis)	28.2±1.2	990±360	1,140±9
Anglojap yew (T. media)	20.5±1.0	1,140±420	1,203±116
Giant sequoia (Sequoiadendron			
giganteum)	19.6±0.9	1,250	1,136±44
American elder (Sambucus canadensis)	19.2±0.5	1,300	1,116±45
Eastern arborvitae (Thuja occidentalis)	16.3±0.9	1,500±550	970±63
Virginsbower (Clematis virginiana) ...	16.0±0.7	1,520±560	1,890±80
Linden viburunum (Viburnum dilatatum)	14.9±0.6	1,600	3,624±698
Northern red oak (Quercus borealis)	5.5±0.3	3,360±1,260	3,653±149
Sugar maple (Acer saccharum)	3.4±0.1	4,800±1,830	4,721±145
White ash (Fraxinus americana)	3.4±0.1	4,800±1,830	7,744±260
Red maple (Acer rubrum)	2.6±0.1	5,850±2,250	5,111±228
Yellow birch (Betula lutea)	2.3±0.2	6,410±2,480	4,281±517
Butterflybushes:			
fountain butterfly bush			
(Buddleia alternifolia)	2.2±0.1	6,630±2,570	7,053±518
orangeeye butterflybush (B. davidii)	0.9±0.05	12,870±5,220	17,500±941

    Sparrow et al. /75, 76/ found that the sensitivity of plants to ionizing
radiation is related to cytogenetic characteristics of the cells, viz. the
nuclear volume and number of chromosomes, and more precisely with the
DNA content in the chromosome.   The degree of radiation damage to a cell
is closely correlated with the number of ionizations in the chromosomal
matter during a single nuclear cycle; for the same number of ionizations,
different species of plants, even differing markedly in their radiosensitivities, display similar radiation effects.   For a given irradiation regime,

the number of ionizations per chromosome is proportional to its volume or its content of DNA, since the choromosomes consist almost entirely of DNA.

Table 9.3 lists the chromosomal volumes of somatic cells for 25 species of coniferous and deciduous arboreal plants. The tabulated data show that the chromosomes of coniferous species are ten times larger on the average than those of deciduous species, and the radiosensitivity of coniferous species is approximately the same number of times higher than that of the deciduous arboreal plants.

When irradiation lasts considerably longer than a single cell cycle, the radiobiological effect is significantly dependent on the fission rate of cells; the longer the cell cycle, the larger is the number of ionizations per chromosome and the stronger the radiobiological effect. This explains the experimentally observed differences in reactions of slowly and rapidly growing plants to short-term and long-term irradiation. Plants are known to react more strongly to short-term (instantaneous) irradiation in periods of active growth (e. g., under favorable weather conditions) than during slow growth periods. This is due to the increase in the relative number of cells in the cell cycle phase that is most sensitive to irradiation with increasing cell fission rate. The pattern is reversed in the case of long-term irradiation; in this case the radiobiological effect is more pronounced in the case of slowly growing plants /75, 76/, because of the increased number of ionization acts in every chromosome with increasing cell cycle duration.

Hence it could be expected that long-term irradiation of trees during the period of their winter rest, when the growth and cell splitting are completely suspended, should produce more severe radiobiological consequences than irradiation during the active growth period. Actually, the situation is the opposite, the radiosensitivity of trees subjected to long-term irradiation in winter being approximately $\frac{1}{3}$ their sensitivity to irradiation during the summer period. The difference can be partially explained by the influence of still another factor. It was established experimentally on a large number of species of arboreal plants /74, 78, 79, 83/ that in the period of slow or totally suspended growth the volume of cell nucleus in the interphase is reduced by an average factor of 1.65, which in turn must increase the stability to irradiation by approximately the same factor. According to Sparrow et al. /78/, seedlings of eastern white pine subjected to long-term irradiation in different seasons of the year proved to be more resistant (approximately by a factor of 3) when irradiated in winter. This means that besides the above-enumerated factors there also exist other, still unknown factors, which augment the radioresistance of arboreal plants in the stage of rest. Under the influence of these factors the actual radiosensitivity may differ by a factor of several units from the value predicted from known size of chromosomes.

Comparison of the radiosensitivities of arboreal and herbaceous plants reveals that experimental data describing the radiation damage to plants as a function of energy absorbed in the cell nucleus do not fall onto a single curve. Thus, the energy absorbed by arboreal plants in short-term irradiation averages $\frac{1}{2}-\frac{1}{3}$ that absorbed by herbaceous plants (with the same reactions to the irradiation). The differences are still more pronounced

in the case of long-term irradiation.   The data listed in Table 9.4 show that
absorbed energy per chromosome producing lethal and sublethal reactions
is approximately one order of magnitude higher in the case of herbaceous
than arboreal plants.   The causes for the differences are not yet clear.
They are possibly due to metabolic differences between the arboreal and
herbaceous plants and the longer duration of the mitotic cycle of cells in
the arboreal plants.   The same causes are responsible for still another
peculiarity of arboreal plants, namely, the pronounced cumulative effects of
the dose accumulated over a prolonged irradiation.   According to Sparrow
et al. /76/, gamma-irradiation of yew and various species of pine with dose
rate 10—20 r/day for three growth periods kills the trees.   Irradiation of
a pine P. rigida with dose rate about 2 r/day for nine years appreciably
suppressed the growth and caused shedding of a considerable proportion of
needles, the total dose not exceeding 4,000 r.

TABLE 9.4. Energy absorbed by a single chromosome in the interphase stage under irradiation causing
damage to herbaceous and arboreal plants /73/

Object	Radiation effect	Short-term irradiation, keV	Long-term irradiation, keV/day	Ratio
Herbaceous plants	Lethal outcome (100%)	3,710±430	450±70	8.3
(mean for	Marked slowing of growth	2,330±370	210.20	11.2
several species)	Slight slowing of growth	1,070±210	100±20	10.2
Arboreal plants	Lethal outcome (100%)	1,450±150	50±5	29.1
(mean for	Marked slowing of growth	1,000±140	27±3	36.9
several species)	Slight slowing of growth	530±70	9±1	59.3

Reaction of arboreal plants to irradiation

The radiobiological effects in arboreal plants may differ widely depending
on the magnitude of the absorbed dose.   Irradiation with relatively small
doses may stimulate the development and growth of certain arboreal species.
With sufficiently high doses there is an increase in the frequency of chromosome breakages and somatic mutations induced by the radiation, manifesting themselves in alterations of certain morphological characters.
Furthermore, irradiation causes a slowing down of growth and development,
sterility of pollen and seeds, and finally death of the plants themselves.
All these effects are used as criteria in assessing the degree of radiation
damage and establishment of a radiosensitivity scale.   Suitable analytical
methods for the radiation effect are used depending on the selective criteria.
Investigations into numerous species of plants have established that in the
majority of cases all methods produce fairly similar radiosensitivity data
/75/.
Radiostimulation.   The phenomenon of radiostimulation has been established only for some species of arboreal plants.   It manifests itself most
distinctly in pines and firs.   Thus, irradiation of Scotch pine, jack pine and

black spruce with doses ranging from a few tens to several hundreds of roentgens accelerates the sprouting of seeds /27,29/. Earlier sprouting and improved field germination were observed for irradiation of the seeds of Austrian pine, oriental abrovitae and bloodtwig dogwood with doses of 500—5,000 r /16/. The germination rate of Scotch pine seeds is improved by irradiation with doses of 300 r /4/; irradiation with doses of 25—300 r increases the length of roots and height of seedlings in some pine species in the first growing season /14,41,44,52,71/. Among deciduous species a similar effect was discovered for irradiation of hedge maple seeds with doses of 500—5,000 r /11/. Radiostimulation effects (increased growth) were also discovered for irradiation of seedlings /29/ and adult pines /92/.

Inhibition of growth and morphological alterations. Plant tissues that are the most sensitive to irradiation are those of the apical meristem. This is because the cell nucleus in the apical meristem of arboreal plants is considerably larger than in other tissues, averaging 1.3 the size of nucleus in the cells of lateral meristem /79/. Irradiation with doses which do not cause any appreciable disturbances in other plant tissues causes a disturbance of mitosis in the apical meristem and temporary suspension of the splitting of cells. Therefore, the first damage is sustained by plants in the terminal bud of their main shoots, while buds on side shoots are somewhat less sensitive. For instance, long-term irradiation of a yew with dose rates 3.7 r/day inflicts damage on the primary meristem of buds on accumulation of a dose of 100 r /76/.

Cambium cells (lateral meristem) and cells of dormant buds of arboreal plants are damaged by higher doses /64/. The relationship between radiation damage to plants and irradiation dose is shown in Table 9.5.

TABLE 9.5. Doses producing different radiation effects in long-term gamma-irradiation of plants for 8—12 weeks during the growing season /75/

Reaction to irradiation	Number of investigated species	Dose with respect to the lethal dose, %
Absence of visible damage of stimulation	14	<10
10% decrease in growth rate	23	25
Loss of capacity for forming full-value seeds	8	30
50% decrease in growth rate	12	35
Pollen sterility	4	40
Marked inhibition of growth	41	60
LD_{50}	17	75
LD_{100}	41	100

When arboreal plants are irradiated with sublethal doses, the growth of the terminal shoot stops first, while the radial growth of plants continues. The result is a low-growing, sprawling plant with relatively broad leaves /34,68/. However, these disturbances are temporary; the mitotic splitting rate in the apical meristem cells is recovered even for prolonged continuous irradiation, and only irradiation with near-lethal doses withers the

terminal bud.   This indicates that the initial cessation of mitosis in meristematic cells is mainly due to some temporary physiological, not irreversible
genetic injuries.

Inhibition of the growth of the terminal shoot is occasionally accompanied
by its death, utter loss of domination and appearance of other morphological
alterations (formation of bundles of shoots from dormant buds or subapical
meristematic tissues, more resistant to irradiation).   In arboreal plants,
disturbance of the domination of terminal shoots may alter their habit,
owing to the appearance of numerous new shoots from axillary buds.   This
effect is ascribed to the decrease in the amount of growth substance auxin
in the terminal shoot under the effect of irradiation, the substance from this
shoot passing into underlying buds, thereby controlling the growth of lower
shoots /37, 64, 68/.   The decrease of auxin in the terminal shoot releases
the lower buds from excess amount of the growth substance, thereby stimulating their growth.   The contents of auxin in plants are decreased by irradiation with doses of 25—1,000 r, although auxin is extremely stable to ionizing
radiation.   It is assumed that the decrease of auxin is caused by a disturbance in the biochemical processes involved in its synthesis from pre-
cursors, proceeding with the participation of several enzyme systems /37/.

Furthermore, irradiation with sufficiently large doses reduces the rate
of lengthwise and radial growth of roots, the terminal shoots and lateral
branches /24, 41, 49, 57/, increases the incidence of tumors in plants /3/,
alters the morphology of different orders of arboreal plants and the number
of needles per bundle; the leaves and needles become thicker and twisted,
their surface becomes coarse and leathery and the flower structure is
affected /55/.   As a result, the branches, leaves and flowers lose their
normal appearance.

The morphological reactions of plants to irradiation are evidently due
to changes on the cellular level, physiological (metabolic disturbances) and
cytogenetic (effect on the chromosomes).   The genetic effect ultimately
manifests itself in the form of a secondary physiological effect /37/.
According to Sparrow et al. /75, 76/, disturbances in chromosomes and
slowing of mitosis are the principal factors responsible for inhibiting the
growth of plants.   Therefore, the reaction to irradiation can be predicted
with a sufficient degree of reliability, provided one knows the number of
chromosomes in somatic cells and the dimensions of their nuclei.   An
appreciable contribution to radiational damage may also be made by physiological alterations due to damage to enzyme systems, disturbances in the
synthesis of DNA and auxin, formation of physiologically active substances
under the influence of irradiation, disturbance in the mineral nutrition,
presence or absence of protective substances in the plants, etc. /7, 22, 37/.
Nevertheless, the changes in the radiosensitivity of plants under the
influence of these factors are apparently limited to a factor of 2—3 in comparison to their radiosensitivity predicted from their known cytogenetic
characteristics.

Effect of irradiation on the reproductive capacity
of arboreal plants

Generative organs of plants are known to be the first damaged by long-
term irradiation. Loss of capacity for forming full-value seeds may be
caused by irradiation with doses that do not yet cause irreversible damage
to maternal plants. Thus, long-term irradiation of pitch pine with dose
rates 2—3 r/day retards fecundation by 7 days /53/, increases the incidence
of chromosomal aberrations in splitting cells (formation of bridges and
micronuclei) /51/, and markedly reduces the amount of formed full-value
seeds (Table 9.6).

TABLE 9.6. Effect of prolonged gamma-irradiation on
development of pitch pine seeds /73/

Dose rate, r/day	Percentage of full-value ripe seeds in cones
Controls	37.8
2.1	23.8
2.4	16.5
2.8	8.0
3.5	3.7

Prolonged irradiation of maternal trees with dose rates 3—20 r/day also
damages the reproductive capacity of oak, impairing the quality of pollen
and appreciably reducing the germination rate of acorns /49, 54, 80/.
Internal irradiation has a much stronger effect on the reproductive
capacity of arboreal plants than external irradiation. According to
Witherspoon /87/, the quality of seeds of the tulip tree was impaired by
irradiation from incorporated ^{137}Cs with dose rate approximately 0.1 rad/day.
The causes of the inferior stability to irradiation of reproductive
organs in comparison to vegetative organs in long-term irradiation were
examined by Sparrow, Woodwell et al. /75, 76/. One such cause is the
considerably larger dimensions of nucleus and chromosomes in the meiosis
prophase, much larger than their dimensions in somatic cells in mitosis,
rendering the nucleus in the meiosis stage relatively more sensitive to
ionizing radiation. Furthermore, although the time necessary for passage
through early meiosis stages varies widely in the higher plants, meiosis is
a more protracted process than mitosis, especially if the comparison is
based on the duration of the prophase stages. In exposure to long-term
irradiation, the longer duration of the nuclear cycle in meiosis increases
the degree of radiation damage in comparison to mitosis, as is confirmed
by experimental data /73, 75, 77/.
Certain arboreal plants (pine, larch, etc.) mature their seeds only in the
second year after fertilization, during which a small number of cell
splittings take place. Therefore the total dose built up in a seed before its
maturation may prove sufficient for irreversible damage to the embryo

even with a relatively low dose rate. A considerable period of time often elapses under natural conditions from the moment of seed maturation to sprouting. Obviously, any delay in the sprouting of seeds during exposure to long-term irradiation will increase the dose absorbed by the seed, thereby intensifying the radiation damage.

When dry irradiated seeds remain at rest for a prolonged time, the radiation damage also increases by itself owing to the so-called storage effect /19, 50/. This effect consists in the appearance of long-lived excitation centers in the seeds under the influence of irradiation, and in the course of time these centers transmit their absorbed energy to vital macromolecules of the cell. If germination of a seed takes place immediately on irradiation, the energy contained in the primary excitation centers is dissipated in the surrounding volume without inflicting any significant damage on the radiosensitive structures of the embryo. Increased storage time of dry irradiated seeds increases the probability that the energy will be transmitted from its primary absorption sites to sensitive centers, thereby intensifying the radiobiological impact.

The radiosensitivity of mature seeds depends on their moisture content. It was demonstrated /50, 58/ that the stability to irradiation of seeds of some coniferous species increases with increasing moisture content, reaching its maximum at a moisture content of 13—16%. Therefore, stratification of seeds irradiated with sublethal doses (in the course of which they become fully saturated with moisture) improve their germination rate in comparison to seeds sown directly after irradiation /50/.

Finally, chromosome breakages induced by radiation in premeiotic somatic cells may result in the formation of unviable haploid daughter cells, since the haploid cells of higher plants are less stable toward a loss of part of their genetic matter in comparison to the diploid somatic cells. Therefore meiosis acts as an efficient barrier in the transmission of different types of induced chromosomal aberrations to offspring, since chromosomal breakages produce underdeveloped pollen and seeds. Besides the lethal outcome, there is also the possibility of the appearance of such radiation-induced harmful mutations with which the seeds develop to the stage of embryo, seedling or even an adult plant. Any increase in the incidence of such mutations will detract from the viability of the populations of irradiated plants, reducing the germination rate of seeds and survival rate of seedlings. Plants grown from such irradiated seeds and attaining maturity may be unable to produce full-value seeds, thereby reducing the reproductive capacity of the population in its entirety. However, the bulk of induced mutations are recessive, and therefore they will manifest themselves only in the second or still later generations of plants. It should be noted, however, that high radiosensitivity of the generative organs of arboreal plants does not manifest itself unless the irradiation is performed during the early spermatogenesis stages, in the period of formation of haploid sexual cells, fecundation and formation of seeds. The stability of seeds to ionizing radiation increases significantly after ripening. For instance, in irradiation of pine stands ripe seeds of the pine Pinus rigida preserve their viability in the presence of doses that prove lethal for the maternal trees /52, 55, 90/. The viability of seeds under the circumstances is obviously due to their low content of oxygen and relatively small size of their chromosomes.

In the case of short-term irradiation the question of the relative stability of generative organs is mostly of purely scientific interest. Under these conditions, the doses causing sterility of pollen and loss of capacity by arboreal plants of producing full-value seeds during irradiation, even in the most radiosensitive stages of meiosis, are approximately the same as the doses killing the maternal trees. Thus, the lethal doses for the pollen of pine and oak are 2,000 and 5,000 r, respectively, but irradiation with the same doses also proves lethal to the trees themselves /54/.

Comparative radiosensitivity of animals

Insects occupy one of the most important places after arboreal plants in the large number of their species and their total biomass, and the role played by them in transformation of matter and energy in the forest bio-geocenosis.

Radiobiological investigations have encompassed a comparatively limited number of insect species, and only a few species of forest insects. Nevertheless, even these limited studies may serve to clarify certain general features of radiation damage to insects in different stages of their development.

The radiosensitivity of insects varies very widely over their developmental cycle. As a rule, embryonal stages are highly sensitive to irradiation. For instance, irradiation with doses of 100—200 r proves sufficient for killing embryos in the initial phase of development (at the age of 1—3 hours) of eggs of ichneumon fly H a b r o b r a c o n   j u g l a n d i s /35/ and fruit fly D. m e l a n o g a s t e r /39/. The stability of eggs toward radiation improves with the passage of time, the lethal dose for fruit fly eggs 7.5 hours old being already 800 r. However, eggs of some insects withstand irradiation with doses of thousands and tens of thousands of roentgens, especially in the late stages of their development /23, 33/. The stability of insects to irradiation increases in the larval stage, and the larvae continue to feed normally and attain the pupa stage even when exposed to sublethal and lethal doses. In this case, radiation damage manifests itself only after the pupation, after the start of the processes of differentiation of imaginal tissues forming anew from the imaginal disks. The damage to these disks, which is initially latent, soon results in the death of pupas formed from the irradiated larvae without reaching the stage of imago. This, however, calls for doses ranging from several thousands to tens of thousands of roentgens. In the pupal stage stability to irradiation increases appreciably with age, along with the development of different tissues and completion of the processes of their differentiation and formation of the imaginal organs. If the absence of yield of adult insects is adopted as the radiation damage criterion, $LD_{50} = 2,800$ r for irradiation of a fruit fly pupa younger than 5 hr. The median lethal dose increases to $LD_{50} = 12,000$ r at the age of 30 hr. Pupas subjected to irradiation at the age of 50 hr, withstand doses of 80,000 r /17/. Imagos usually possess high stability to ionizing radiation and their destruction calls for doses of tens and hundreds of thousands of roentgens. However, eggs laid by irradiated insects become unviable and do not yield larvae at much

lower doses. For instance, male and female forest pests P i s s o d e s
s t r o b i, L i m a n t r i a   d i s p a r and C o l e o p t e r a   s c o l i t i d a e are
rendered sterile by doses of 5,000—10,000 r /42, 67, 88/.

In view of the specific diversity of forest insects and the broad range of
their radiosensitivity, it may be assumed that many of the species are much
more stable toward ionizing radiation than the majority of species of arboreal plants.

Mammals and birds are highly radiosensitive in all stages of their development. Irradiation with doses of a few hundred roentgens proves lethal
for adult individuals of the majority of species (Table 9.7), while embryos
and young mammals and birds are damaged by even smaller doses /2, 13/.
For instance, a dose of 25 r has a detrimental effect on fetal development
of mammals /2/.

TABLE 9.7. $LD_{50/30}$ for some mammals and birds

Species	$LD_{50/30}$, r	Source	Species	$LD_{50/30}$, r	Source
Mammals			Birds		
Small rodents ...	300—600	/2/	Linnet	400	/8/
Rabbit	750—820	/2/	Canary	500	/8/
Racoon	580	/30/	Goldfinch	600	/8/
Gray fox	710	/30/	Greenfinch	600	/8/
Lynx	580	/30/	Sparrow	625	/8/
Goat	350	/2/	Dove	920	/81/

Radiation damage to forest

Information concerning the radiosensitivities of organisms makes it
possible to draw rough conclusions concerning the expected primary effects
in an irradiated forest in relation to the irradiation dose.

In view of the high radiosensitivity of mammals and birds it may be
expected that for an irradiation of a forest these classes of animals will be
affected by the ionizing radiation in the first place. Coniferous arboreal
plants are somewhat more stable to irradiation. However, doses that are
lethal for mammals and birds will also have an appreciable inhibiting effect
on the principal forest-forming coniferous species (pine, spruce, fir, stone
pine and larch). Deciduous arboreal plants are considerably more stable
toward irradiation, and doses one order of magnitude larger are required
to inflict damage on them. The same and even larger irradiation doses are
withstood by many insects, herbaceous plants, fungi, mosses, lichens and
microorganisms.

This conclusion in its general features is confirmed by experimental data.
Experiments on irradiation of forest were conducted in the USA, at the Brook-
haven National Laboratory for Nuclear Research under Sparrow and at
Atlanta University under Platt.

The irradiated biogeocenosis at Brookhaven is /73, 89, 92/ an oak-pine
stand homogeneous with respect to species and soil conditions. The

principal forest-forming species are white oak, scarlet oak and pitch pine. A ^{137}Cs gamma source of 9,500 curies installed directly in the forest was in operating position for 20 hr daily for several months beginning on 22 November 1961. The dose rate at a distance of several meters from the source was several thousand roentgens daily, decreasing to 2 r/day at 130 m.

TABLE 9.8. Predicted and actual radiosensitivities of arboreal plants (with 24 chromosomes) subjected to long-term irradiation for 9 months /92/

Species	Nuclear volume, μ^3	Dose rate killing 90% of plants, r/day		Ratio of predicted to actual dose rate
		predicted	actual	
Trees				
Pinus rigida	880	15	23	0.7
Quercus alba	155	70	110	0.6
Quercus coccinea .	85	150	110	1.4
Shrubs				
Quercus ilicifolia.	105	80	90	0.9
Subshrubs				
Vaccinium	45—60	500	100	5.0
Gaylussacia	32	500	70	7.1
Ratio $\dfrac{LD_{max}}{LD_{min}}$	—	33	5	—

The irradiation consequences for the arboreal plants were forecast even before the beginning of the experiment from the known relationship between their radiosensitivity on the one hand and the size of their chromosomes on the other. The forecast and experimental results obtained after 9 months of irradiation are listed in Table 9.8. They show fairly satisfactory agreement between the predicted and actual reactions to irradiation in the case of forest-forming species making up the dominant story. However, there was a considerable discrepancy between the predicted and observed results in the case of arboreal plants in the understory (Vaccinium and Gaylussacia subshrubs), these plants proving to be 5—7 times more sensitive to irradiation than expected.

At dose rates exceeding 45 r/day the killing effect of ionizing radiation also extends to a considerable part of herbaceous species growing under the canopy of the irradiated forest. There is a distinct selection manifesting itself in better survival of perennials, possessing underground perennial buds protected against irradiation. The position in the community vacated by radiosensitive species is taken over by more radioresistant species, such as sedge (Carex pensilvanica), horseweed fleabane (Erigeron canadensis) and American burnweed (Erechtites hieracifolia), which are uncharacteristic of the forest community under ordinary conditions /84/.

The loss of radiosensitive species by the forest community led to the appearance of four annular zones with different plant covers in place of the former oak-pine stand around the radiation source. A brief description of these zones is provided in Table 9.9.

TABLE 9.9. Plant zones formed around the radiation source after six months' irradiation /89/

Predominant vegetation in the irradiated zone	Species lost by the community	Dose rate, r/day	Dose, r
Mosses, lichens	All higher plants	350	62,000
Sedge	All arboreal and most herbaceous plant	150—350	27,000—62,000
Subshrubs	Oak, pine, some herbaceous plants	60—150	11,000—27,000
Oak stand	Pine	20—60	3,600—11,000

Platt et al. /32, 48, 65, 66, 86/ studied the effect of neutrons and gamma-quanta produced in the active zone of a reactor on an oak-hickory stand with an admixture of pine. The reactor was installed directly in the forest and operated periodically, unshielded. According to observations, pine needles exposed to irradiation with doses of 7,500 r acquired a bright orange color after one week, and the tree crowns dried up. Pine seedlings were killed by a cumulative dose of 6,000—6,500 r. Discoloration of needles and death of apical buds in pines were caused by a dose of 1,000 r.

In the case of hard-wooded broadleaved species (oak, hickory) irradiation with doses 15,000 r caused defoliation seven weeks ahead of the natural leaf fall, while irradiation with dose 4,000 r accelerated the leaf fall by one week. In April the following year, when the foliage in surrounding forests was fully developed, the irradiated trees within a radius of up to 1 km from the reactor were still in a state of hibernation. The delay in developing the foliage was proportional to the dose, being 1—2 weeks after a dose of several hundred roentgens, 7—8 weeks after doses of 10,000—15,000 r. All the apical buds were killed in deciduous trees within a radius of 300 m from the source. Subsequently, many trees irradiated with doses 10,000—15,000 r developed only 15—20 buds, mainly on the bole and lateral branches. These obser-vations confirm the previously noted considerable differences in the radio-sensitivities of the apical and lateral meristems. Owing to the high resistance of dormant lateral buds to irradiation some trees retained their viability even after a cumulative dose of 35,000 r. Observations of the development of radiation damage to the foliage in direct proximity to the reactor, where neutrons made a considerable contribution to the absorbed dose /32/, made it possible to estimate the RBE of neutron radiation for arboreal plants as 12 with respect to RBE of gamma-radiation from ^{60}Co /86/.

Besides the damage to arboreal plants, the irradiation of the forest like-wise modified the specific composition of herbaceous cover under the forest canopy. Similar to the foregoing experiment, the herbaceous cover mainly lost radiosensitive annual species, while perennial species capable of vege-tative regeneration survived /65/.

Still another experiment with irradiation of a forest was carried out in Puerto Rico /46/. A tropical forest irradiated from a ^{137}Cs source with activity 10,000 curies proved to be much more stable toward ionizing radiation than forests of the temperate zone. The 5,000 r doses did not have any appreciable effect on the tropical arboreal plants in the forest community, the seedlings of many species of these plants surviving under irradiation with doses of 25,000—30,000 r.

In addition to recording radiation effects on the plant cover in an irradiated forest, observations were also conducted of the activities of insects and birds /25, 70, 89/.

Studies were made /25/ of the effects of irradiation on the numerical strength of populations of ipids (Ips), aphids (Myzocallis discolor), tortricids (Argyrotata semipurpurana), chrysomelids (Chrysomelidae), ants (Formica integra) and insects feeding on organic residues (Psocoptera). The results enabled the community of insects to be divided into two groups, those ecologically radiosensitive and radioresistant. The first group included herbivorous insects, their parasites and predators, the numerical strength of this group of insects near the radiation source being close to zero. Normal incidence of a majority of the species occurred at distances at least 60 m from the source where the dose rate did not exceed 20 r/day. However, a marked increase in the numbers of aphids Myzocallis discolor was observed in 1963 in the zone of partial damage to oak (dose rate 20—60 r/day), whereas the same pest was absent from the intact forest. The oak trees damaged by ionizing radiation likewise exhibited an increase in the amount of prematurely shed foliage due to the activities of tortricids and chrysonelids.

The ecologically radioresistant group included insects feeding on organic residues and those with a well-developed capacity for areal migration, as well as ants. Their numerical strength remained at a fairly high level even at a distance of a few meters from the radiation source, where the dose rate amounted to several hundred roentgens/day. For instance, at the dose rate 300 r/day the fraction of cambium utilized by ipid larvae in dead trees and trap logs reached 80%. The numerical strength of Psocoptera was not lower (possibly even higher than usual) in the zone where all the arboreal plants including shrubs and subshrubs had been killed by the irradiation. The ants Formica integra escaped irradiation with high doses by constructing a subsurface path leading from the anthill away from the source and subsequently did not appear on the irradiated side of the stump which contained their nest.

Irradiation produced significant alterations in the populations of forest birds in the zone around the reactor /70/. The songbirds Passerina cyanea, Geothlypis trichas, Colinus virginiana, Thryothorus ludovicianus, Contopus virens, Vireo griseus and Vireo olivaceus, occupying permanent feeding grounds, disappeared from the area one week after the irradiation. The doses suffered by the eliminated species reached 20,000 r. The abundance of these species diminished with the passage of time also in the control, but not to complete disappearance. Since, however, no dead birds were found on the irradiated area, it remained a moot point whether the decrease in the numbers of songbirds resulted from their high mortality under the influence of irradiation or to migration from the irradiated area.   The numerical strength of birds without permanent feeding grounds was approximately the same in the irradiated as in the control area.

Influence of the structure of forest cenosis and
environmental factors on the effect of irradiation

The above-described experiments on irradiation of forest with long-range radiation provides only a very averaged general pattern of the possible consequences of irradiation.   They show that irradiation with doses starting at several thousand roentgens and higher injured and destroyed many species of arboreal plants which were eliminated from the forest cenosis within a short time.

At the same time, the available experimental data revealed that in the case of sublethal irradiation doses there is the possibility of considerable variations in the reactions to irradiation of the entire cenosis and also its constituent groups of organisms, on account of differences in environmental conditions.   The fact is that under natural habitation conditions the organisms are continuously exposed to biological pressure from other organisms and also from their environment.   Combined effect of ionizing radiation and unfavorable environmental factors may significantly modify the stability of ecological systems subjected to irradiation.   The pathological effect produced by the irradiation (such as shortening of the lifespan, weakening of resistance to infections, parasites, predators and unfavorable geophysical environmental factors, weakening of the viability of offspring, etc.) and generally any disturbance in vital functions may have a significant effect on the organisms' capacity for withstanding the outside pressure.   Therefore, under extremely unfavorable environmental conditions the ultimate effect of ionizing radiation may be expected to manifest itself in an apparent decline in the stability to irradiation of the arboreal plants and the forest cenosis as a whole.   This surmise is borne out by the following known experimental facts.

Woodwell and Miller /91/ studied the effect of climatic conditions on the growth of arboreal plants exposed to long-term irradiation for a period of many years with dose rate 0.1—5 r/day.   The irradiation reduced the radial increment of bole, especially in its lower portion.   The decline was most pronounced in years with unfavorable climatic conditions.   In these years the irradiated trees often yielded no radial increment altogether, although the increment of controls (nonirradiated trees) amounted to at least 50% of its maximum possible value.   Furthermore, the investigators discovered that the effect of irradiation was strongest in the case of trees exposed to competitive pressure from surrounding trees.   The increment of irradiated trees measured with reference to that of nonirradiated trees proved to be considerably smaller in dense forest than in the case of free standing trees not subjected to competition from their neighbors.   Trees growing in a dense stand react to irradiation at dose rate 2 r/day, yielding no increment in their lower bole.

McCormick and McJunkin /45/ studied the combined effect of radiation and various artificially produced environmental conditions (shading, drought, high temperature) on the seeds and sprouts of longleaf and slash pines. The reaction of seeds and seedlings to these combined influences was found to depend on the irradiation dose. Thus, irradiation of seeds with doses of 2,000—4,000 r significantly improved the drought-resistance of sprouts. The lifespan of pine seedlings, shaded and irradiated with doses of 1,000—5,000 r, averaged 1.5—2 months longer than the lifespan of irradiated seedlings left to grow in sunlight. The survival of seedlings irradiated with doses exceeding 500 r was considerably lower during exposure to high temperatures and drought than during exposure to only one of these factors or irradiation.

Similar results were obtained also by other authors, who studied the variations in the radiosensitivity of seed and seedlngs of certain species of pine under conditions of drought /56/ and acute thermal exposure /28/. All these results testify that a forecast of the consequences of irradiation of organisms and their communities under natural conditions cannot be based solely on radiosensitivity data obtained in laboratory experiments.

Unfavorable environmental conditions, such as lack of food or inclement weather, are liable to have a significant effect also on the consequences of irradiation of populations of higher animals and birds. Consequently, the lethal doses for irradiation under natural conditions are found to be con- siderably smaller than in laboratory experiments. For instance, irradiation of a population of the rodents S y g m o d o n h i s p i d u s in their natural habitats altered the animals' behavior: they lost their watchfulness and agressiveness, their reflexes and feeling of equilibrium became weaker, and their overall motor activity declined /69/. Specimens with these symptoms were never caught a second time and were presumed dead.

The consequences of irradiation of organisms in a forest cenosis are also significantly dependent on its structure. When mature stands are sub- jected to irradiation, the irradiation doses on different trees and their parts are found to differ at the same distance from the source /47/. Some trees, protected by the trunks of neighboring trees, receive considerably smaller doses and may survive even though surrounding trees have received a lethal dose. Some portions of the crown and bole on the opposite side of the radiation source are likewise appreciably screened from the radiation by the other portion of crown and bole. In this case the injured crowns may be rehabilitated by shoots sprouting from uninjured dormant buds on the screened portion of the tree.

Attenuation of the intensity of radiation is most pronounced in the case of an irradiation dose on subterranean organs of arboreal plants and edaphic organisms. Soil acts as a screen attenuating the irradiation dose by a factor of several units /57/, and this attenuation may have a decisive effect on the survival of populations of irradiated organisms. For instance, deciduous stands damaged by irradiation recover their former structure by vegetative renovation, i. e., by means of root suckers and stool shoots arising from protected subterranean buds /31/.

The protective effect due to screening is liable to manifest itself in irradiation not only of plants but also of animals. This effect was detected in irradiation of populations of small rodents, which included mice

Peromyscus leucopus and P. polionotus and cotton rat Sigmo-
don hispidus in their natural habitats /69/. The gamma-radiation doses
at ground surface needed to produce considerable reduction in the numbers
of populations of irradiated rodents were found to be nearly an order of
magnitude larger than the lethal doses for the irradiation of the animals in
the laboratory. Thus, irradiation of the populations of both species of
Peromyscus for 15 days with dose 5,690 r at ground surface reduced
their numbers by only 38%, although the lethal dose for these species of
rodents does not exceed 500 r. The discrepancy is explained by the presence
of natural protection, the rodents spending only a small fraction of their total
time on the ground surface, while their burrows lie so deep beneath the surface that the irradiation does not reach them, being almost completely absorbed by the overlying layer of soil.

Disturbances caused by alteration of biocenotic relationships
in an irradiated forest cenosis

    Changes directly produced by irradiation in the structure of a forest
cenosis are inevitably followed by further alterations, due to disturbances
in the former and establishment of a new biocenotic situation in the irradiated community. We shall now examine in more detail the nature of such
alterations and their consequences.
    Every species in a forest biocenosis occupies a definite ecological place
and is in a state of relative ecological equilibrium with the other cenosis
components. Elimination of any species from the cenosis, irrespective of
the causes, disturbs the equilibrium and interrelationships that have become
established among the different populations of organisms and their habitats
and gives rise to further rearrangement of the cenosis. This explains the
damage to subshrubs from irradiation doses that should not have caused
any appreciable consequences, in line with the forecast of the above-
described experiment on irradiation of forest stands /89, 92/. The actual
damage must probably be explained by alteration of the microclimatic
conditions under the canopy of irradiated forest. Damage to the crowns
and death of trees in the upper story due to irradiation produce very significant modifications in the photic, thermal and water regimes under their
canopy, i. e., alteration of the formerly optimum conditions for the subshrubs.
The combined effect of altered microclimatic conditions and irradiation
resulted in the elimination of subshrubs during irradiation with approximately the same doses as those damaging the crowns of trees in the upper story.
Their place is taken by more light-demanding and more radioresistant
species.
    Analogous consequences are brought about by phenological changes in
the foliage unfurling dates in irradiated trees. The delay in unfurling
appreciably prolongs the time during which solar radiation penetrates un-
hindered to the lower layers of the forest, producing favorable conditions
for the development of light-demanding herbaceous plants, especially
ephemers, the entire developmental cycle of which takes place in spring
before the unfurling of foliage. The result is a marked increase in the

biomass of herbaceous plants under the forest canopy. In its turn, the luxuriant growth of herbaceous cover has an appreciable effect on the arboreal layer, causing a marked deterioration in the conditions for rehabilitation of damaged stands by seeding.

Furthermore, every species in a forest community usually constitutes a link in the food chain and occupies its proper place in the corresponding trophic level. A disturbance in one or several links of the food chain inevitably produces alterations throughout the chain, with a variety of consequences.

Damage to arboreal plants, irrespective of its causes, is usually accompanied by a deterioration in their resistance to forest pests (herbivorous insects, pathogenic fungi and bacteria). The same consequences are produced also by damage to arboreal vegetation caused by irradiation. Thus, in an experiment /71/ pine seeds irradiated with dose 1,000 r had germination rate 30%, but the sprouts were weak and were suddenly killed by the parasitic fungus Fusarium, whereas control seedlings growing alongside proved resistant to this parasitic fungus. A similar phenomenon was described in /60/.

Brower /25/ reported that arboreal plants irradiated with sublethal doses and having lost part of their foliage suffered from a mass attack of aphids, whereas the number of the same pests in a nonirradiated forest were insignificant. The number of ipids (bark beetles) were likewise considerably larger on trees that had been weakened by ionizing radiation. In some cases the damage sustained from insect pests is relatively heavier in the case of trees damaged by irradiation than in the case of healthy trees even for the same numbers of a pest, since the biomass of foliage and needles (i.e., the insect pests' food resources) is smaller in the irradiated than in the nonirradiated forests. The injurious effect of ionizing radiation is enhanced by the weakening of resistance of irradiated trees with respect to insect pests, and this circumstance is liable to destroy a forest stand with doses considerably below the lethal dose.

A characteristic instance of a rearrangement of a community due to biocenotic unbalance caused by an alteration of alimentary relationships among species under the influence of irradiation was described in /20/. The investigation was performed on a community of anthropods from the hollow of a beech decomposing organic substances from animal and plant residues. Among the 155 species belonging to several classes the most common species were those of springtails Collembola (31 species). For a one-time irradiation of the community with doses up to 15,000 r there was an increase in the numerical strength of Collembola populations, the numbers of some species increasing 2—2.5 times in comparison to controls. This stimulational effect persists for at least three months after irradiation. However, the increase in the numbers of Collembola was apparently not due to radiostimulation but to alteration of alimentary relationships between these insects and predators, the Collembola being devoured by five species of predatory ticks, which are probably more radiosensitive than their prey. The growth of the numerical strength of Collembola could be explained by the decrease in the numbers of predators in the irradiated samples.

The above examples suggest that irradiation of a forest cenosis can often alter the numerical strength of populations of its constitutent organisms under the influence of doses that are too small to cause direct destruction of organisms by irradiation, but merely alter their specific interrelationships and their relationship with the environment. Generally speaking, ionizing radiation should be recognized as one ecological factor to which universally recognized principles of modern ecology can be applied. One of the basic principles proclaims that if an organism is situated under optimum environmental conditions, it is capable of more successfully withstanding the effect of one or several injurious factors, including irradiation. On the other hand, the appearance of new adverse factors may reinforce the detrimental effects to such a degree that the ultimate effect may prove stronger than the simple sum total of effects produced by every factor separately. This explains, among other things, the apparent increase in the radiosensitivity of certain organisms when irradiated in their cenosis, in comparison to their radiosensitivity determined in the laboratory, where all other detrimental agents have been excluded. There have been reports to the effect that under natural conditions lethal consequences for many organisms may be produced by doses considerably smaller than those established in laboratory experiments. This provides grounds for introducing in radioecology the concepts of relative ecological efficacy of radiation. This we shall express as the ratio of doses producing identical reactions to irradiation in populations of a given species when these organisms are insulated against environmental influences (irradiation of isolated organisms) and in their natural habitat (at the community level). For the majority of organisms constituting the forest cenosis (arboreal plants, insectivorous birds, mammals) the relative ecological efficacy of radiation should be greater than unity. For instance, irradiation of populations of forest insectivorous birds under natural conditions may reduce their numbers in the case of doses amounting to only 0.1—0.2 of the lethal doses established by measurements of their radiosensitivity /85/.

Thus, alterations in a forest cenosis and its structure produced by irradiation may be due to many causes. They may be due either to the direct effect of ionizing radiation on the organisms which may manifest itself differently for different species, for a variety of reasons, or to the rearrangement of the internal structure of the cenosis. The ultimate results of irradiation of a forest cenosis with sufficiently large doses consists in a simplification of its structure, modification of its specific composition, and weakening of its overall stability toward unfavorable environmental conditions.

Radiation damage to forest in the case of
radioactive contamination

In the case of a fallout of radioactive aerosols on a forest, the distribution of absorbed doses in a forest stand is very nonuniform and irradiation doses for different groups of organisms differ markedly among themselves /1, 15/.

The nonuniform distribution of doses is primarily due to the nonuniform distribution of radionuclides deposited on the forest. At first they are principally intercepted in the tree crowns of the upper story. Therefore the crowns of arboreal plants receive the largest irradiation doses, whereas the organisms living under the forest canopy are largely protected against the effect of ionizing radiation. This specific distribution of the fields of ionizing radiation in forest stands contaminated by radioactive aerosols must result in a considerably broader spectrum of possible reactions to irradiation from the different components of forest cenosis than under the influence of gamma-radiation from point sources.

In the presence of radioactive fallout from the atmosphere, radiation burns are mostly inflicted on the organs and parts of trees exposed to falling-out radioactive particles, i. e., the shoots, leaves, needles and buds situated on the windward portions of crowns or in the treetops /36/. The leeward and lower portions of crowns, as well as crowns of trees in the lower stories, are largely protected against radioactive fallout and are subjected to lesser irradiation. Part of advanced growth and self-sown seedlings protected against fallout by the crowns of adult trees may be expected to survive and to serve as a base for the regeneration of the structure of the forest community even in the case of heavy radiational damage and destruction of trees in the upper story. For the same reasons plants of the ground cover, mammals, edaphic (soil-inhabiting) insects and other organisms living under the forest canopy absorb relatively smaller irradiation doses in the initial period. Nevertheless, the principal features involved in radiational damage to a forest remain the same, notwithstanding certain particular differences in the effect of ionizing radiation on forest irradiated from a point source or by radioactive fallout, namely, the effect of ionizing radiation in both cases manifests itself in the first place on arboreal plants, especially coniferous species.

In the following period, i. e., several months after the radioactive fallout, when the bulk of radionuclides shifts from the crowns down under the forest canopy, the organisms in the lower stories are subjected to more intensive irradiation than those in the upper stories. The radionuclides become concentrated in the relatively thin layer of forest floor and the adjacent soil layer, where rather high irradiation levels may be generated. This is the layer harboring the seeds of arboreal and other plants from their ripening to sprouting. Therefore radioactive contamination of a forest and studies into the radiational effect on arboreal plants give rise to yet another task: examination of possibly regenerating the arboreal plants from seeds under conditions of radioactively contaminated soil.

Karaban' and Tikhomirov /5, 6/ conducted experimental studies into the effect of beta radiation from $^{90}Sr-^{90}Y$ on the seeds and sprouts of coniferous species (pine, spruce and larch) for a superficial addition of radioactive solution to the soil. According to their observations, radioactive contamination of soil in the range 0.05—15 mcuries/m^2 does not affect the field germination of the seeds of coniferous species, the absorbed dose accumulated by the seeds over their germination period being below lethal.

Since beta particles belong among short-range radiation, their effective range is limited to a relatively thin contaminated layer of forest floor and soil, the absorbed dose rate declining rapidly with increasing distance from the surface. For this reason, the most critical period in the life of seedlings growing on contaminated soil is the initial period. During this period,

practically all the vitally important organs of seedlings are exposed to
strong irradiation, including the terminal bud, stems and a considerable part
of their root system. As the seedlings grow taller and their root system
becomes better developed, the terminal bud and an increasing portion of the
roots emerge beyond the irradiation zone. According to experiments, the
effect of radioactive contamination of young sprouts manifests itself already
in the first growing season. In the case of the absorbed dose rate being at
least 70 rad/day at the soil surface, corresponding to contamination density
40 mcuries/m^2, the damaging effect of irradiation on the sprouts of coniferous species manifests itself two months after sowing, and most of the
seedlings die toward the end of the growing season. Irreversible damage
to seedlings of investigated arboreal species occurred at absorbed dose
rate 10—15 rad/day, corresponding to contamination density 6—9 mcuries/m^2.
At these contamination levels, the seedlings practically cease their growth in
the third year of their life, and appear sickly and unviable. At absorbed
dose rate 4 rad/day, irradiation, although reducing the increment, does not
produce irreversible damage, at least in the first 4 years. Moreover, in
the 4th year of their life the seedlings recover from the damage and begin
to yield satisfactory height increments. The experimental results indicate
that normal regeneration by seed of a forest contaminated by long-lived
radionuclides is possible at radioactive contamination levels up to
3 mcuries/m^2.

Radioactive contamination levels one order of magnitude smaller stimulate
the growth of seedlings, the latter's height being in some cases double that
of the control. Characteristically, the specific radioactive contamination
levels producing stimulating or inhibiting irradiation effects depend on the
environmental conditions, shifting toward higher values under favorable
growth conditions. This confirms the previous suggestion, that the radiobiological effect is controlled by combined effects of irradiation and other
environmental factors.

Tasks in the field of forest radioecology

The fundamental task of forest radioecology consists in forecasting the
possible consequences of irradiating forest with ionizing radiation over a
broad range of conditions based on investigated features. The fulfilment
of this task calls for more complete information concerning the radiosensitivities of organisms constituting the forest biocenosis. An important
source of such information is provided by direct experiments on the irradiation of organisms under a variety of regimes of the time distribution of
the dose and for different physiological states of the irradiated organisms.
For this purpose it is necessary to enlarge the range of species used in
radiobiological investigations by including the principal arboreal and
frutescent species, forest insects, birds and mammals.

Furthermore, for the same purposes it is possible to utilize the relationship between the radiosensitivity of plants and the size of chromosomes
established by Sparrow and co-workers, but this requires a knowledge of
the cytogenetic characteristics of at least the most important forest-forming

arboreal species.   The determination of these characteristics is of considerable interest in connection with the radiosensitivity of plants.

A variety of criteria are used for assessing the effect of radiation.   The
criteria depend on the investigation method, including changes in the biomass increment of irradiated plants, changes in the color of leaves and
needles, decrease in the quantity and deterioration in the quality of formed
seeds, increase in the incidence of induced mutations and chromosome
breakages, necrosis of different organs and tissues, and finally, death of
irradiated organisms.   Owing to the comparatively narrow variation range
of the ratios of doses producing the above-enumerated effects in plants
from one species to another, it is possible to predict the consequences of
irradiation with any doses, provided we know the doses at which the major
effects manifest themselves.   Therefore, establishment of the most characteristic radiation damage criteria and the corresponding scale of doses is
likewise an important task in studying the problem of radiosensitivity of
organisms and entire communities.

However, the problem of the stability of a community of organisms
toward irradiation is not limited to measurements of the radiosensitivity of
individual isolated groups of organisms under conditions excluding the
influence of all other environmental agents.   Irradiation is always combined with effects of other ecological factors at the level of the community
and even population.   Therefore determination of the resulting effect calls
for more detailed studies into the combined effect of irradiation and other
possible influences, including variations in the conditions of the external
geophysical environment (temperature, light and water regimes), compe-
tative and alimentary relationships among different groups of organisms,
modifications (under the influence of irradiation) of the reactions of
organisms to incoming signals from the geophysical environment concerning imminent changes in it, and so on.   Examination of the major
factors modifying the organisms' reactions to irradiation and quantitative
assessments of the roles played by these factors should take their due
place in radioecological investigations.   In order to fulfill this task it is
feasible to use relatively simple models in which the number of variable
input parameters does not exceed two, the irradiation dose being one
variable, while the other variable is one of the ecological factors modifying
the irradiation effect.   The output quantities should be such characteristics
of the irradiated system as the numerical strength of populations, shift of
the ecotone, reproductive capacity and mortality of irradiated species, the
biomass, i. e., the characteristics used in descriptions of biological systems
at the level of a population or community of organisms.

In order to obtain radioecological information concerning the reactions
to irradiation of a forest cenosis in its entirety, it is feasible to utilize the
experience amassed by scientists working outside the USSR in setting up
model experiments on irradiation of relatively small forest areas from
powerful gamma-radiation sources.   From the practical standpoint experiments simulating contamination of forests by radioactive aerosols deposited
from the atmosphere are of still greater interest.   However, due to the
difficulties involved in simulating these phenomena for radiation safety
considerations, experimental work of this kind is confined to small-scale
experiments on radioactive dusting or spraying (under a protective hood)

of small areas occupied by trees or seedlings. Although information elicited by experiments of this kind refers to a rather limited territory and cannot be unequivocally extrapolated to the forest cenosis as a whole, in the main it nevertheless permits representation of the consequences of irradiation of forests over a broad range of doses as well as climatic, seasonal and other natural conditions.

The final stage consists in theoretical generalization of amassed experimental data and construction of a general scheme of the effect of ionizing radiation on the forest biogeocenosis. In constructing such a scheme one must resort not only to radioecological information, but also to modern concepts of the structure and functioning of the forest biogeocenosis. In this case one encounters a large number of interacting factors and it is far from a simple task to determine the specific part played by each of them with ordinary methods. The solution of this problem can be significantly facilitated by designing analog computers to simulate the functioning of forest cenosis. Devices of this kind have been successfully used, for instance, to construct the circulation scheme of radioactive substances in a forest artificially contaminated with radionuclides /63/.

References

1. Aleksakhin, R. M. et al. — Ekologiya, No. 1:27. 1970.
2. Bacq, Z. M. and P. Alexander. Fundamentals of Radiobiology. Pergamon. 1961.
3. Gunckel, J. E. and A. H. Sparrow. — In: International Conference on the Peaceful Uses of Atomic Energy, Geneva. 1955.
4. Karaban', R. T. — Lesnoe Khozyaistvo, No. 7:36. 1966.
5. Karaban', R. T. and F. A. Tikhomirov. — Radiobiologiya, Vol. 7:275. 1967.
6. Karaban', R. T. and F. A. Tikhomirov. — Lesovedenie, No. 2:91. 1968.
7. Kuzin, A. M. — In: "Osnovy radiatsionnoi biologii," edited by A. M. Kuzin and N. I. Shapiro, p. 247. Moskva, "Nauka." 1964.
8. Kushnuruk, V. A. Radiochuvstvitel'nost' ptits. Biologicheskoe deistvie radiatsii (Radiosensitivity of Birds. Biological Effect of Radiation), No. 1. Izdatel'stvo L'vovskogo universiteta. 1962.
9. Morozov, G. F. Uchenie o lese (Silvics). Moskva, Sel'khozgiz. 1931.
10. Sukachev, V. N. and N. V. Dylis (editors). Osnovy lesnoi biogeotsenologii (Principles of Forest Biogeocenology). Moskva, "Nauka." 1964.
11. Podol'skaya, O. I. — Trudy Tashkentskogo sel'skokhozyaistvennogo instituta, No. 16:243. 1964.
12. Privalov, G. F. — Radiobiologiya, Vol. 3:770. 1963.
13. Radiobiology [Russian translation from English. 1960].
14. Sparrow, A. H. and G. M. Woodwell. — In Sbornik: "Voprosy radioekologii," p. 57. Moskva, Atomizdat. 1968.

15. Tikhomirov, F. A. et al.— In: "Simpozium po migratsii radioaktiv-
    nykh elementov v nazemnykh biogeotsenozakh. Tezisy dokladov,"
    p. 26. Moskva. 1968.
16. Turakhodzhaeva, M.— Trudy Tashkentskogo sel'skokhozyaistvennogo
    instituta, No. 16:273. 1964.
17. Fritz - Niggli, Kh. Radiobiologiya, ee osnovy i dostizheniya (Radio-
    biology, Its Principles and Achievements). Moskva, Gosatomizdat.
    1961.
18. Khil'mi, G. F. Osnovy fiziki biosfery (Principles of Physics of the
    Biosphere). Leningrad, Gidrometeoizdat. 1966.
19. Zimmer, K. G. Quantitative Radiation Biology. Hafner. 1961.
20. Auerbach, S. I.— Ecology, Vol. 39:522. 1958.
21. Bowen, H. J. M.— Radiation Botany, Vol. 1:223. 1961.
22. Bowen, H. J. M. and J. Thick.— Radiation Res., Vol. 13:234. 1960.
23. Bletchly, J. D. and R. Fisher.— Nature, Vol. 179:670. 1957.
24. Brandenburg, M. K.— Radiation Botany, Vol. 2:251. 1962.
25. Brower, J. H.— Nucl. Sci. Abstr., Vol. 19:40307. 1965.
26. Capella, J. A. and A. D. Conger.— Radiation Botany, Vol. 7:137. 1967.
27. Clark, G. M. et al.— Health Phys., Vol. 11:1627. 1965.
28. Clark, G. M. et al.— Radiation Botany, Vol. 7:167. 1967.
29. Clark, G. M.— Radiation Botany, Vol. 8:59. 1968.
30. Colley, F. B. et al.— Health Phys., Vol. 11:1573. 1965.
31. Cotter, D. J. and J. H. McGinnis.— Health Phys., Vol. 11:1663. 1965.
32. Cowan, J. J. and R. B. Platt.— In: Radioecology, p. 311, Ed. by
    V. Schultz and A. W. Klement. New York, Reinhold Publ. Corp.,
    1963.
33. Davis, A. N. et al.— J. Econ. Entomol., Vol. 52:868. 1959.
34. Donini, B.— Radiation Botany, Vol. 7:183. 1967.
35. Erdman, H. E.— Radiation Res., Vol. 12:433. 1960.
36. Fosberg, F. R.— Nature, Vol. 183:1448. 1959.
37. Gunckel, J. E. and A. H. Sparrow. In: Encyclopedia of Plant
    Physiology, Vol. 16:555, Ed. by W. Ruhland. Berlin, Springer
    Verlag. 1961.
38. Gustafsson, A. and M. Simak. Medd. Statens Skogsforsningsinst.,
    Vol. 48:5. 1958—1959.
39. Hasett, C. C. et al.— Nucleonics, Vol. 10:42. 1952.
40. Heaslip, M. B.— In: Recent Advances in Botany, Vol. 2:1372, Univ.
    Toronto Press. 1961.
41. Herbst, W.— Atompraxis, Vol. 10:361. 1964.
42. Jagnes, H. H. and P. A. Godwin.— J. Econ. Entomol., Vol. 50:393.
    1957.
43. May, J. T. and H. J. Posey.— Forestry, Vol. 56:854. 1958.
44. McCormick, J. F.— Ass. Southeast. Biol. Bull., No. 11. 1964.
45. McCormick, J. F. and R. E. McJunkin.— Health Phys., Vol. 11:1643.
    1965.
46. McCormick, J. F.— Nucl. Sci. Abstr., Vol. 20:42853. 1966.
47. McCormick, J. F. and F. B. Colley.— Health Phys., Vol. 12:1467.
    1966.
48. McGinnis, I. T.— In: Radioecology, p. 282, Ed. by V. Schultz and
    A. W. Klement, New York, Reinhold Publ. Corp. 1963.

49. M e r i c l e, L. W. et al.— Radiation Botany, Vol. 2:265. 1962.
50. M e r g e n, F. and J. G u m m i n g s.— Radiation Botany, Vol. 5:39. 1965.
51. M e r g e n, F. and T. S. J o h a n s e n.— Radiation Botany, Vol. 3:321. 1963.
52. M e r g e n, F. and T. S. J o h a n s e n.— Radiation Botany, Vol. 4:417. 1964.
53. M e r g e n, F. and S. S i m p s o n.— Radiation Botany, Vol. 7:247. 1967.
54. M e r g e n, F. and G. R. S t a i r s.— Health Phys., Vol. 19:37. 1970.
55. M e r g e n, F. and B. A. T h i e l g e s.— Radiation Botany, Vol. 6:203. 1966.
56. M i l l e r, L. N.— Health Phys., Vol. 11:1653. 1965.
57. M o n k, C. D.— Radiation Botany, Vol. 6:329. 1966.
58. O h b a, K.— Hereditas, Vol. 47:283. 1961.
59. O h b a, K.— Nippon Kungaku Kaishi, Vol. 48:12. 1966, Nucl. Sci. Sbstr., Vol. 21:22796. 1967.
60. O h b a, K. and M. S i m a k.— Silvae Genetica, Vol. 10:64. 1961.
61. O k u n e w i c k, J. P. et al.— Nature, Vol. 204:391. 1964.
62. O k u n e w i c k, J. P. and S. E. H e r r i c k.— Nature, Vol. 214:514. 1967.
63. O l s o n, J. S.— Health Phys., Vol. 11:1385. 1965.
64. P e d i g o, R. A.— In: Radioecology, p. 295, Ed. by V. Schultz and A. W. Klement, New York, Reinhold Publ. Corp. 1963.
65. P l a t t, R. B.— Discovery, Vol. 23:42. 1962.
66. P l a t t, R. B.— In: Radioecology, p. 243, Ed. by V. Schultz and A. W. Klement. New York, Reinhold Publ. Corp. 1963.
67. P o p a, A. and G. M i h a l a c h e.— Rev. Padulor, Vol. 80:59. 1965.
68. S a x, K. and L. A. S c h a i r e r.— Radiation Botany, Vol. 3. 1963.
69. S c h n e l l, J. H.— In: Radioecology, p. 339, Ed. by V. Schultz and A. W. Klement. New York, Reinhold Publ. Corp. 1963.
70. S c h n e l l, J. H.— Auk., Vol. 81:528. 1964.
71. S i m a k, M. and A. G u s t a f s s o n.— Hereditas, Vol. 39:458. 1953.
72. S n y d e r, E. B. et al.— Silvae Genetica, Vol. 10:125. 1965.
73. S p a r r o w, A. H.— Radiation Botany, Vol. 6:377. 1966.
74. S p a r r o w, R. S. and A. H. S p a r r o w.— Science, Vol. 147:1449. 1965.
75. S p a r r o w, A. H. and G. M. W o o d w e l l.— Radiation Botany, Vol. 2:9. 1962.
76. S p a r r o w, A. H. et al.— Radiation Botany, Vol. 1:10. 1961.
77. S p a r r o w, A. H. et al.— Radiation Botany, Vol. 3:169. 1963.
78. S p a r r o w, A. H. et al.— Radiation Botany, Vol. 5:7. 1965.
79. S p a r r o w, A. H. et al.— Radiation Botany, Vol. 8:149. 1968.
80. S t a i r s, G. R.— Forest Sci., Vol. 10:397. 1964.
81. S t e a r n e r, S. R. and S. A. T y l e r.— Intern. J. Rad. Biol., Vol. 5:205. 1962.
82. T a y l o r, F. G.— Radiation Botany, Vol. 6:307. 1966.
83. T a y l o r, F. G.— Radiation Botany, Vol. 8:67. 1968.
84. W a g n e r, R. H.— Ecology, Vol. 46:517. 1965.
85. W i l l a r d, W. K.— In: Radioecology, p. 345, Ed. by V. Schultz and A. W. Klement. New York, Reinhold Publ. Corp. 1963.
86. W i t h e r s p o o n, J. P.— Health Phys., Vol. 11:1637. 1965.
87. W i t h e r s p o o n, J. P.— Radiation Botany, Vol. 8:45. 1968.
88. W o o d, D. L. and R. W. S t a r k.— Canad. J. Entomol., Vol. 98:1. 1964.
89. W o o d w e l l, G. M.— Science, Vol. 138:572. 1962.
90. W o o d w e l l, G. M.— Scient. Amer., Vol. 208:40. 1963.
91. W o o d w e l l, G. M. and L. N. M i l l e r.— Science, Vol. 139:222. 1963.
92. W o o d w e l l, G. M. and A. H. S p a r r o w.— Radiation Botany, Vol. 3:231. 1963.

*Section 3. General Radioecological Aspects
of Land Biogeocenoses*

Chapter 10

TASKS OF DOSIMETRY IN RADIOECOLOGY

Introduction

The progress of radioecology over the past decade has been largely due
to experimental investigations conducted under natural conditions with a
broad range of ionizing radiation sources. Irradiated objects included
separate populations (the radioecology of populations) as well as the more
complex communities of organisms — biogeocenoses (radiation biogeocenology).

The foremost task of such investigations is to obtain quantitative dosimetric information that is essential for correct interpretation of phenomena
occurring in ecological systems under the influence of irradiation. However, unlike laboratory experiments in which the irradiation conditions and
doses are strictly controlled, dosimetric monitoring in natural communities
involves certain difficulties.

These difficulties may be due to different causes, depending on the
choice of dosimetric method:

1) nonuniform distribution of radiation sources in the irradiated system;

2) simultaneous exposure to external and internal irradiation;

3) simultaneous irradiation with different types of ionizing radiation, of
complex spectral composition;

4) nonuniformity of ambient medium with respect to density and atomic
composition;

5) diversity of forms, dimensions and other morphological characters of
organisms and nonuniformity of their distribution in the irradiated community.

Consequently, a correct understanding of biophysical phenomena calls
for a large volume of dosimetric information. As a rule, eliciting such
information involves large expenditures of labor and means. One could
name several other factors interfering with the procurement of necessary
quantitative information concerning absorbed doses. Analysis of these
factors and methods for estimating the absorbed doses in irradiation of the
most characteristic communities forms the subject of this chapter. At
first we shall concern ourselves with interesting radiation sources and
some of their characteristics.

Radiation sources

Radioecological investigations make use of point and distributed radiation sources.

Point sources are usually ^{60}Co and ^{137}Cs or unshielded atomic piles /24—26, 31, 32, 35, 37, 38/.

In the case of point sources only strongly penetrating radiation components (gamma quanta and neutrons) are of interest from the standpoint of dosimetry in an ambient biotic environment. Most of the energy of weakly penetrating radiation (alpha and beta particles) is absorbed in the source itself and its immediate surroundings; the contribution of alpha and beta particles to the absorbed dose is negligibly small a few meters from the source. The dosimetry of weakly penetrating radiation in radioecological investigations does not present any fundamental difficulties /9/.

Distributed sources are used in investigations of radioactive fallout and of the behavior of radionuclides within a community, and when recording their biological effect. Sources of this kind are fallout of radioactive fission products from the atmosphere or simulated fallout over small areas. In the latter case the radioactive substances are introduced by spraying experimental areas with radioactive solutions from special sprayers or by introducing them into the soil.

Radioactive fallout comprises mainly fission products of heavy nuclei. The characteristic property of a mixture of fission products is that the radiation emitted by it is mixed, consisting of gamma quanta and beta particles. In a fresh mixture, not over 10 days old, there is approximately one beta particle for every gamma quantum. In older mixtures there may be several beta particles per gamma quantum /12/. The average energy of a beta particle approximates the average energy of a gamma quantum (0.4—0.6 MeV) /1/. Consequently, the total energy of beta particles emitted by a fresh mixture of fission products is approximately equal to the total energy of gamma radiation, but in the case of an older mixture the total energy of beta particles is several times that of gamma radiation.

The above properties of radioactive fallout are very important from a radiological standpoint. Gamma radiation is a strongly penetrating radiation, the half-value thickness for gamma radiation from a mixture of fission products being about $10 \, g/cm^2$ for biological tissues. In contrast, the penetrating power of beta particles is very weak, the maximum not exceeding $1 \, g/cm^2$ for the majority of fission products. We shall now consider the consequences that may result from these differences in the penetrating powers of beta and gamma radiation for radioactive contamination of a community of plants and animals, and the influence of the properties and structure of the community and its constituent organisms on the distribution of absorbed doses.

Role of ecological factors

A considerable part of radioactive aerosols (40μ and finer) when deposited on the earth's surface are intercepted on plants and the exterior integuments of animals and is liable to remain there for fairly prolonged periods /19—22, 27—30/. The remainder of the radioactive particles are deposited on the soil surface. For relatively large animals, with bodily dimensions far exceeding the range of beta particles, external irradiation due to the beta activity of particles does not constitute any significant danger. For animals with small body surface to body mass ratio the amounts of intercepted radioactive particles per unit body weight, and consequently also the absorbed dose turned out to be relatively small. Moreover, the beta particles intercepted on the body surface are mainly absorbed in the skin-hair integument and in the subcutaneous fatty layer. The vital internal organs of large animals are reliably protected against external beta radiation by their external tissues, and the contribution of beta radiation to the absorbed dose is insignificant. An even smaller contribution to the absorbed dose is made by beta particles emitted by radionuclides deposited on the soil surface and surrounding objects. The main contribution to external radiation in this case comprises gamma radiation of radionuclides deposited on the animal's body and in the ambient environment.

The bodily dimensions of small animals (rodents), small birds and insects are commensurate with the range of beta particles. For such animals the contribution of external beta radiation to the irradiation dose of vital organs may be comparable to, or even greater than the contribution of gamma radiation.

The role of beta radiation is still more important in the case of radioactive contamination of plant cover. Plants usually possess a very ramified shape and a large surface of assimilating organs; in some plants these organs are pubescent (hairy). The ratio of surface area to biomass is very large in the majority of plant species — at least a hundred times greater than in mammals and approximately ten times greater than in large and medium-sized insects (Table 10.1).

TABLE 10.1. Ratio of the surface area of organisms (or organs) to their weight

Species	Ratio, cm^2/g
Man	0.25
Mouse	2.0
Insects Plant leaves	$n \cdot 10 - n \cdot 10^2$

Moreover, the vital organs of plants (reproductive and assimilating tissues) are as a rule open and unprotected against dusting by radioactive aerosols and consequently also against external irradiation. The dissection and pubescence of leaves and also the large surface of plants are conducive to their retaining a considerable proportion of the radioactive particles settling on the plant cover. These particles and associated radioactive substances are distributed with varying degrees of uniformity throughout the volume of plant cover. With this distribution of radioactive emitters, most of the energy of emitted beta radiation is absorbed directly in the biomass of plants and in the surface air layer occupied by them.

The situation is quite different in the case of gamma radiation. The fraction of energy of gamma radiation absorbed in the plant cover biomass does not usually exceed a few percent, by the most generous estimates, reaching 10—15% only in the case of forests /4/. Therefore beta radiation constitutes the principal radiation danger for radioactively contaminated plants. The contribution of beta radiation to the absorbed dose accumulated by assimilating and reproductive organs of plants is approximately one order of magnitude, and in some cases even several tens of times, larger than the contribution of gamma radiation.

However, the consequences of radioactive pollution are not limited to external irradiation of organisms. Radiation of incorporated nuclides may play an important role from the radiobiological standpoint. The internal irradiation dose depends on the species of organisms, the capacity of critical organs for the buildup of different radionuclides, seasonal conditions, the type and energy of radiation, etc. The behavior of radionuclides in organisms and their migration along food chains are considered in detail in Section 1 of Part I and in Part II. Nevertheless, irrespective of this behavior, in the case of radioactive contamination of natural communities by a mixture of fission products the internal irradiation of organisms will be mainly due to beta radiation from incorporated radionuclides.

In estimating absorbed doses built up by organisms in different layers of a community, one must take into account the nonuniform distribution of deposited radioactive particles among the vegetation layers. This nonuniformity is most pronounced in complex multilayer communities. Characteristic of such a community is forest biogeocenosis. The forest has a very high retentive capacity with respect to particles of atmospheric impurity, which are first intercepted mainly by tree crowns. In densely closed coniferous stands, especially spruce stands, the deposited aerosols are liable to be almost completely intercepted by the tree crowns. For instance, in periods of intensive global fallout the specific activity of pine needles was 10—20 times that of the herbaceous cover /33/. In such densely closed stands the surface of the forest floor and soil, the arboreal plants of lower layers and animals living under the forest canopy are protected by the tree crowns of the upper story against contamination by radioactive substances. Consequently, the arboreal plants of the upper story are subjected to the most intensive irradiation in the initial period, during which the tree crowns protect the advance growth, seedlings and other organisms under their canopy against irradiation. With time the radioactive substances migrate into the lower forest layers under the influence of atmospheric precipitation and wind, contaminating the underwood, herbaceous cover, forest floor and soil and their inhabiting animals. As a result, the irradiation of the upper story of the forest biogeocenosis becomes weaker, and if the fallout is represented by long-lived radionuclides, there may be simultaneous intensification of irradiation in the lower layers.

Mechanisms of protection against irradiation in communities of organisms are not confined to the screening effect of the upper story with respect to the lower layers. For instance, many plants have subterranean organs (dormant buds, rhizomes, bulbs, etc.) protected against irradiation by a layer of soil. Many animals likewise spend much of their time underground (in burrows, lairs) or in tree hollows. It has also been reported that many animals (insects, fish, rodents) possess the capacity for determining the location of ionizing radiation sources, enabling them to avoid irradiation /8, 14/. In cases of intensive radioactive fallout such animals can leave the contaminated territory or hide temporarily in burrows and shelters.

It follows that irradiation doses of organisms due to radioactive fallout depend on many factors. It is not always possible to determine precisely the part played by every particular factor. However, this may not even be necessary in certain cases, everything being dependent on the dosimetric-data accuracy required. The solution of this problem may be approached in different ways depending on the required accuracy.

Dosimetric assessment methods

Calculation method. This is based on calculation of absorbed doses from a known distribution of ionizing radiation sources in the irradiated system. Such a calculation does not involve any fundamental difficulties in the case of relatively simple types of distribution, with point sources, or with sources distributed uniformly throughout a known volume or with sources that can be represented as an entity of sources of relatively simple geometry with uniform distribution of the emitting substance throughout the volume. The corresponding calculation formulas can be found in books on the dosimetry of ionizing radiation /9/.

As an example, consider the calculation of gamma radiation doses from ^{60}Co and ^{137}Cs point sources situated above the soil surface /17/. For a point source situated in an infinite homogeneous medium the dose rate is

$$M(r) = P_\gamma \frac{A}{r^2} e^{-\mu_0 r} B(r, E_\gamma),$$

where $M(r)$ is the dose rate at distance r from the source; P_γ is the gamma constant of the source; A is the effective activity of the source (corrected for self-absorption of radiation in the source itself); μ_0 is the absorption coefficient of gamma radiation for a "good" geometry (for a narrow beam); $B(r, E_\gamma)$ is the buildup factor accounting for the contribution of scattered radiation to the dose; E_γ is the energy of gamma quanta (tabulated function /9/).

For 1 curie of ^{60}Co and ^{137}Cs, respectively, the gamma constant referred to distance 1 m in air is 1.32 and 0.356 r/hr. The mean free path in air over which the primary radiation intensity is attenuated by factor e is 146 m for ^{60}Co and 108 m for ^{137}Cs.

In the case under consideration, when the radiation source is situated near the interface of two different media (soil and air), the results calculated by the above formula differ significantly from actual observations. Berger studied gamma radiation spreading through matter along the interface of two media of different densities but of the same atomic composition. Since soil and air consist mostly of light elements (H, O, N, Si), Berger's theory is applicable to the problem in hand in the first approximation. Berger introduced still another coefficient in the formula for an infinite homogeneous medium, the numerical values of this coefficient being provided graphically in /17/. In this case calculated and experimental results are in satisfactory agreement.

However, in irradiation of communities of organisms occupying a considerable proportion of the surface air layer and distinguished by a large biomass per unit area, it is necessary to take into account radiation attenuation due to absorption and scattering of radiation in the biomass itself. For instance, the dose rate in a forest at distances ranging from several tens to several hundreds of meters from the source may differ from the dose rate on an unforested area by approximately a factor of 2, other conditions being equal.

Consider now the method for estimating absorbed doses when irradiation of a community and its constituent organisms is due to radiation from distributed sources. In this case it is necessary to know the distribution of emitters in space and the time variation of this distribution.

Dose from passing radioactive clouds. In the first approximation, the clouds may be assumed to be of infinite extent horizontally and vertically, while the distribution of radioactive admixture throughout the clouds may be assumed to be uniform. Then the absorbed dose of gamma radiation in air near the ground can be calculated by the approximate formula

$$D \approx 0.25 \bar{q} t \bar{E} \text{ (rad)},$$

where $\bar{q}$ is the mean concentration (μcuries/cm^3) in a cloud during duration t (sec) of the passage of clouds through a given point, while $\bar{E}$ is the mean radiation energy per disintegration (MeV).

For more accurate estimates it is necessary to make allowance for the finite dimensions of the cloud (emergence of radiation beyond its boundaries) and nonuniform distribution of radionuclides throughout its volume, as well as the contribution to the dose of scattered radiation from the ground. However, these factors give rise to corrections that are of opposite signs and therefore cancel each other. Therefore the error in the dose calculated by the approximate formula is usually small. The same formula is also applicable to calculation of the absorbed dose of beta radiation in air, the only difference being that the numerical factor increases with increasing distance above the ground from 0.25 (at the surface) to 0.5 (at a height equal to the maximum free path of beta particles).

The above relationships can be used to estimate absorbed doses accumulated by animals and plants enveloped by a radioactive cloud. However, this is not permissible unless their dimensions (or the dimensions of their critical organs) are within the free path of radiation quanta, i.e., it is only permissible in the absence of an appreciable attenuation of the radiation flux due to its absorption in the organism. This condition is usually fulfilled for gamma radiation, whereas the use of these relationships for assessing absorbed doses from beta radiation is confined to assimilating organs of plants with small leaves and for very small animals.

External radiation dose due to radioactive fallout. In the case of a fallout of fresh fission products, the external irradiation dose can be determined from an empirical relationship between the radioactive contamination density σ and the dose. The dose rate of gamma radiation above an infinite, ideally smooth surface for $\sigma = 10^6$ curies/km^2 at height 1.0 m is

10–15 r/hr /3/. Microrelief is taken into account by means of a correction factor, which is approximately 0.8 for the surface of level virgin land /5/.

Calculation of the absorbed dose from beta radiation calls for an individual approach in every specific case, taking into account the dimensions of irradiated objects, their geometry, and their spatial arrangement.

Internal irradiation dose. In the case of a buildup of fission products in an organism, the internal irradiation dose is mainly due to beta particles. The absorbed dose can be calculated from the known specific activity of radionuclides in the tissues. If the dimensions of the irradiated organ exceed the free path of beta particles and the radionuclide distribution throughout the tissue is uniform, then the dose rate is calculated by the formula $M = 2.13 \, \bar{q} \bar{E}_\beta$ (rad/hr), where $\bar{q}$ is the specific activity in μcuries/g while $\bar{E}_\beta$ is the mean energy of beta radiation per disintegration (MeV).

If the dimensions of an organ are less than the free paths of the beta particles, a correction must be made for the emergence of some radiation beyond the organ. In accordance with calculations /23/ the fraction of energy of beta radiation from ^{90}Sr and its daughter isotope ^{90}Y incorporated in the skeleton that is absorbed in the bone tissues and marrow of a human subject weighing 70 kg amounts to about 85%; this fraction is 62% in a rabbit weighing 2 kg and does not exceed 30% in a mouse weighing 30 g.

The relevant formulas are cited in /9/, together with formulas for calculating the internal irradiation doses from gamma emitters when their distribution within the organism is known.

Some examples of assessing irradiation doses for organisms and communities. The required degree of detail concerning the distribution of radiation sources may differ in accordance with the task. For practical purposes it is often only necessary to estimate the irradiation doses for major components of a specific community for given radioactive contamination levels. In other words, the task is reduced to finding the relationship between radioactive contamination density and irradiation doses. Solving such a problem in the first approximation sometimes calls only for the most general information concerning the distribution of radionuclides in the community in question.

The most elementary case from the standpoint of dosimetric estimates is a closed water body polluted by radioactive substances together with its inhabiting organisms. Shortly after the influx of radioactive substances into the water body, the latter achieves a state of ecological equilibrium in which the ratio of concentrations of radionuclides in the aqueous space, bottom sediments and biomass of aquatic organisms remains constant.

Consider the components of an irradiation dose on an aquatic organism living at a constant depth and of spherical body shape. Suppose the radionuclides accumulated within the organism are distributed uniformly throughout its volume. Then the system in question can be represented as the sum of the following sources with known uniform distribution of emitters: external irradiation sources — two plane layers I and II (bottom sediments and aqueous phase), "negative" source of spherical shape III (water volume displaced by organism together with radioactive substances in it), and spherical internal irradiation source IV adequate for the emitters incorporated within the organism. The irradiation dose at any point in the organism can be calculated by the formula

$$D = D_I + D_{II} - D_{III} + D_{IV}.$$

The corresponding analytical expressions for the terms of this formula can be found in manuals on the dosimetry of ionizing radiation /9/. A specific example of such a calculation is provided in /7/.

A similar calculation scheme can also be used for land communities of organisms. Consider, for instance, a meadow community of plants. We shall attempt to estimate the irradiation dose for herbaceous plants at radioactive fallout density 1 mcurie /m^2 for a mixture of fission products of 10 days' age. According to Russell /30/, a herbaceous cover typical of the temperate zone primarily intercepts an average of some 25% of radioactive fallout. It is also known that the rate of field losses of different radionuclides under the influence of wind and atmospheric precipitation is approximately the same. The field half-life, i.e., the period during which the radioactivity of herbaceous plants is halved (disregarding radioactive decay), amounts to some two weeks under average weather conditions /21, 22, 30/. It may likewise be roughly assumed that radionuclides deposited on the soil surface directly in the course of fallout or while migrating from the plant cover become uniformly distributed in the top soil layer 0.5 cm thick. Suppose the plant cover height is 20 cm, its aerial biomass 5 tons/ha, and the radioactive substances intercepted by the plants are uniformly distributed throughout the volume occupied by the plant cover. Then the absorbed dose at any point in the meadow community can be calculated as the sum of absorbed doses generated at the point by radiation of two plane sources of infinite extent: the top soil layer which is 0.5 cm thick (corresponding approximately to $h_1 = 0.5 \, g/cm^2$) and the 20 cm surface air layer occupied by the plants, together with their biomass corresponding to $h_2 = 0.08 \, g/cm^2$. For instance, the absorbed dose accumulated in the organs of plants situated directly on the soil surface is determined by the formula

$$D = 8.0 \cdot 10^{-9} \left\{ \sum_i 0.75 \alpha_i \sigma \bar{E}_{\beta_i} v_i g_i \left(v_i h_1 \right) + \right.$$

$$\left. + \sum_i 0.25 \alpha_i \sigma \bar{E}_{\beta_i} v_i g_i \left(v_i h_2 \right) \right\} \, (rad),$$

where σ is the surface density of radioactive contamination (disintegrations /cm^2); α_i is the fraction of the i-th emitter of decayed fraction of the mixture of fission products; $\bar{E}_{\beta_i}$ is the mean energy of beta particles emitted by the i-th emitter (MeV); v_i is the absorption coefficient for beta particles of the i-th emitter (cm^2/g) and may be assumed to be the same for soil, air and biological tissues; g_i are geometrical factors which depend on dimensionless quantities $v_i h_1$ or $v_i h_2$, respectively.

The percentage content α_i of different emitters in a mixture of fission products for different dates is provided in /2/, and the corresponding values of g_i in /9/. These relationships enable one to calculate the absorbed dose from beta radiation at any desired point in the system under consideration. It has been calculated, for instance, that for the fallout of a mixture of fission products of 10 days' age on a community of meadow plants with radioactive contamination density 1 mcurie/m^2, the absorbed dose rate at the initial moment will be 3 rad/day in the upper portion of plant cover, approximately 3.5 rad/day in its middle portion, and 2 rad/day at the soil surface.

The absorbed dose rate generated by gamma radiation is practically constant throughout the plant cover height and can be estimated from curves in /3/. When $\sigma = 1$ mcurie/m^2 the absorbed dose rate is about 0.25 rad/day, which is $\frac{1}{10}$–$\frac{1}{15}$ the absorbed dose rate generated in the plant by beta radiation.

The time variation of absorbed dose rate depends on the radioactive decay rate and the field losses rate. If the field half-life is assumed to be two weeks and the radioactive decay of a mixture of fission products is $A(t) \sim t^{-1.34}$, then in the case under consideration the absorbed rate from beta radiation in the meadow community two weeks after radioactive fallout of a mixture of fission products of 10 days' age will be reduced by approximately a factor of 5, while the dose rate due to gamma radiation will be reduced by a factor of 3.

This assessment method is also applicable to more complex ecological systems, such as a multilayer forest community, the only difference being that the tree crowns possess a significantly larger intercepting capacity than the aerial portions of herbaceous plants /33/. Tree crowns in a middle-aged coniferous stand are capable of intercepting 70–90% of radioactive particles from a fallout. By means of the above-described calculation method it is seen that the absorbed dose rate from beta radiation at the growing points of accumulating organs of arboreal plants when $\sigma = 1$ mcurie/m^2 will initially amount to about 1.5–2 rad/day. The absorbed dose rate of gamma radiation will be 0.25 rad/day as before, remaining approximately constant throughout the height of the tree crowns.

If the radiosensitivities of the major species in the community are known, then estimates of this kind may be used to portray, in its general features, the pattern of radiological consequences of the irradiation of the community for the given radioactive contamination density. In this manner it is possible, for instance, to determine the radioactive contamination level producing heavy damage or destruction of the bulk of the population of various species of organisms of known radiosensitivities. However, the technique of averaging the distribution of radioactive emitters throughout the volume occupied by every layer of the community used in this case is a very rough approximation to possible real situations. It ignores the nonuniform distribution of the fields of ionizing radiation in the community due to the nonuniform distribution of emitters themselves even within relatively small areas of contaminated territory or screening of some community members by other members against radiation. It also ignores the protective effect of the subterranean shelters of animals or their ability to avoid the irradiation zone.

The nonuniform distribution of absorbed doses accumulated by different community members may have a decisive effect on its further destiny. Consider a case in which the radioactive contamination density is so high that the mean absorbed dose considerably exceeds the lethal level for the given species. Then, according to the prognosis based on the above-described approximation, all the individuals belonging to the given species must perish, and the species itself must be eliminated from the community. However, this need not actually happen. Owing to the nonuniform spatial distribution of radioactive substances and the shielding effect, some individuals may be exposed to a dose smaller than lethal and survive. In this case there is a possibility of a fairly rapid recovery of the numerical strength of the population of the given species in the community to its original level. As an

example one may refer to a census of the population of rat Rattus exu-
lans on Engebi islet in Eniwetok Atoll, where thermonuclear explosions
occurred in 1952 and 1954 /18/. The dose of gamma radiation at the earth's
surface due to the powerful explosion of 1952 was 2,800—6,700 r. The irradiation dose for rats in their burrows was considerably lower (250—2,500 r);
irradiation from the second explosion (1954) was much less intensive. In
spite of such a high irradiation level at the ground surface, the population
of rats reached approximately 10,000 already in 1965, although these animals
had almost totally disappeared after the 1952 explosion. A population core of
up to 50 individuals had apparently survived in their shelters. Calculations
revealed that 50 rats were capable of producing a progeny of about 10,000
after only two years, although in this case one cannot preclude the possibility
of repopulation of the island by rats migrating from nearby islands.

Another instance is provided by forest biogeocenosis. In the case of
destruction of trees on the upper story due to radioactive contamination of
a forest, the former structure of the forest stand may be rehabilitated quite
rapidly by the surviving advanced growth that was protected by the crowns
of maternal trees against beta radiation.

Hence it becomes evident that a complete description of the radiation
situation in a contaminated community calls for a knowledge not only of
the mean value of the absorbed dose, but also of the range of possible
deviations from these means under the influence of various ecological and
other factors. Furthermore, it must be remembered that in the course of
time a steadily increasing contribution is supplied by internal irradiation
due to the intake of radionuclides from soils through roots in the case of
plants or through food chains in the case of animals. These irradiation
sources were ignored by us. Allowance for all these factors is far from
simple. A knowledge of only the general distribution and migration of radioactive emitters in a community is insufficient for this purpose. There is
a need for more detailed information on their microdistribution, the buildup
of radionuclides in different organisms, the behavior of animals, and so on.
The collection of such extensive information is very labor-consuming. However, even if such information were obtained, difficulties would inevitably
arise due to the cumbersome and complex calculations. These difficulties
can be avoided when determining external irradiation doses by directly
gathering dosimetric information under natural conditions.

The devices and techniques used for this purpose will be examined in
more detail.

Experimental methods. Dosimeters must satisfy rather rigid requirements in order to be suitable for use in field radioecological investigations.
They must be portable, reliable and simple in operation, and suitable for use
under any weather conditions. They must also be capable of recording
radiation doses of different kinds over a broad range of energies and
dosages. These requirements are best satisfied by thermoluminescence
dosimeters, which are being increasingly used in practice outside the USSR
/10, 11, 13, 15, 16/.

The phenomenon of thermoluminescence is based on the following
mechanism. In the presence of impurity atoms or disturbances in the
crystal lattice of the working medium of the dosimeter it develops metastable electron states with energies exceeding the ground state energy by

several electron-volts. Under the influence of ionizing radiation part of the electrons are converted from the ground to metastable state. Heating of an irradiated crystal to a temperature of several hundred degrees causes the reverse transition accompanied by emission of light. From the intensity of emitted light, measured with a photomultiplier, it is possible to determine the energy absorbed by the crystal and hence the absorbed dose /10/. Thermoluminescence dosimeters are usually made with crystalline LiF or CaF_2:Mn, introduced in the form of fine powder into glass or teflon. Dosimeters based on LiF added to teflon are most suitable for work with biological objects. Dosimeters of this type possess several important advantages.

1. They can be made of any desired shape and size, for instance, in the shape of disks, rods and strips. Platelets as thin as 15μ $(3\,mg/cm^2)$ can be obtained with a microtome /11/. Such dosimeters are suitable for measuring absorbed doses from beta radiation; the thinness of their platelets precludes the possibility of large gradients of absorbed doses for small energies of beta particles, while absence of a can (cassette) makes it possible to abut the radiation-sensitive volume of the dosimeter on the object to be investigated.

2. The yield of photic energy during heating of a dosimeter irradiated with neutrons, gamma quanta or electrons depends solely on the amount of absorbed radiation energy and is independent of the dose rate. Therefore, thermoluminescence dosimeters can be used for recording mixed radiation /16/.

3. LiF and CaF_2:Mn crystals possess a linear or near-linear relationship between the yield of photic energy and the dose for doses from fractions of a rad to several thousand rads /16/. The measurement error does not usually exceed 5% for dosages of $10-10^5$ rad /11,13/. Spontaneous losses of energy stored in the dosimeter do not exceed 5% per year at room temperatures. However, prolonged exposure of dosimeters to sunlight during or after irradiation reduces the yield of thermoluminescence by 20—25% /24/. These losses are independent of humidity and mechanical factors.

4. Thermoluminescence dosimeters can be used repeatedly.

These qualities of the dosimeters as well as their being portable and simple in operation offer extensive possibilities for radioecological applications. In field experiments, dosimeters based on LiF in the form of thin disks or rods of teflon are currently used mainly for recording gamma radiation. They were used, for instance, to obtain detailed information on the distribution of gamma-radiation doses in forest stands and field communities of herbaceous plants irradiated by point sources. Detailed studies were made of the shielding effect due to the crowns and boles of trees and soil at different distances from the source and at different levels with respect to the soil surface /24/. Results have also been obtained indicating the possibility of using thermoluminescence dosimeters to determine internal irradiation doses for organisms without significantly interfering with their functions. For instance, LiF dosimeters introduced into the organism were used to obtain experimental data on internal irradiation doses for the gastro-intestinal tract of a sheep due to gamma radiation from ^{137}Cs ingested with the ration /34/ and on irradiation doses for a rabbit's lungs /10/.

There is no doubt that thermoluminescence dosimeters will occupy a fitting place in radioecology and the dosimetry of beta radiation. However, further methodological studies will be necessary for this purpose, including the choice of the optimum dosimeter geometry, search for the best calibration standards (sources) for each specific task, etc.

Among dosimeters of other types, film dosimeters possess many valuable properties from a radioecological standpoint. The recording of doses in this case is based on measuring the blackening density of photographic film under the influence of irradiation. This method had long been used in radiation monitoring of personnel working with radioactive substances and ionizing radiation sources. The principles of the method can be found in appropriate manuals /6/.

Film dosimeters are suitable for measuring doses from gamma and beta radiation under natural conditions. For this purpose one should use thin film holders made of lightproof polyethylene film capable of protecting the photodosimeter against sunlight and distinguished by a relatively weak absorption with respect to the flux of beta particles. Such film holders can be made fully moistureproof.

In order to preclude energy dependence of the dosimeter, the photographic films used to construct a calibration curve must be irradiated with standard sources, the beta radiation of which is identical or similar with respect to composition and energy spectrum to the corresponding characteristics of the radiation doses being measured.

Film dosimeters in holders with walls $0.02\,g/cm^2$ thick were used for recording doses of beta radiation from ^{90}Sr at the surface of contaminated soil. A radioactive solution of ^{90}Sr was introduced on the soil surface by sprinkling. The photographic film was calibrated against a thick-walled flat ^{90}Sr source made from finely ground soil with the addition of a known amount of the radionuclide. Simultaneously with the measurements, the dose at the same points was calculated from the known radionuclide distribution over the soil profile. The calculated and experimental results proved to be in close agreement, discrepancies between them not exceeding 30%.

One advantage of film dosimeters (similar to thermoluminescence ones) lies in the possibility of placing several dozen or even hundreds of such dosimeters within an irradiated ecological system with a minimum time expenditure. Such a procedure would yield simultaneously information on the dosages at practically all points of the system in which the researcher is interested.

Method of assessing doses from irradiation effects. The above discussion refers to a task that can be designated as the direct problem of dosimetry with respect to radioecological problems, its purpose being the determination of doses and forecasting biophysical consequences of irradiation according to a given radiation source distribution (or radioactive contamination density). The inverse problem (determination of radioactive contamination levels on a territory from radiological effects) may likewise possess practical significance. This problem may arise, for instance, when it is necessary to draw the isopleths of radioactive contamination density over large territories. Visually observed reactions to irradiation of certain species sufficiently common in the given district may be utilized as radioactive contamination indicators in this case. For example, the pine, the reaction of which to irradiation has been thoroughly studied, is the most suitable indicator among tree species.

It is known that irradiation with doses of 5,000—6,000 rad accumulated in he course of a few days causes reddening of needles in two or three weeks, .ollowed by death of the pine /25, 26, 31, 32, 37, 38/. The change in the color of needles allows one to visually demarcate the zone of radiation damage to coniferous stands (by aircraft and surface observations) and to distinguish areas in which the irradiation doses for the pine proved to be above lethal. Once the radioisotope composition of contaminating mixture (in the case of a mixture of fission products, a knowledge of its age is sufficient) and the coefficient of interception of radioactive aerosols by tree crowns are known, the radioactive contamination density corresponding to the boundary of radiation damage can be determined for the pine from the absorbed dose, and the time variation of contamination density can be computed.

Tasks of dosimetry in radioecological problems

We now examine possible experimental approaches to the solution of the above-formulated task of dosimetery with respect to radioecological problems, i. e., establishment of relationships between radioactive contamination levels in natural communities of organisms and irradiation doses on their components. The starting data necessary to solve this problem can be obtained in studies of the fallout of radioactive substances.

If contamination is due to global radioactive fallout, direct measurement of its doses is difficult owing to the presence of a natural radioactive background and the inadequate sensitivity of existing dosimeters. Nevertheless, studies of the distribution of global fallout may yield the necessary information for assessing irradiation doses by computation. For approximate calculations it is important to know the following characteristics of radioactive fallout:

1) radioactive contamination density, i. e., the amount of radioactive substances deposited on unit area;

2) radionuclide composition of fallout;

3) interception coefficients for radioactive aerosols by different layers of every specific community (i. e., the fraction of radionuclides intercepted on the surface of organs of animals and plants in every layer);

4) rate of subsequent radionuclide migration under the influence of atmospheric precipitation and wind (usually stated in terms of field half-lives) and the rate of attenuation of specific activity of tissues due to the effect of "dilution" (i. e., the increment in the biomass of organisms);

5) buildup factors for individual radionuclides in different organs of animals and plants.

Whenever more complete dosimetric information is needed it is likewise necessary to determine the possible scatter of these parameters, due to nonuniform distribution of radioactive substances in the ambient medium, to the nonuniform structure of the contaminated community, and to the effects of shielding and protection of organisms from radiation.

Direct measurement of doses is only possible at relatively high radio-active contamination levels which either occur in local fallout of radioactive substances or can be produced artificially in model experiments. In these cases thermoluminescence and film dosimeters permit experimental establishment of the relationship between the radioactive contamination density and dose rate, and also its time variation.

A large number of experiments have been described concerning the simulated radioactive contamination of communities of organisms under laboratory and field conditions. However, the majority of these experiments were aimed at solving particular problems: studying the distribution of radionuclides in the ambient medium and in different organs of animals and plants, examining the mechanisms governing radionuclide penetration into these organs and their further redistribution. On the other hand, radiation monitoring aspects were practically ignored in experiments of this kind. However, such monitoring would permit the establishment of the relationships (between radioactive contamination levels and irradiation doses on different community components) in which we are interested, provided the distribution of radioactive substances among the contaminated system components were monitored simultaneously. Moreover, since a knowledge of the distribution and composition of radioactive emitters is a necessary and sufficient condition for the computational determination of doses, the experimental results make possible also incidental assessment of the accuracy of the calculation method, i. e., the boundary of its applicability.

Dosimetric information received directly from an experiment is valuable in that it may be extrapolated, to some degree, also to other possible cases of radioactive contamination. This is so, since an empirically determined correlation between the radioactive contamination levels of different community components and their irradiation doses permits determination of the doses in practically any likely situation (including global and local fallout), provided one knows the distribution of radionuclides among the components of the system under consideration. In order to place such information on a strong experimental base it is necessary to solve the following two problems.

First, it is necessary to examine the measurement techniques pertaining to doses for mixed beta and gamma radiation under field conditions over a broad range of doses and energies. Thermoluminescence dosimeters can be usefully utilized for this purpose. Second, it is desirable that the problems of experimental investigations into the behavior of radioactive substances in communities of organisms and their components should provide for the gathering of necessary dosimetric data by directly measuring absorbed doses at the most important points in the irradiated community.

References

1. Bykov, A. G. et al. — Atomnaya Energiya, Vol. 10:362. 1961.
2. Grechushkina, M. P. Tablitsy sostava produktov mgnovennogo deleniya U^{235}, U^{238}, Pu^{239} (Tables of the Composition of Instantaneous Fission Products of ^{235}U, ^{238}U, ^{239}Pu). Moskva, Atomizdat. 1964.

3. Deistvie yadernogo oruzhiya (Effect of Nuclear Weapons).   Moskva, Voenizdat. 1964.

4. I z r a e l ', Yu. A. et al. — Izvestiya Akademii Nauk SSSR. Seriya geofizicheskaya, No. 8:1126. 1962.

5. I z r a e l ', Yu. A. and E. D. S t u k i n.   Gamma-izluchenie radioaktivnykh vypadenii (Gamma Radiation of Radioactive Fallout).   Moskva, Atomizdat. 1967.

6. K o z l o v, V. F.   Fotograficheskaya dozimetriya ioniziruyushchikh izluchenii (Photographic Dosimetry of Ionizing Radiation).   Moskva, Atomizdat. 1967.

7. P a r c h e v s k a y a, D. S. — Radiobiologiya, Vol. 9:281. 1969.

8. P r a v d i n a, G. M.   Povedenie zhivotnykh organizmov v polyakh ioniziruyushchikh izluchenii (Behavior of Animal Organisms in Fields of Ionizing Radiation).   Thesis. Moskva. 1967.

9. H i n e, J. E. and G. L. B r o w n e l l (editors).   Radiation Dosimetry, Academic Press. 1956.

10. B e n n e r, S. et al. — In: Solid State and Chemical Radiation Dosimetry in Biology and Medicine, p. 65. Vienna, IAEA. 1966.

11. B j a r n g a r d, B. E. and D. J o n e s. — In: Solid State and Chemical Radiation Dosimetry in Biology and Medicine, p. 99. Vienna, IAEA. 1966.

12. B j o r n e r s t e d t, R. — Arkiv for Phys., Vol. 16:293. 1959.

13. B r o o k e, C. and R. S c h a y e s. — In: Solid State and Chemical Radiation Dosimetry in Biology and Medicine, p. 31. Vienna, IAEA. 1966.

14. B r o w e r, J. H. — Nucl. Sci. Abstr., Vol. 19:403. 1965.

15. C a m e r o n, J. R. et al. — Health Phys., Vol. 10:25. 1964.

16. C a m e r o n, J. R. et al. — In: Solid State and Chemical Radiation Dosimetry in Biology and Medicine, p. 77. Vienna, IAEA. 1966.

17. C o w a n, F. P. and C. B. M e i n h o l d. — Radiation Botany, Vol. 2:241. 1962.

18. F r e n c h, N. R. — Health Phys., Vol. 11:1557. 1965.

19. K r i e g e r, H. L. et al. — In: Radioecological Concentration Processes, p. 59, Ed. by Aberg, B. and F. P. Hungate. Pergamon Press. 1967.

20. L o u t i t, J. F. and R. S. R u s s e l l. — In: Progr. in Nucl. Energy. Ser. VI, Biol. Scie., Vol. 3:126, Ed. by Loutit, J. F. and R. S. Russell. Oxford, Pergamon Press. 1961.

21. M a r t i n, W. E. — Health Phys., Vol. 11:134. 1965.

22. M a r t i n, W. E. — Radiation Botany, Vol. 4:275. 1964.

23. M a y s, C. W. — U. S. Atomic Energy Commission Report, COO-218, p. 113. 1959.

24. M c C o r m i c k, J. F. and F. B. C o l l e y. — Health Phys., Vol. 12:1467. 1966.

25. P l a t t, R. B. — Discovery, Vol. 23:42. 1962.

26. P l a t t, R. B.   Problems of Radioecology [Russian translation, 1968.]

27. R o m n e y, E. M. et al. — Ecology, Vol. 44:343. 1963.

28. R u s s e l l, R. S. and J. V. P o s s i n g h a m. — In: Progr. in Nucl. Energy, Ser. VI, Biol. Sci., Vol. 3:2, Ed. by Loutit, J.F. and R. S. Russell. Oxford, Pergamon Press. 1961.

29. R u s s e l l, R. S. — Nature, Vol. 182:834. 1958.

30. R u s s e l l, R. S. — Health Phys., Vol. 11:1305. 1965.

31. Sparrow, A. H. et al. — Radiation Botany, Vol. 5:7. 1965.
32. Sparrow, A. H. — Radiation Botany, Vol. 6:377. 1966.
33. Szepke, R. — Atompraxis, Vol. 11:391. 1965.
34. Watson, C. R. — Nucl. Sci. Abstr., Vol. 21:1539. 1967.
35. Witherspoon, J. P. — Health Phys., Vol. 11:1637. 1965.
36. Woodley, R. — Radiation Botany, Vol. 7:1. 1967.
37. Woodwell, G. M. — Science, Vol. 138:572. 1962.
38. Woodwell, G. M. and A. H. Sparrow. — Radiation Botany, Vol. 3:231.
       1963.

Chapter 11

RADIOECOLOGY OF WILD ANIMALS

The radioecology of animals examines mainly the life of populations of animals and zoocenoses exposed to ionizing radiation as an abiotic factor of their habitat. To this end it is necessary to study the accumulation of radionuclides by animals, the migration of radioactive substances along food chains in the biogeocenosis, and the distribution of doses in the different zoocenosis components.

Accumulation of radioactive fission products
in animal populations

The general buildup of radionuclides in animals depends on several conditions, the major ones being the contamination levels and distribution patterns of nuclides in the population's habitat, the biological features of animal species, the trophic level occupied by animals in the food chains of the biogeocenosis, and the chemical properties of radionuclides.

TABLE 11.1. Accumulation factors and biological half-lives of some radionuclides in rodents /96/

Parameter	Albino rats		White mice	
	^{90}Sr	^{106}Ru	^{137}Cs	^{60}Co
Accumulation factor	0.15	0.03	0.8	0.3
Biological half-life, days	350	85	6.5	9.4

In the accumulation of radionuclides by land animals, an important part is played by the latter's chemical properties and the physiological characteristics of animal species (Table 11.1). The concentration quotients of ^{90}Sr (the ratio of its concentrations in the skeleton and in the ration) for some wild rodents are as follows: hare (L e p u s   c a l i f o r n i c u s) 0.20; rabbit (S y l v i l a g u s   a u d u b o n i) 0.22; kangaroo rat (D i p o d o m y s   m e r r i a m i) 0.16 /56/. The accumulation rate of fission products following territorial contamination has been studied by many authors /33, 98, 100, 102, 134/. Only nine days after the fallout of fresh fission products during nuclear tests, equilibrium is attained between the content of fission products in the animals and the degree of territorial contamination; this equilibrium becomes stable toward the 20th day.

In contrast to the aquatic environment, contamination of land areas by radioactive fission products proceeds nonuniformly /15, 18/, due primarily to the nonuniform fallout of radioactive aerosols on account of uneven relief, presence of plant cover, and so on. Nonuniform contamination of the soil cover results in nonuniform contamination of plants assimilating radionuclides from the soil and serving as food for animals.

The intake of artificial radionuclides by animals on land mainly proceeds by the alimentary circuit with food or by respiration of aerosols or gases containing the radionuclides. Differences among land animals with respect to behavior, feeding, the use of territory and other biological characteristics likewise play an important role in the accumulation of radioactive substances by different species of animals. An equally important role in the accumulation of nuclides in the populations of different species and ecological groups of animals is played also by the area of contaminated territory, controlling the degree of contact of the animals with the radioactive substances.

Studies into the accumulations of some radionuclides (^{137}Cs, ^{65}Zn, ^{90}Sr, ^{60}Co and ^{106}Ru) by rodents on sites of local burial of radioactive wastes showed that small animals (rats and mice) accumulated considerably larger amounts of radionuclides than larger rodents, such as hares. According to some authors /96/, these differences are controlled by the size of individual feeding territories of the investigated animals. Animals with smaller individual feeding areas come in more frequent contact with contaminated territory. For instance, Willard /141/ reported different degrees of contact of several bird species with a radioactive area, resulting in their unequal accumulation of radionuclides.

Differences in the habitat distribution of populations over a nonuniformly contaminated terrain may likewise affect the accumulation of nuclides by animals. Thus, the ^{90}Sr buildup and irradiation doses for three small rodent species inhabiting the site of emptied Lake White Oak, previously used for disposal of radioactive wastes, are controlled by the habitat distribution of the populations of these animals /109/. The white-footed mice (P e r o m y s - c u s   l e u c o p u s) inhabits the elevated portion of the lake site, where the contamination density is lowest. Cotton rats (S i g m o d o n   h i s p i d u s) live in the depressed humid portion of the lake basin with a comparatively high contamination level, while rice rats (O r y t o z o m y s   p a l u s t r i s) occupy the lowest, bogged portion of the valley along the lake shore, where the contamination density attains its maximum. The accumulation of ^{90}Sr by these rodent species was found to be proportional to the contamination density on the territories occupied by their populations.

As a rule, concentration of fission products in the body of land animals is on the average higher than in aquatic animals, for the same radioactive contamination density on land and water areas. For instance, dry land birds of Alaska are distinguished by a higher concentration of fission products than aquatic birds /76/.

In studies of the accumulation of radionuclides by animals from the environment, an important role is played by an examination of specific differences, which are controlled by the animals' physiology, morphology, alimentation, distribution within the biogeocenosis, and so on. Specific differences in the concentration of radionuclides by animals may ultimately control the internal irradiation doses due to radioactive substances incorporated in their tissues. Consequently, the biological effect of ionizing radiation may manifest itself differently in different species.

Specific differences in the accumulation of radionuclides are especially distinct in the case of monophagous insects, which accumulate ^{90}Sr and ^{137}Cs in amounts that are proportional to their contents in the plants on which the insects feed /74, 75/. Thus, there are different contents of ^{137}Cs in different species of edaphic arthropods inhabiting the forest floor /115/ and there are different contents of several radionuclides from global fallout in certain species of insects, fishes and amphibians /69/. Specific differences in the accumulation of radionuclides by populations of birds in contact with locally contaminated terrain (bottom of former Lake White Oak) were pointed out by Willard (Figure 11.1) /141/. Different concentrations of ^{90}Sr in wild rodents were detected only 9 days after a nuclear explosion: 7.65 ± 0.5 Sr units in kangaroo rats and 22.1 ± 0.4 Sr units in hares.

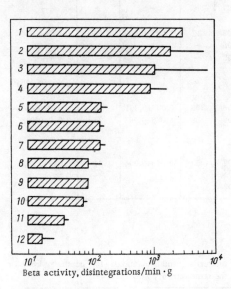

Beta activity, disintegrations/min · g

FIGURE 11.1. Total beta activity of the body of birds in contact with a territory locally contaminated by radionuclides /141/. The bars and straight lines respectively signify mean and maximum beta activities:

1—4 — birds that are permanent inhabitants of the contaminated territory; 5—11 — birds nesting outside the contaminated zone but often visiting the latter; 12 — hummingbirds, only seldom visiting a contaminated zone.
(1 — Seiurus motacilla; 2 — Melospiza melodia; 3 — Icteria virens; 4 — Spizella pusilla; 5 — Spinus tristia; 6 — Geothlypis trichas; 7 — Passerina cyanea; 8 — Dumatella carolinensis; 9 — Oporonis formosus; 10 — Vireo griseus; 11 — Richmondena cardinalis; 12 — Archilochus colibris.)

Allied species of small rodents and insectivores inhabiting the same territory but subsisting on different kinds of food differ in the degree of accumulation of ^{90}Sr and ^{137}Cs (Tables 11.2 and 11.3). Large differences in the accumulation of ^{90}Sr deposited in global fallout were detected in different animal species in Alaska /138/ and in North America /82/ (Table 11.4). Furthermore, the concentration of this nuclide was found to be higher in the skeleton of mountain animals (deer) than in species occurring on plains /118/. A similar pattern in the content of global ^{137}Cs in the skeleton was observed also in deer Odocoileus hemionus in the State of Colorado, USA. The body of deer living at an altitude of 250 m was found to contain larger amounts of ^{137}Cs than the body of deer living at lower altitudes /140/, the difference being explained by the larger quantity of fission products on elevated territories. The lowest ^{90}Sr concentration is found in the skeleton of the coyote occupying a higher trophic level in the food chain.

TABLE 11.2. Strontium-90 content in plants and skeletons of different species of small mammals, μcuries/g /17/

Object of investigation	Level of ^{90}Sr contamination on experimental areas, mcuries/m^2					
	2.5		1.0		0.6	
	number of samples	^{90}Sr content	number of samples	^{90}Sr content	number of samples	^{90}Sr content
Plants	45	0.16±0.013	–	–	–	–
Rodents						
Field vole (Microtus agrestis) ..	10	0.43±0.07	–	–	–	–
Northern redbacked vole (Clethrionomys rutilus)	157	0.42±0.02	32	0.37±0.003	26	0.027±0.004
Common vole (M. arvalis)	14	0.43±0.05	–	–	–	–
Narrow-skulled vole (M. gregalis) .	28	0.29±0.03	–	–	–	–
Common hamster (Cricetus cricetus)	–	–	–	–	21	0.006±0.0001
Common field mouse (Apodemus sylvaticus)	30	0.25±0.06	308	0.023±0.001	45	0.018±0.002
Striped field mouse (A. agrarius)...	40	0.18±0.02	28	0.038±0.005	26	0.017±0.003
Insectivores						
Common shrew (Sorex araneus) ..	40	1.0±0.07	69	0.049±0.004	34	0.039±0.003
Middle-sized shrew (S. caecutiens)	108	0.49±0.03	–	–	–	–

TABLE 11.3. Cesium-137 content in plants (μcuries/kg wet matter) and small mammals (10^{-2} μcurie/kg body weight without the digestive tract) inhabiting three areas in the same territory /45/

Object	^{137}Cs contamination level of experimental areas, μcuries/m^2		
	7.85±0.22	5.30±0.14	4.45±0.10
Plants	3.9	4.0	2.9
Northern redbacked vole	2.2	–	2.0
Field vole	–	–	2.7
Narrow-skulled vole	–	1.0	–
Common field mouse	0.7	0.9	0.4
Striped field mouse	–	1.1	0.7
Common shrew	–	3.0	–

TABLE 11.4. Strontium-90 concentration in the skeletons of large animals inhabiting the same territory /82/

Object	Number of samples	Concentration, Sr units
Cattle	125	12.7±1.4
Hare	195	14.8±1.6
Antelope (A n t i l o c a p r a a m e r i c a n a)	41	14.7±3.3
Deer (O d o c o i l e u s h e m i o n u s, C e r v u s c a n a d e n s i s)	28	21.7±4.5
Coyote (C a n i s l a t r a n s)	43	10.3±1.3

Among bone tissue samples from cattle, deer (C e r v u s sp.), elks (A l c e s   a l c e s) and roe deer (C a p r e o l u s   c a p r e o l u s) the highest ^{137}Cs concentration was determined in the roe deer and the lowest in the farm animals /23/.

Differences in the contents of radionuclides have been reported with respect to fish, amphibians, reptiles and mammals /26/. Distinct differences in accumulation levels have also been detected in other animals belonging to different classes /121/.

Seasonal variations in the accumulation of radionuclides in animal populations

Investigations of seasonal differences in the accumulation of radionuclides by mammals related to changes in alimentation have been carried out on reindeer (R a n g i f e r   t a r a n d u s) and caribou (R. c a r i b o u) in the arctic tundra /90, 92, 113, 114, 132, 138/. The maximum concentration of radionuclides in venison occurs in winter, when the deer mainly feed on lichens which sorb the fission products from air. In summer, when the deer switch over to herbaceous plants and twigs, the content of nuclides in their body decreases. Analogous variations in ^{137}Cs accumulation in the skeleton of large herbivorous animals (elks, deer and roe deer) in the forest zone were discovered by Kruglikov /23, 24/. Seasonal variations in ^{90}Sr concentration in the skeletons of rodents are characteristic for American hares /82/ and field voles in the USSR /16/. Hanson /88/ observed a variation in the content of ^{131}I in the thyroid gland of hares by seasons of the year at Hanford.

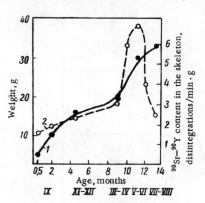

FIGURE 11.2. Age-contingent and seasonal features of the accumulation of [90]Sr–[90]Y in the population of field voles:

1 – body weight; 2 – [90]Sr–[90]Y concentration in the skeleton.

Seasonal variations in the diet of field voles control the variations in the content of ^{90}Sr in the skeleton of field voles born in different seasons

TABLE 11.5. Seasonal and age-contingent differences in the accumulation of ^{90}Sr—^{90}Y in a population of field voles inhabiting a contaminated territory, 10^3 disintegrations/min · g skeleton /16/

Age, months	March–April		May		June		July		August		September		December	
	number of samples	^{90}Sr—^{90}Y	number of samples	^{90}Sr—^{90}Y	number of samples	^{90}Sr—^{90}Y	number of samples	^{90}Sr—^{90}Y	number of samples	^{90}Sr—^{90}Y	number of samples	^{90}Sr—^{90}Y	number of samples	^{90}Sr—^{90}Y
~0.5	—	—	7	370.8±14.9	36	342.3±10.5	16	169.5±16.0	—	—	—	—	—	—
1–1.5	—	—	—	—	29	438.0±6.6	27	285.2±2.1	26	163.0±7.3	51	199.0±7.7	—	—
2–2.5	—	—	—	—	—	—	15	397.5±30.2	17	244.8±15.1	—	—	—	—
4–4.5	—	—	—	—	—	—	—	—	—	—	25	348.6±19.2	91	240.3±3.6
6–7	—	—	—	—	—	—	—	—	—	—	—	—	9	263.0±5.5
9–10	70	315.8±4.8	—	—	—	—	—	—	—	—	—	—	—	—
10–11	—	—	19	599.4±8.6	—	—	—	—	—	—	—	—	—	—
11–12	—	—	—	—	39	638.8±5.0	—	—	—	—	—	—	—	—
12–13	—	—	—	—	—	—	22	397.2±13.3	—	—	—	—	—	—
13–14	—	—	—	—	—	—	—	—	13	246.0±7.2	—	—	—	—

(Table 11.5). In the voles born in the second half of summer the accumulation of this nuclide proceeded at only half the rate in those born in spring and early summer. The accumulation rate of ^{90}Sr in the field voles is considerably higher in summer than in winter. The accumulation of this nuclide is especially intensive in old field voles which have survived winter. The ^{90}Sr accumulation dynamics is not strictly correlated with the seasonal variation in the weight of these animals (Figure 11.2). The variation in the accumulation of ^{90}Sr in the rodents is mainly due to seasonal variations in their diet /16/. The variation of ambient temperatures may also play some part /89/.

In connection with seasonal variations in the buildup of radionuclides in the body of animals, one must point out the relationship between accumulation of radionuclides and intensity of global fallout. Fineman et al. /82/ discovered yearly differences in the buildup of global ^{90}Sr in animals caught in the same locality. The increase of ^{90}Sr concentration in the skeleton of several species of wild and domestic animals proceeded synchronously in one and the same year following the increase in the density of soil contamination by global fallout after nuclear tests. A similar increase of ^{90}Sr content was observed in the skeleton of Virginia deer (O d o c o i l e u s   v i r g i - n i a n u s) in the State of Maryland, USA, over the period 1953–1959 /119/.

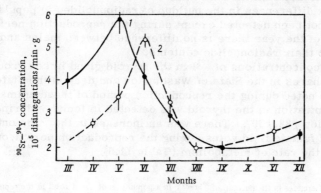

FIGURE 11.3. Differences in ^{90}Sr–^{90}Y contents in the skeleton of field voles in 1964 (1) and in 1965 (2)

A population of field voles displayed statistically significant differences from one year to another in the accumulation of ^{90}Sr in their skeletons in open air cages /16/ with this nuclide artificially added to the soil (Figure 11.3). The differences were apparently due to dissimilar climatic conditions in the compared years.

Age, sex and individual differences in the buildup of
radionuclides in animal populations

Differences in the buildup of radionuclides in different age and sex groups in populations of animals have been reported by several authors. In natural populations of small animals inhabiting areas locally contaminated by ^{90}Sr, they have exhibited distinct age-contingent differences in the buildup of this radionuclide related to seasonal variations in the diet and the duration of the reproduction period of animals in the population.

Studies of the accumulation by animals of radionuclides from global fallout revealed differences in the accumulation of ^{90}Sr by different age groups in a population of undulate animals. Schultz /117/ reported that the average ^{90}Sr content in the antlers of Virginia deer were 11.2 and 8.8 pcuries/g ash, respectively, in animals aged 4.5 and 2.5 years. Studies of the age-contingent regularities of the buildup of these nuclides in the lower jaw of deer revealed the average ^{90}Sr concentration to decline steadily, from 13.2 to 7.0 pcuries/g ash, on the average, between the ages of 4 months and 6.5 years /118, 119/.

Contrasting data were obtained for certain other species of animals. Thus, ^{90}Sr concentration in the bones of 4-year-old reindeer proved to be considerably higher than in 2-year-olds /12/.

Studies of the buildup of ^{90}Sr in small wild mammals on a territory that had been contaminated by fallout following nuclear tests (in the State of Nevada, USA) revealed higher ^{90}Sr concentrations in old kangaroo rats and hares than in young ones /102/. Young field voles likewise had lower ^{90}Sr contents in their skeleton than old ones.

Distinct sex differences in the buildup of radionuclides in populations of animals have not been detected except during the reproduction period. In other seasons of the year there is no difference between males and females with respect to their radionuclide contents.

Dissimilar concentrations of ^{131}I in the thyroid gland in sex groups of populations of hares in the State of Washington and deer in the State of Colorado were noted during the reproduction period of these animals. A high ^{131}I concentration in the thyroid was detected in females during gestation and lactation /88, 139/. There was an increase in the ^{90}Sr content in the skeleton of female field voles during the reproduction period, on account of a change in the rate of metabolism (Table 11.6).

TABLE 11.6. Sex differences in the buildup of ^{90}Sr–^{90}Y in a population of field voles in the reproduction period, 10^9 disintegrations/min · g skeleton /17/

Sex	March–April		May–June		July	
	number of animals	^{90}Sr–^{90}Y concentrations	number of animals	^{90}Sr–^{90}Y concentrations	number of animals	^{90}Sr–^{90}Y concentrations
Females....	38	312.2±3.9	40	641.4±7.3	16	415.8±19.0
Males	32	283.0±17.0	18	564.5±9.8	6	348.0±2.5

Individual differences have been noted with respect to the buildup of ^{90}Sr by animals in a population. For instance, such differences (due to nonuniform contamination of the soil surface by fission products, distribution of plant cover microgroups and the situation of individual areas in a population) were discovered for a population of small rodents /16/.

Variations in the buildup of radionuclides
in animal populations

The nonuniform contamination of terrain by radioactive aerosols, as well as seasonal, age-contingent and other peculiarities in the buildup of radio-nuclides in organisms, produce variability of their content in a population. In the case of local contamination by fission products this variability in-creases, owing to migrations of animals between clean and contaminated territories.

The coefficient of variation of ^{137}Cs content in a single herd of caribou generally amounts to 30% /81/. Appreciable fluctuations of ^{90}Sr concentra-tion in the skeleton of small mammals were detected in populations of rodents and insectivores in locally contaminated areas. The degree of variability in the buildup of radionuclides in the body of animals increases with diminishing areal contamination density /17/. The variability of ^{90}Sr concentration in the skeleton of animals depends on the variability of its content in the ration, which is in turn controlled by the soil contamination pattern /18/.

Studies of the variability in the radionuclide content in populations of animals show that different individuals constituting the population may receive different irradiation doses from emitters incorporated in their bodies. Therefore different intrapopulational groups of animals (age, sex, etc.) differ with respect to the effects of internal irradiation, which is liable ultimately to affect the biological processes occurring in the population occupying territories with a high level of contamination by radionuclides.

Migration of radionuclides in zoocenoses

The majority of investigations concerned with radionuclide migration in biogeocenoses are limited to an investigation of the migration of radioactive substances along food chains leading to man. As a rule, wild land animals do not play an important role in these food chains, nevertheless their parti-cipation in the transport of radionuclides along food chains and the latter's redistribution among the biogeocenosis components is very great. The buildup level of radionuclides in populations of animals may be significantly affected by specific relationships in the biogeocenosis. Similar groups of animals occupying the same trophic level in alimentary relationships within a biocenosis may concentrate different amounts of radionuclides in different landscape zones, depending on their habitat conditions. The behavior of radionuclides in food chains and their distribution among trophic levels differ according to the chemical properties of radionuclides and their meta-bolism.

In the zoocenosis inhabiting the forest floor (consisting mainly of decaying leaves of trees) and represented by various species of invertebrates, the ^{137}Cs concentrations diminish along the chain, organic residues — saprophages — predators, being respectively 25, 4.9 and 2.5 pcuries per mg dry matter. The buildup factors in the same links are as follows: forest floor — saprophage, 0.2; saprophage — predator, 0.5 /115/. The

concentration of radionuclides likewise diminishes along the chain, plant —
— herbivores (insects) /74, 75/. As ^{137}Cs and ^{90}Sr migrate from plants
to these animals, their concentrations become 80% of the concentration in
plants for ^{137}Cs and 10% for ^{90}Sr /75/.

In higher vertebrates in the chain, plants — herbivorous mammals,
the ^{90}Sr concentration as a rule increases, while that of ^{137}Cs decreases.
The muscles of deer contain considerably less ^{137}Cs per unit weight than
the plants on which they feed /140/.

The behavior of these radionuclides undergoes a change in the next link
of the food chain. Strontium-90, which is mainly incorporated in the bone
tissue not consumed by predators, is not concentrated in the link herbivores — predators, and its content per unit weight of skeleton diminishes
on the following trophic levels. The behavior of ^{137}Cs exhibits an opposite
pattern, since this nuclide is deposited in soft tissues, which are consumed
by carnivores. Consequently, migration of ^{137}Cs in the link herbivores —
— predators is accompanied by concentration of this radionuclide. Thus,
as the food chain in a zoocenosis along which the radionuclides migrate
becomes longer, the concentrations of ^{90}Sr in the following trophic levels
decrease, while those of ^{137}Cs increase.

The migration of nuclides along food chains and their distribution in
zoocenosis primarily depend on the type of feeding and the modes of consumption of their food by animals.

Table 11.2 provides data on the buildup of ^{90}Sr in the skeletons of several
species of small mammals differing in their mode of feeding. The contents
of this nuclide differ in the bone tissue of different rodents subsisting on
different parts of plants. Voles subsisting on vegetative parts of plants
accumulate relatively more ^{90}Sr than mice feeding on seeds. The content of
^{90}Sr is higher in the skeleton of the insectivorous shrews occupying the next
trophic level in the zoocenosis than in the skeleton of rodents. The difference is evidently due to the shrews consuming whole edaphic animals and
insects. Similarly, the content of ^{90}Sr in the skeleton in insectivorous birds
is higher than in insects on which they feed /141/. Predatory insects have
the same contents of radionuclides as the herbivorous insects that are their
prey /74/. The migration coefficient of ^{137}Cs in this chain is 0.8 /115/. It
should be noted in this connection that as a rule predatory insects do not
completely devour their prey; they only suck their haemolymph which contains more Ca than the chitinous integument, but considerably less than in
the intestinal epithelium and the excretory system. This distribution is
characteristic of all the metamorphosis stages of insects /72/. It may be
assumed that the distribution of ^{90}Sr in these animals will be the same as
that of Ca.

The concentration of ^{90}Sr in the muscles of grizzly bear (U r s u s   h o r -
r i b i l i s) is two orders of magnitude lower than in most predatory mammals /138/. The content of this radionuclide in the bones of coyotes in
the state of Colorado is considerably smaller than in herbivorous animals
(see Table 11.4).

In studies of radionuclide migration along food chains leading up to man,
in the arctic zone, an increase in the concentration of ^{137}Cs was noted along
the chain, lichens — reindeer — man. The specific property of lichens

and mosses of sorbing nuclides directly from air /90, 99, 106, 107, 111, 132, 138/ leads to high concentration of radionuclides in the body of reindeer (R. tarandus) and caribou feeding on these plants. Over 50% of the deer's ration is composed of ground-surface lichens /120/. The concentration ratio ^{137}Cs/^{90}Sr is 7 in the lichens but 800 in venison /89/. The muscles of wolverines (Gulo luscus), wolves (Canis lupus) and foxes (Vulpes fulva) in Alaska feeding mainly on caribou contain 2—3 times more ^{137}Cs in winter than venison /90—92/. The concentration of these nuclides in the link prey — predator is also known for pumas (Felis concolor) feeding on deer, their ^{137}Cs content being 3—4 times the concentration in venison (1,736 and 504 pcuries/kg, respectively) /109/.

Information concerning the migration of other radionuclides along food chains in which one of the links is represented by animals is very limited. The migration of ^{210}Po in the terminal links of a food chain is accompanied by a decrease in its concentration. The content of ^{210}Po in the meat of seals (apparently Phoca sp.) is 0.2 μcurie/kg, as against 0.008 μcurie/kg in white bears (Ursus maritimus) feeding mainly on pinnipeds /61/. Similar data were obtained for ^{55}Fe and ^{131}I: the concentration of ^{55}Fe in the body of reindeer is 728 pcuries/kg, but its concentration in the body of eskimos, whose ration contains up to 40% venison, is 61 pcuries/kg /107/. The thyroid gland of herbivores is known to contain more ^{131}I than that of predators and insectivores /88/.

Transport of radionuclides by migratory animals

Some researchers ascribe considerable importance to the transport of radionuclides from water bodies to land by insects with aquatic metamorphosis /36, 37/. However, Nelson /103/ studied the transport of ^{90}Sr by insects from bottom sediments of Clinch River and found that the spreading of ^{90}Sr by insects had little effect on contamination of dry land. It was likewise demonstrated that the transport of natural radionuclides by Teteraonidae birds (whose biomass is small and which inhabit forest biocenoses) within a radius of up to 20—30 km from a site with high radionuclide concentration does not have any significant effect on territorial contamination /5/.

The reverse process (concentration of radionuclides on a small area by populations of animals) is quite possible. The nesting of colonial birds every year on the same small area may increase the content of radioactive substances per unit area by contamination with droppings, which concentrate the radionuclides from food /27/.

In edaphic (soil) zoocenoses the transport of radionuclides by edaphic animals, even such relatively large and numerous ones as earthworms (Lumbricidae), does not have any significant effects on the distribution of radioactive substances in soil /37, 38/. Larger burrowing animals may apparently play some part in this respect, especially in years when they are abundant /5/.

Biological effect of territorial radioactive contamination
on animal populations

Studies into the effect of radioactive contamination on animals are
currently being performed experimentally, mostly under laboratory conditions.   There are extremely few analogous investigations performed on
populations of animals under natural conditions, and the results are frag-
mentary.

Investigations of this kind are important for explaining certain aspects
related to the existence of populations of animals (especially mammals) in
areas contaminated by artificial radionuclides.   Ionizing radiation emitted
by radioactive fission products polluting the environment is a constituent
of the complex of abiotic habitat factors affecting populations of animals.
Therefore a rise of the radiation level in a biogeocenosis may produce
considerable alterations in the structure of populations and the entire community of animals, leading to disturbances in the biogeocenosis and modification of its further evolution.   Owing to dissimilar radiosensitivities
of different organisms, the effect of ionizing radiation may manifest itself
differently in populations of different species of animals.   Especially important is the investigation of aspects related to the effect on animal populations of ionizing radiation originating from radionuclides accumulated in
the biogeocenosis.

Mortality and life-span in animal populations on a territory
with high content of radioactive substances

A laboratory investigation of a group of narrow-skulled voles taken from
the population of a region with a high natural radiation level revealed the
life-span of this group to be considerably longer than in the controls;
changes were also noted in other parameters /29/.

Studies were made /71, 78—80, 90/ of the biological effect of ionizing
radiation on populations of small rodents inhabiting a territory contaminated
by fission products (150 curies/km^2 ^{90}Sr, 270 curies/km^2 ^{137}Cs) /75/.   No
increase in mortality or shortening of the population life-span were detected
under these conditions.

Effect of radioactive pollution on life-span and mortality was studied in
experiments in large open-air cages to which ^{90}Sr was artificially introduced (1.8—3.4 mcuries/m^2) /16/.   The observations comprised two species
of rodents: field and northern redbacked voles differing in the degree of
accumulation of ^{90}Sr.   The total ^{90}Sr content in the skeletons of the field and
northern redbacked voles were 0.11 ± 0.01 and $0.05 \pm 0.01 \mu$curie/g, respectively.

The survival of wintered populations of these species of voles did not
proceed in a similar manner (Table 11.7).   In the population of field voles
kept in the cage that was ^{90}Sr-contaminated, the mortality was higher and
dying-off of its population proceeded more rapidly than in the clean cage.
This feature was less distinct in the population of northern redbacked voles,
because there were differences in the irradiation dose due to skeleton-in-
corporated ^{90}Sr.

TABLE 11.7. Survival of wintering population of voles in a clean area and an area contaminated by ^{90}Sr,% /16/

Species	Territory	Months				
		May	June	July	Aug.	Sept.
Field voles	^{90}Sr-contaminated	76.1	57.1	41.2	28.5	0
	Control	100	84.5	–	46.1	3.8
Northern redbacked voles	^{90}Sr-contaminated	88.8	77.7	33.3	22.2	0
	Control	76.9	46.1	–	30.7	0

Note. The number of wintering voles in April was taken as 100%.

The difference in the life-spans of wintered field voles in the investigated populations was insignificant. The average life-span of these animals was 11.6 ± 0.8 and 12.5 ± 0.8 months in the ^{90}Sr-contaminated territory and clean cages, respectively.

Seasonal populations of voles are exposed to different irradiation doses /20/. Therefore, the survival on contaminated territory differs in voles born at different dates during the reproduction period (Table 11.8). The majority of voles born at the beginning and middle of summer do not survive until spring of the following year. The animals born at the end of the reproduction period account for 98% of the spring population. Only half of the voles in the ^{90}Sr-contaminated cage survived through winter and until spring of the following year, even among those born at the end of summer, in contrast to the low mortality of the control population.

TABLE 11.8. Ratio of age groups in the population of field voles in winter and spring and their survival in the ^{90}Sr-contaminated cage and in the control population,% /20/

Territory	Age groups	December	April	Survival in the second half of winter
Controls	Those born at the beginning and middle of summer	46.7	50.0	100
	Those born at the end of summer	53.3	50.0	87.5
^{90}Sr-contaminated	Those born at the beginning and middle of summer	14.8	2.0	6.6
	Those born at the end of summer	85.2	98.0	55.6

The mortality in the population of small rodents inhabiting an ^{90}Sr-contaminated territory depends not only on the accumulated level of these nuclides in their skeleton and different radiostabilities of different age groups, but also on the simultaneous effect of adverse climatic factors of the population. Animals that have been weakened by exposure to ionizing radiation perish more rapidly in periods that are critical in the life of their population (Table 11.9). The most critical moments in the life of rodents

are the periods of establishment of a permanent snow cover and spring
snowmelt /19, 32, 53, 54/. In winter lower numbers of these animals are
killed by the climatic conditions, but the effect of these conditions differs
with different species.   The data listed in Table 11.9 show that the death
of field voles on ^{90}Sr-contaminated territory was due to the effect of
ionizing radiation in all periods of the year.   The mortality in the population
of northern redbacked voles, exposed to a smaller dose of radiation, is more
strongly effected in autumn and spring by climatic conditions, while the
deadly effect of radiation predominates only in winter.

TABLE 11.9. Mortality in vole populations on ^{90}Sr-contaminated territory /16/

Species	Experimental territory	Percentage of perished voles with reference to the total population in the preceding season			
		autumn	winter	spring	total over the autumn-spring period
Field voles	^{90}Sr-contaminated	42.8	50.0	7.7	60.9
	Controls	11.7	6.3	3.4	17.6
Northern redbacked voles	^{90}Sr-contaminated	46.2	47.0	11.1	72.9
	Controls	50.2	10.6	21.4	70.2

Variation in the fecundity of populations
exposed to ionizing radiation

The effect of ionizing radiation on the fecundity of animals is studied by
different methods, but mostly under laboratory conditions with X-ray or
gamma-irradiation of experimental groups of animals and also by introducing radionuclides into the animals.   The fecundity in experimental groups
of animals was reported to be lower /6/. Data on the variation of animal
fecundity for exposure to ionizing radiation under natural conditions are
very limited.

X-ray and gamma-irradiation of populations of birds under natural conditions reduce their fecundity /104, 137/.   The incubation period in female
Sialia sialis irradiated with 200—600 r was longer than in the control
birds (14.4 and 13.9 days, respectively).   The number of hatched nestlings
in the experimental females was 20% reduced in comparison to the controls.

Irradiation of only the male mice in the experimental groups reduces
the numerical size of litter in females /63, 64, 95/.   X-ray irradiation with
500—700 r of male hamsters Peromyscus leucopus and P. mani-
culatus reduced the average numerical size of litter in the population
from 4.46 ± 0.25 to 3.40 ± 0.22 and from 5.03 ± 0.18 to 3.86 ± 0.41, respectively.
A striking effect of ionizing radiation on the gonads of male mammals is
also observed under natural conditions.   Verkhavskaya et al. /4/ discovered
partial sterility in male root voles (Microtus oeconomus) caught on
areas with high radium and uranium concentrations.   Long-term exposure
of bleaks (Alburnus alburnus) and Leucaspius delineatus to

small amounts of ^{238}U retards the fission of cells in testes, causes transformation of ovaries into testes, and produces intersexes /47, 48/.

Studies of reproduction patterns for populations of small rodents on a
contaminated territory established a decrease in the fecundity of animals
subjected to long-term irradiation. The number of embryos per female
decreased and the reproduction period of the population became shorter
/78, 80/. Incorporated radionuclides have different effects on the fecundity
of different species. The number of embryos per female decreased with
increasing ^{90}Sr concentration in the skeleton of northern redbacked voles,
but the same fecundity index remained substantially unchanged in common
field mice (Table 11.10).

TABLE 11.10. Number of embryos per female northern redbacked voles and common field mice as a
function of ^{90}Sr–^{90}Y concentration in the skeleton /45/

^{90}Sr–^{90}Y concentration in skeleton, μcuries, /g	Northern redbacked voles		Common field mice	
	number of animals	number of embryos	number of animals	number of embryos
$1 \cdot 10^{-1} - 9 \cdot 10^{-1}$	16	5.63±0.49	–	–
$1 \cdot 10^{-2} - 9 \cdot 10^{-2}$	24	7.04±0.32	95	6.58±0.2
$1 \cdot 10^{-3} - 9 \cdot 10^{-3}$	11	7.24±0.42	33	6.51±0.3
Control	14	7.85±0.32	22	6.48±0.14

Several investigations of rodents revealed the possibility of some increase in the numerical size of litter following irradiation with small doses.
According to Monastyrskii and Polovinkina /29/, there were an average of
10.5 embryos per female narrow-skulled vole on an area with natural
radiation background 5—7 times the normal value, as against 9.0 embryos in
the control population, while the numerical sizes of litters were, respectively,
8.3 and 7.2. Furthermore, on the territory with high radiation level females
with 13 or 14 embryos accounted for about 25%, whereas no females with
such large litters were discovered on the control area.

The litter of experimental northern redbacked voles with small content of
^{90}Sr–^{90}Y in their skeleton (($63.2 \pm 1.5) \cdot 10^3$ disintegrations/min · g) was
6.5 ± 0.6, as against only 4.5±0.5 in the controls /15/. In some cases the
females have an increased number of embryos even with high ^{90}Sr concentration in their skeleton. In the case of field voles maintained in an open-
air cage with addition of ^{90}Sr, the average litter of females with
$(959.6 \pm 1.2) \cdot 10^3$ disintegrations/min · g ^{90}Sr–^{90}Y in their skeleton was
6.2 ± 0.3, while females with $(434.8 \pm 5.1) \cdot 10^3$ disintegrations/min · g ^{90}Sr–^{90}Y
had average litter of only 4.0 ± 0.5 /16/. The apparent "stimulation" in the
case of field voles was probably their reaction to the disturbance introduced
into their hormonal or sexual system by exposure to ionizing radiation.
Unfortunately, the fate of these females' litters could not be traced.

Effect of ionizing radiation on development of animals
under natural conditions

The effect of ionizing radiation on the development of animals in their
natural habitat has been inadequately studied, altough its examination is of
practical importance.

The effect of long-term irradiation from a ^{137}Cs source on the growth of
animals was studied on an isolated population of iguanas Uta sansbu-
riana. Irradiation of the experimental area occupied by the lizards was
conducted continuously from July through November. The lizards were
irradiated with a dose of 100—1,200 r over the investigation period /136/.
An appreciable lag in the growth of young iguanas behind the controls was
discovered only at the end of October and in November, i.e., 5 months after
the beginning of the population's irradiation. In an experiment carried out
in this investigation, one-time doses up to 1,200 r did not produce any
changes in the lizards' growth, but irradiation with doses 1,300—1,600 r
slowed their growth. According to the experimenters' interpretation of results, the slower growth of lizards on the irradiated area could result from
the reduction in the amount of available food.

Studies into the effect of a one-time gamma-irradiation with ^{60}Co in
different doses on Sialia sialis nestlings aged 2 and 9 days revealed
that their growth and plumage development lagged behind the nonirradiated
control nestlings /142/. The difference between the experimental and
control groups was especially pronounced when the nestlings were irradiated
at the age of two days (Figure 11.4).

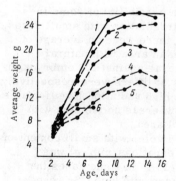

FIGURE 11.4. Effect of ionizing radiation on the growth of Sialia sialis nestlings irradiated at the age
of two days /142/:

1 — control; 2 — 300 r; 3 — 600 r;
4 — 1,500 r; 5 — 2,000 r; 6 — 3,000 r
(perished).

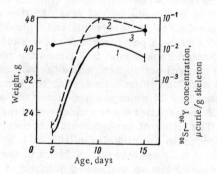

FIGURE 11.5. Average weight of Sturnus vulgaris nestlings (without the stomach contents)
as a function of the buildup of ^{90}Sr—^{90}Y in their
skeleton:

1 — experimental population; 2 — control population; 3 — ^{90}Sr—^{90}Y buildup in the skeleton.

The postembryonal development of animals inhabiting areas contaminated by radioactive fission products was studied in the development of
young starlings (Sturnus vulgaris). Nestlings aged 5, 10 and 15 days

were taken from artificial nesting sites placed on an area that had been
artificially contaminated with ^{90}Sr. Comparison of the average weight of
nestlings of different ages from the contaminated and control areas revealed
differences in their development. The growth of nestlings in the population
occupying the ^{90}Sr-contaminated area lagged behind the controls (Figure 11.5).
Filling of the nestlings' stomachs with food on both territories was compared
and revealed a difference (though not always statistically significant) in the
quantity of food received by the nestlings (Table 11.11).

TABLE 11.11. Average filling of the stomach of starling nestlings with food at various ages, g

Territory	Age, days					
	5		10		15	
	number of specimens	filling of stomach	number of specimens	filling of stomach	number of specimens	filling of stomach
Territory contaminated with ^{90}Sr to the level 1.8–3.4 mcuries/m^2 ...	19	8.7±0.4	18	13.6±0.5	14	11.1±0.3
Control	15	9.7±0.4	14	15.4±0.3	14	12.2±0.3

Effect of incorporated radionuclides on immunity, blood status and appearance of pathological phenomena in animals in a population

Studies into the effect of animal irradiation on their natural immunity
and susceptibility to endo- and ectoparasites have recently attracted increasing attention on the part of researchers. An answer to these questions
will explain the nature of radiation damage produced by irradiation of animal
populations with different doses and the animals' reaction to exposure to
ionizing radiation as a factor of the habitat.

Ionizing radiation is known to impair the natural resistance of an irradiated organism to infectious and parasitic diseases /9, 22, 51, 122/. The
increase in susceptibility of irradiated animals is due to a decrease in
their resistance to infections, not to any increase in the virulence of microorganisms in their body /51/. Irradiation increases the propagation rate
and activity of any infection that might remain latent under normal conditions /13, 66/. Thus, X-ray irradiation of goldfish (Carassius aura-
tus) with a dose of 10,000 r increases their susceptibility to pathogenic
bacteria Aeromonas salmonicida and ectoparasites Gyrodacti-
lus sp. /123/. Irradiation of chickens infected with Plasmodium
iophurae and P. gallinaceum with doses of 400 and 500 r exhibited a
marked increase in parasitemia /62, 133/. An analogous decline of immunity was observed in the case of laboratory mice irradiated with a dose of
550 r with respect to trypanosomas (Trypanosoma duttoni) and albino
rats with respect to both T. lewisii /101/ and pathogenic Leptospira
/9/.

Long-term penetration of radionuclides into the organism of laboratory animals produces the same phenomena as external X-ray and gamma-irradiation of animals. Long-term poisoning of laboratory animals with various amounts of ^{210}Po, ^{232}Th or ^{90}Sr produced a decline of their natural immunity to gastrointestinal and pulmonary infections /1, 7, 33, 34, 39, 94/.

An investigation of wild populations of rodents inhabiting areas contaminated by fission products, for incidence of endo- and ectoparasites, did not produce positive results. The degree of infestation of experimental animals did not differ from that of the controls, while a slight increase in the incidence of endoparasites in rodents (larvae of parasitic flies) on contaminated territory (discovered by Childs and Cosgrove /71/) could be attributed to natural populational differences.

Rodent populations inhabiting contaminated areas sometimes display multiple abscesses, mainly on lips and forepaws. These abscesses occur as a result of fights between the animals and appear on the sites of wounds inflicted with their incisors. The incidence of animals with abscesses in a population may serve as an index of the population's stability toward bacterial diseases. Statistical data processing did not reveal differences between animals inhabiting the contaminated territory and controls /80/.

High ^{90}Sr concentration in the skeleton of animals inhibits the ossification of cartilageous tissues even to the extent of producing strontium rickets /20/, which is the cause of frequent fractures of tubular bones in mammals and bone tumors /40, 46/. Strontium-90 concentrations of $10-500\mu$curies/kg produce carcinogenic transformation of bone tissue in mice and rats /30, 31, 55/.

There is one known case of osteosarcoma in wild rodents. A muskrat (Ondatra zibethica) was caught in March 1953 in the region of Lake White Oak, contaminated with fission products, and was found to suffer from a bone tumor on its right hind paw in the region of the distal portion of tibia with metastases into lungs and kidneys. The ^{90}Sr—^{90}Y concentration in its skeleton averaged 1.1μcurie/g bone, and the total content of these nuclides in the skeleton was 100μcuries. With this content of ^{90}Sr in the bones, the dose rate on bone tissues amounted to 37.2 rad/day /97/.

No osteosarcomogenic effect was detected in small rodents inhabiting ^{90}Sr-contaminated areas for a prolonged time, although the recorded buildup levels of ^{90}Sr in the skeleton of small animals considerably exceeded the minimum concentrations of this nuclide producing bone tumors in laboratory rodents. The maximum ^{90}Sr—^{90}Y concentration in the skeleton of small animals for which no tumor-formation was observed, was as follows (μcuries/g skeleton):

Common field mice	1.37
Northern redbacked voles	1.6
The middle-sized shrew (Sorex caecutiens)	1.43
Common shrews	up to 2.0—2.24

The absence of osteosarcomas in natural populations of small mammals may be due to their higher radiostability in comparison to laboratory animals, or is a reflection of the process of rapid elimination of sick animals from the population. Furthermore, the animals' life-span must also be taken

into account, since formation of tumors of visible dimensions requires a
fairly prolonged period of time. In the case of rats osteosarcomas usually
appear 200—500 days after their poisoning with carciogenic amounts of ^{90}Sr
/14/. The life-span of large rodents (such as muskrats) reaches 3 years
/44/, but the life-spans of voles and mice are 0.7 and 1 year /21/, respectively, only rare individuals living beyond one year.

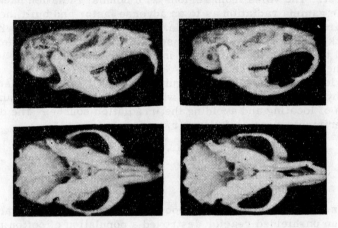

FIGURE 11.6. Disturbed growth of incisors in a field vole that lived one year
in an ^{90}Sr-contaminated area. The skull of a normal vole is shown at left.

In working with isolated populations of voles in an ^{90}Sr-contaminated
open-air cage /16/, some field voles aged 12—13 months exhibited irregularities in the growth of front teeth. Since the incisors of rodents are subject to continuous wear, they grow continuously, compensating the wear.
The irregularity in the growth of incisors in the ^{90}Sr-poisoned animals
manifested itself in their intensified growth in the upper jaw and falling-out
from the lower jaw (Figure 11.6). These disturbances were of a regular
nature and were discovered in 16% of the voles surviving to the age of one
year. No such phenomena were observed in the control population. Such
deviations in the development of incisors were also observed in laboratory
rats poisoned with ^{90}Sr /34/. The accelerated growth of incisors is a
reaction to irradiation of the animal's continuously growing tissues. As
yet, no disturbances in the blood system have been reliably detected in
populations of animals inhabiting contaminated territories under natural
conditions /71/.

Effect of ionizing radiation on the behavior
of animals in a population

The effect of ionizing radiation on animals and their distribution over a
territory with a high content of radioactive substances were studied by
many investigators. Raushenbakh and Monastyrskii /43/ demonstrated that

narrow-skulled voles, sampled from regions with a radiation background 5—7 times the natural value and from a territory with a normal background, reacted differently to additional gamma irradiation from a ^{137}Cs point source (dose rate 0.001 r/sec). The experiments were conducted in a cage, and the criterion of the voles' behavior toward the gamma radiation source was provided by the distance from the source to the site on which the animals built their nest. The voles from regions with normal radiation background reacted to the gamma radiation as to an adverse factor and built their nest as far away from the source as possible. Animals from regions with high radiation background proved to be less sensitive to the effect of gamma radiation and the site of their nest was independent of the location and distance of the gamma source.

The behavior of monkeys, rats and guinea pigs in a gamma radiation field was studied by Darenskaya et al. /10/, who demonstrated that the animals were capable of locating and avoiding the irradiation source. The dose rates in these experiments were as follows (r/sec):

For guinea pigs	0.0017
For rats	0.0127
For monkeys	0.008
	(irradiation of head)

Irradiation of a natural forest biogeocenosis with neutron and gamma quanta from an unshielded reactor destroyed a population of cotton rats and considerably reduced a population of hamsters Peromyscus leucopus and P. polionotus when a cumulative dose of about 4,300 rads was reached over a period of 21 days /125/. Irradiation of the biogeocenosis reduced the aggressiveness and caution of animals, impaired their motor reflexes and equilibrium, and caused loss of orientation and overall reduction of activity. Studies into the effect of lower irradiation doses (dose rate 3 r/day) on a forest biogeocenosis did not reveal any changes in the distribution of P. leucopus population due to the effect of ionizing radiation /67/.

The mobility of animals and of the distribution of populations of insects and rodents of eight species over a territory affected by a nuclear explosion was studied. Results showed that changes in the animals' mobility and the distribution pattern of populations depended on certain physical factors (including the height of plants and the overall plant coverage) and were not due to the effect of primary or residual radiation /57—59/.

Studies into the distribution of lizard populations before and after an atomic explosion on a testing ground in the State of Nevada, USA, showed that the number of lizards hatched from eggs after the explosion increased with increasing distance from the epicenter, but the change in the population density was not controlled by the effect of ionizing radiation but apparently by changes in the plant cover /135/.

In studies of the distribution of mammal populations over ^{90}Sr-contaminated territories, it was noted that the numerical strength of four rodent species did not depend on the areal contamination density, but obeyed the usual habitat distribution of species. Northern redbacked and common voles are more numerous on the outskirts of a forest, while common field-mice are more numerous in thickets of tall weeds. The distribution of the rodent species was controlled by the specific features of their feeding and by the forage capacity of their habitats /45/.

Reaction of zoocenoses to ionizing radiation and animals'
adaptation to ionizing radiation as an environmental factor

The accumulation of radionuclides by animals and the distribution of
radioactive substances in populations were discussed above. The characteristic features of radionuclide migration along food chains in a biogeocenosis, as well as age, sex, seasonal and specific differences in the accumulation of radionuclides by animals, result in the accumulation of different
amounts of nuclides in animals of different species and intrapopulational
groups. This in turn results in the formation of dissimilar dose loads
received by the animals and ultimately produces different biological effects.
The biological effect of ionizing radiation from incorporated radionuclides on
animal populations manifests itself in reduced life-span, survival, altered
fecundity, rates of development, and so on. Different species and groups of
animals are affected differently by ionizing radiation, and the reaction of
seasonal population compositions to this environmental (habitat) factor
varies.

A dissimilar reaction of organisms, even belonging to closely related
species, during exposure to ionizing radiation is controlled by their different
radiostabilities, which are liable to vary widely. Even closely allied species
of animals may differ significantly in their radiosensitivities. Owing to
these differences irradiation of a zoocenosis kills those species that are
less resistant to irradiation, whereas more highly radioresistant forms survive and thrive. These developments gave rise to variations in ecological
links in the biogeocenosis, specific relationships, and relationships between
plant and animal groups. These were noted in experimental investigations
with irradiation of freshwater and forest phytocenoses /49, 50, 112/. However, many aspects related to the course of biocenotic processes in land
zoocenoses during exposure to lethal and sublethal doses of ionizing radiation have been inadequately studied, especially when the radiation source is
represented by radionuclides accumulated by the cenosis. In particular
one must emphasize the limited nature of information concerning the range
of radioresistances of animals belonging to different species.

At present, the possibility of animals' adaptation to ionizing radiation
remains unsolved. According to some authors, the general biological
adaptation of highly organized animals to the effect of ionizing radiation is
improbable /2/. Others, for instance Raushenbakh and Monastyrskii /43/,
established that narrow-skulled voles living in the presence of a high natural
radiation background possess a higher radioresistance in comparison to
populations from habitats with normal radiation background. After many
generations have remained on a territory with high radiation background,
the animals acquire a genetically conditioned, high radiostability evolved
by these generations. The mechanism of such adaptation has not been yet
clarified. It may be related to natural selection and preservation of the
more radiostable individuals within the population /11/ or to modification
of intrapopulational phenomena (in the first place, altered ratio between
life-span and fecundity). It is known that a population of animals responds
to adverse environmental (habitat) factors by increasing the variability of
organisms. Prolonged exposure of a zoocenosis to ionizing radiation may
be expected to produce such an increase in the variability of animal populations. The enhanced variability must facilitate acceleration of the natural
selection and, possibly, the population's adaptation to radiation.

TABLE 11.12. LD$_{50}$/$_{30}$ for wild vertebrates

	Forest and steppe zones	Deserts	Reference
Pisces			
Carassius auratus..	2,315 (-20—25° C)	–	/124/
	3,500 (-15° C)		/68/
Amphibia			
Rana pipiens	700	–	/130/
Diemictylus viri- descens	1,460	–	/93/
Reptilia			
Elaphe obsoleta, Coluber constric- tor, Natrix sipedon	300—400	–	/73/
Terrapena carolina	1,250*	–	/60/
	850	–	/73/
Aves			
Chloris chloris	600	–	/25/
Carduelis carduélis	600	–	/25/
C.cannabina	400	–	/25/
Passer domesticus..	625	–	/25/
Serinus canarina....	500	–	/25/
Mammalia			
Marsupialia Didelphus mar- supialis	900*	–	/86/
Insectivora Cryptotis parva	795	–	/65/
Blarina brevicauda	716	–	/65/
Primates Tamarinus nigri- collis	200	–	/84/
Rodentia Cricetidae Reithrodontomys humulis	–	1,200	/85/
		874	/65/
Peromyscus leucopus	972	–	/65/
P.nuttali	976	–	/65/
P.polionotus	–	1,125	/85/
P.gossypinus	–	1,130	/85/
Microtus pinetorum	870	–	/65/
Oryzomys palustris	498	–	/65/
Sigmodon hispidus	893	–	/65/
Meriones unguicu- latus	–	1,175	/70/
Lagurus lagurus	760	–	/35/

TABLE 11.12 (contd.)

	Forest and steppe zones	Deserts	Reference
Heteromyidae			
Perognathus for-			
mosus	–	1,300	/83/
P. longimembris	–	1,520	/83/
Muridae			
Mus musculus	700	–	/85/
	744–795	–	/65/
Rattus norvegicus ..	818	–	/65/
Sciuridae			
Citellus tridecem-			
lineatus	700	–	/77/
C. major	600	–	/41/
Marmota monax	360*	–	/128/
Carnivora			
Procyon lotor	580	–	/86/
Urocyon cinereo-			
argentatus	710	–	/86/
Lynx rufus	600*	–	/86/

* According to the author's calculations.

Radioresistance of animals

The radioresistance of animals depends on many factors, including the physiological status, and ecological and evolutionary characteristics. Lethal doses of X-ray and gamma irradiation in the animal kingdom vary broadly, from a few hundred to several tens of thousands of roentgens /3, 42, 110/. $LD_{50}/_{30}$ for mammals, which are the most sensitive animals, lies in the range 200–1,000 r /8, 110/. The radiosensitivity of animals varies with their age /8, 110, 116, 129–131, 141, 142/, season of the year /20, 128/, ambient temperature /105, 108, 126/, and various other factors.

Table 11.12 provides information concerning $LD_{50}/_{30}$ values for different species of vertebrates. Even closely allied species belonging to the same order and family differ in their radiosensitivities. For example, rodents living in arid zones have a higher radioresistance than closely related species from other zones. Consequently, the degree of radiostability of animal populations and possibly entire zoocenoses may be expected to depend as a whole on the nature of the landscape. If this conclusion should prove correct, then one should expect significant differences in the consequences of irradiation for animal populations and zoocenoses, for the same contamination density of biogeocenoses in different geographical zones.

Variability in populations of animals inhabiting
a territory contaminated by radioactive fission products

The increased variability of plants under the influence of ionizing radiation is well known and widely used in radiational selection of plants /11/.
The particulars of the effects of ionizing radiation on the variability of
animals have been less fully studied than for plants. In 1933, Ushatinskaya-
Dekalenko /52/ established a direct relationship between the degree of
ionization of air due to high concentration of radioisotopes of Ra and other
natural radionuclides in the soil-plant cover and the bodily dimensions of
the mountain field-mice Sylvimus sylvaticus fulvipectus
(Table 11.13).

TABLE 11.13. Variability of mountain field-mice in the North Caucasus (calculated according to data in /52/)

Index	Georgian Military Highway, village of Gvilety			Mountain Ingushetia, village of Bisht and the Salgi R.		
	number of specimens	mean, mm	coefficient of variation	number of specimens	mean, mm	coefficient of variation
Length of:						
body	20	95.6±0.8	3.7	14	106.5±1.9	6.5
tail	17	99.9±1.4	5.6	14	111.3±1.6	5.5
foot	20	22.3±0.2	2.9	14	22.1±0.4	6.1
ear	20	15.3±0.2	4.4	14	18.5±0.2	4.8
nasal bones	20	9.0±0.6	3.0	13	10.2±1.3	4.7
dental row	20	3.9±0.1	2.1	14	4.2±0.1	4.3
Zygomatic cranial width .	19	12.9±0.1	2.9	9	13.5±0.2	3.9

Our statistical processing of the data reported by Ushatinskaya-Deka-
lenko showed that, besides an increase in the linear indices in the population
of mice in the gorge of the Salgi River (experimental area with high radiation background level), there was also a considerable increase in the degree
of their variability in comparison to the controls (Village of Gvilety).

Changes in the morphophysiological indices in a population of rodents
inhabiting an area with high natural radiation level was also noted by
Monastyrskii /28/. However, studies of a population of black rats (Rattus
rattus) inhabiting a territory with a high natural radiation level did not
detect any increase in the variability of the skeletal morphological characteristics of these rodents /87/.

Table 11.14 lists data on the variability of some morphological characters in populations of northern redbacked voles, field mice and common

TABLE 11.14. Variability of morphological indices in populations of mammals inhabiting a small area contaminated with ^{90}Sr /45/

Index	Northern redbacked voles				Field mice				Common shrews			
	control		contaminated area		control		contaminated area		control		contaminated area	
	mean	coefficient of variation	mean	coefficient of variation	mean	coefficient of variation	mean	coefficient of variation	mean	coefficient of variation	mean	coefficient of variation
Number of embryos	7.85±0.32	15.2	6.63±0.1	25.2	6.48±0.1	9.7	6.54±0.23	26.0	—	—	—	—
Body weight, g	22.6±0.3	7.5	23.4±0.6	12.2	—	—	—	—	6.51±0.07	10.4	6.52±0.07	10.4
Body length, mm	98.4±0.1	4.6	97.4±1.0	7.2	—	—	—	—	—	—	—	—
Length of ear, mm	—	—	—	—	14.2±0.1	3.2	14.3±0.2	5.4	—	—	—	—
Length of hind foot, mm	—	—	—	—	20.2±0.2	3.9	20.6±0.2	3.7	—	—	—	—
Length of tail, mm	—	—	—	—	80.2±0.6	4.8	80.5±1.1	6.9	—	—	—	—
Relative weight of spleen, %	0.38±0.02	28.9	0.2±0.02	41.0	—	—	0.35±0.02	41.5	0.41±0.02	20.0	0.33±0.03	48.7
Relative weight of liver, %	6.6±0.2	12.5	6.5±0.2	16.6	—	—	5.7±0.2	13.4	5.9±0.1	10.0	5.7±0.1	11.2
RBC, 10⁶	11.9±0.2	10.5	13.0±0.2	10.4	—	—	—	—	—	—	—	—
WBC, 10³	4.5±0.2	35.2	4.2±0.4	66.6	—	—	—	—	—	—	—	—
Agranulocytes, %	77.0±1.6	15.1	65.5±2.3	26.2	71.7±2.9	18.6	70.7±2.6	24.1	—	—	—	—
Total length of cranium, mm	23.0±0.1	1.1	23.3±0.1	2.4	—	—	—	—	—	—	—	—
Condylobasal length, mm	22.7±0.1	1.1	23.1±0.1	2.2	—	—	—	—	19.1±0.1	1.3	19.1±0.1	1.3
Diastema length, mm	7.01±0.04	3.9	6.9±0.1	4.2	—	—	—	—	—	—	—	—
Length of dental row, mm	4.56±0.02	3.0	4.63±0.02	3.4	—	—	—	—	—	—	—	—
Interorbital width, mm	3.71±0.02	2.4	3.76±0.1	5.3	—	—	—	—	—	—	—	—
Cranial height, mm	8.7±0.02	1.3	8.7±0.1	3.3	—	—	—	—	—	—	—	—
Life-span, months*	12.54±0.83	34.7	11.62±0.8	50.3	—	—	—	—	—	—	—	—

* For field voles.

shrews occupying an area artificially contaminated with ^{90}Sr
(0.6—3.4 mcuries/m^2). Under these conditions, there is an increase in the
coefficient of variation for the majority of populational characteristics,
indicating increased variability of animals in the population as a result of
exposure to external and internal irradiation from ^{90}Sr.

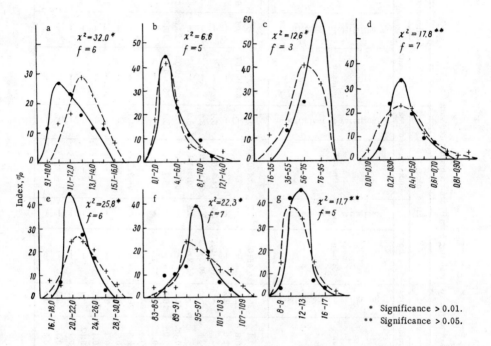

FIGURE 11.7. The variability of certain indices in a population of northern redbacked voles and of the life-
span of field voles. Distribution of size classes on "clean" (background) territory (——) and on ^{90}Sr-contami-
nated area (— — —):

f — number of degrees of freedom; x^2 — significance test; a — RBC; b — WBC; c — agranulocyte count;
d — relative weight of spleen; e — absolute weight of body; f — body length; g — life-span.

Figures 11.7 and 11.8 pertaining to northern redbacked voles show that,
under the influence of ionizing radiation, there is a change in the kurtosis
of these curves and an increase in the variability range in a population subjected to long-term irradiation. A certain number of individuals appear
among the population, the parameters of which exceed the corresponding
variability range in animals from the control population. In some cases
the curves are asymmetrical. The difference between the distribution
curves in the compared populations was proved to be statistically significant by the chi-squared (x^2) test.

A notable feature is the absence of any change in the variability of
certain indices (body weight and condylobasal cranial length) in the population of common shrews. This may indicate possible unequal biological

effects of ^{90}Sr on populations of animals of different species occupying different stages on the evolutionary ladder.

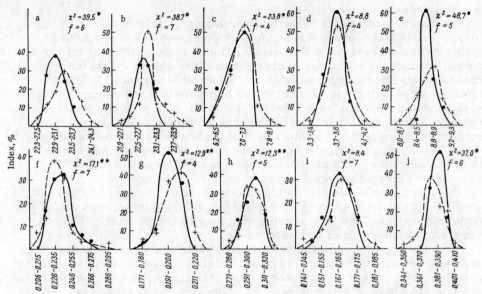

FIGURE 11.8. The variability of craniological indices of northern redbacked voles. Notation is the same as in Figure 11.7:

a — total length; b — condylobasal length; c — rostral length; d — interorbital width; e — cranial height. Ratios: f — cranial index; g — dental row length to condylobasal length; h — diastemal length to condylobasal length; i — interorbital width to condylobasal length; j — cranial height to condylobasal length.

The increase in the variability of morphological indices in a population is undoubtedly related to the change in the degree of variability of the animals' physiological and biochemical features.

The increase in the coefficient of variation of morphological indices in a population of rodents inhabiting a ^{90}Sr-contaminated area and the change in the distribution pattern of the features indicate that the effect of radiation as an extreme environmental factor (including those cases in which the territory with high radiation level occupies a large area) may possess evolutionary significance for the irradiated populations.

Current tasks of the radioecology of animals

Living organisms appear to have adapted to the natural radiation background as an environmental factor in the course of their evolution. Artificial heightening of the radiation level by injecting radionuclides into the environment is liable to create extreme habitation conditions for populations and for those zoocenoses characterized by a shift of the irradiation levels to extreme values at which, nevertheless, the animals' existence is still possible.

Local pollution of land and drainless inland water bodies with radioactive wastes may increase the radiation background markedly. Under these conditions animals are exposed to prolonged irradiation from internal and external sources.

The dissimilarity in the behavior of even closely related animal species toward a radioactive environment, the ecological features of organisms giving rise to differences in the accumulation of radionuclides by different animal species, the migration of different nuclides along food chains in a biogeocenosis and the seasonal status of a habitat significantly affect the distribution of ionizing irradiation doses in populations of different animal species within a zoocenosis. However, the dissimilar radioresistances of different animal species formed in the course of their evolution, and seasonal differences in the radiostabilities of populations of organisms may result in different consequences from the same irradiation doses applied to populations of animals of different species in different seasons of the year and in different geographical regions.

This gives rise to one task of the radioecology of animals: to study the specific, seasonal and geographical effects of different irradiation doses (from both external and incorporated emitters) on the structure of populations, animal communities and zoocenoses.

It has been proved experimentally that ionizing radiation increases the degree of variability of indicators in a population. This is liable to possess a definite evolutionary significance for the animals, because variability is a factor capable of affecting the course of the evolutionary process.

The increase of variability in animal populations may apparently create favorable conditions for natural radioselection of organisms. The selection in a population possibly proceeds toward preservation of radioresistant forms. Artificially created extreme habitat conditions for animals in the case of radioactive pollution of their habitats may lead to the formation of several adaptive mechanisms in populations. Adaptation of rodent populations in large natural regions with a high radiation level was detected by Raushenbakh and Monastyrskii /43/.

Cases are known of a rise in bacterial populations, which are stable toward high-power ionizing radiation in the water of reactors /11/.

The emergence of populations of microorganisms that are stable to antibiotics and ionizing radiation, and of insect populations that are stable to many poisons (for instance, development of populations resistant to DDT) suggests that adaptation of animals to various environmental factors may obey general biological laws, but that the adaptation mechanisms operating in animal populations belonging to different taxa and ecologic groups may be different. Detailed studies into the adaptation of animal populations in their natural environments to a radioactive habitat, and an examination of the adaptation mechanisms of populations constitute an important task of the radioecology of animals. This involves an investigation into the degree of variability of animals and populations exposed to ionizing radiation, and the natural selection mechanisms facilitating the adaptation of animals to this abiotic factor.

Since animal organisms are, on the whole, less stable to ionizing radiation than plants, one may expect in the first place disturbances under the influence of ionizing radiation in animal populations. Such changes may produce considerable shifts in the productivity and evolution of an entire biogeocenosis.   Consequently, studies into the effect of changes in a zoo-cenosis under the influence of ionizing radiation on changes in the entire biogeocenosis is another important task of radioecology.

References

1. Alekseeva, O. G. — In Sbornik: "Vliyanie radioaktivnogo strontsiya na zhivotnyi organizm," p. 156. Moskva, Medgiz. 1961.
2. Arbuzov, S. Ya. and A. M. Stashkov. — Voenno-meditsinskii Zhurnal, Vol. 4:26. 1966.
3. Bacq, Z. M. and P. Alexander.   Fundamentals of Radiobiology, Pergamon. 1961.
4. Verkhovskaya, I. N. et al. — Radiobiologiya, Vol. 5:720. 1965.
5. Verkhovskaya, I. N. et al. — Izvestiya Akademii Nauk SSSR, Seriya biologicheskaya, No. 2:270. 1967.
6. Golovinskaya, K. A. and D. D. Romashov. — Voprosy Ikhtiologii, Vol. 11:16. 1958.
7. Goloshchakov, P. V. and V. A. Nikiforova.   Raspredelenie i biologicheskoe deistvie radioaktivnykh izotopov (Distribution and Biological Effect of Radioactive Isotopes). Moskva, Atomizdat. 1966.
8. Gorizontov, P. D. et al. — In Sbornik: "Voprosy obshchei radio-biologii," p. 63. Moskva, Atomizdat. 1966.
9. Grigor'ev, I. I. — Meditsinskaya Radiologiya, No. 3:46. 1958.
10. Darenskaya, N. G. et al. — In Sbornik: "Voprosy obshchei radiobiolo-gii," p. 90. Moskva, Atomizdat. 1966.
11. Dubinin, N. P.   Problemy radiatsionnoi genetiki (Problems of Ra-diation Genetics). Moskva, Atomizdat. 1961.
12. Durkina, V. V.   Voprosy eksperimental'noi i klinicheskoi rentgeno-logii i radiologii (Problems of Experimental and Clinical Roent-genology and Radiology). Leningrad. 1966.
13. Dzhikidze, E. K. and L. S. Aksenova. — Meditsinskaya Radiologiya, No. 4:44. 1959.
14. Zakutinskii, D. I. — Meditsinskaya Radiologiya, No. 2:22. 1957.
15. Il'enko, A. I. — In Sbornik: "Ekologiya mlekopitayushchikh i ptits," p. 122. Moskva, "Nauka." 1967.
16. Il'enko, A. I. — Ibid., p. 126.
17. Il'enko, A. I. — Zoologicheskii Zhurnal, Vol. 67:1695. 1968.
18. Il'enko, A. I. — Radiobiologiya, Vol. 10:151. 1970.
19. Il'enko, A. I. and E. V. Zubchaninova. — Zoologicheskii Zhurnal, Vol. 42:609. 1963.
20. Il'enko, A. I. and G. N. Romanov. — Radiobiologiya, Vol. 7:90. 1967.
21. Karaseva, E. V. — Byulleten' MOIP. Otdelenie biologicheskoe, Vol. 60: 32. 1955.
22. Kiselev, P. N. and P. A. Buzin. — Meditsinskaya Radiologiya, No. 4:36. 1959.

23. Kruglikov, B. P. — In: Materialy I Respublikanskoi konferentsii po radiatsionnoi gigiene LatvSSR, p. 99. Riga. 1966.

24. Kruglikov, B. P. — In Sbornik: "Voprosy eksperimental'noi i klini-cheskoi rentgeno-radiologii," p. 160. Leningrad. 1966.

25. Kushniruk, V. A. — In Sbornik: "Biologicheskoe deistvie radiatsii," No. 1:81. Izdatel'stvo L'vovskogo gosudarstvennogo universiteta. 1962.

26. Lambrev, Zh. and N. Balabanov. — Nauchna tr. Vyssh. ped. in-ta, Vol. 3(2):119. 1965.

27. Mironov, O. G. — Okeanologiya, Vol. 5:715. 1965.

28. Monastyrskii, O. A. — In Sbornik: "Voprosy ekologii," p. 223. Tomsk. 1966.

29. Monastyrskii, O. A. and R. A. Polovinkina. — In Sbornik: "Voprosy zoologii," p. 224. Tomsk. 1966.

30. Moskalev, Yu. I. Biologicheskoe deistvie radiatsii i voprosy raspredeleniya radioaktivnykh izotopov (Biological Effect of Radiation and Problems of the Distribution of Radioactive Isotopes). Moskva, Gosatomizdat. 1961.

31. Moskalev, Yu. I. and V. N. Strel'tsova. — In Sbornik: "Raspredelenie, biologicheskoe deistvie i migratsiya radioaktivnykh izotopov." Moskva, Gosatomizdat. 1961.

32. Naumov, N. P. Ocherki sravnitel'noi ekologii myshevidnykh gryzunov (Comparative Ecology of Muridae). Moskva, Izdatel'stvo MOIP. 1948.

33. Nikiforova, V. A. and P. V. Goloshchakov. — In Sbornik: "Raspredelenie, biologicheskoe deistvie i uskorenie vyvedeniya radioaktivnykh izotopov," p. 35. Moskva, "Meditsina." 1964.

34. Novikova, A. P. and L. I. Burykina. Vliyanie radioaktivnogo strontsiya na zhivotnyi organizm (Effect of Radioactive Strontium on a Living Organism). Moskva, Medgiz. 1961.

35. Pegel'man, S. G. Vliyanie gamma-oblucheniya na organizmy (Effect of Gamma Irradiation on Organisms). Tallin. 1965.

36. Peredel'skii, A. A. and I. O. Bogatyrev. — Izvestiya Akademii Nauk SSSR. Seriya biologicheskaya, No. 2:186. 1959.

37. Peredel'skii, A. A. and I. O. Bogatyrev. — Byulleten' MOIP. Otdelenie biologicheskoe, Vol. 64:149. 1959.

38. Peredel'skii, A. A. et al. — DAN SSSR, Vol. 135:185. 1960.

39. Pigalev, I. A. Paper Submitted to the First International Conference on the Peaceful Uses of Atomic Energy. Geneva. 1955.

40. Pigalev, I. A. — Meditsinskaya Radiologiya, No. 3:80. 1958.

41. Pomerantseva, M. D. — Zhurnal Obshchei Biologii, Vol. 18:194. 1957.

42. Raevskii, B. Dozy radioaktivnykh izluchenii i ikh deistvie na organizm (Doses of Radioactive Radiation and Their Effect on an Organism). Moskva, Atomizdat. 1959.

43. Raushenbakh, Yu. O. and O. A. Monastyrskii. — In Sbornik: "Vliyanie idoniziruyushchikh izluchenii na nasledstvennost'," p. 165. Moskva, "Nauka." 1966.

44. Sludskii, A. A. Ondatra i akklimatizatsiya ee v Kazakhstane (Muskrat and Its Acclimatization in Kazakhstan). Alma-Ata. 1948.

45. Sokolov, V. E. and A. I. Il'enko.— Uspekhi Sovremennoi Biologii, Vol. 67:235. 1969.
46. Strel'tsova, V. N. and Yu. I. Moskalev.— Meditsinskaya Radiologiya, No. 11:39. 1957.
47. Telitchenko, M. M.— Nauchnye Doklady Vysshei Shkoly. Seriya biologicheskikh nauk, No. 1:114. 1958.
48. Telitchenko, M. M. and E. N. Levitova.— Vestnik MGU. Seriya biologicheskaya, No. 1:45. 1959.
49. Timofeev - Resovskii, N. V.— In Sbornik: "Problemy kibernetiki," p. 201. Moskva, "Nauka." 1964.
50. Timofeev - Resovskii, N. V. et al.— Trudy Instituta biologii Ural'skogo filiala Akademii Nauk SSSR, Vol. 9:202. 1957.
51. Troitskii, V. L. et al.— Radiatsionnaya immunologiya (Radiational Immunology). Moskva, "Meditsina." 1965.
52. Ushatinskaya - Dekalenko, R. S.— Izvestiya 2-go Sev.-Kavkazskogo pedagogicheskogo instituta, Vol. 10:83. 1933.
53. Formozov, A. N.— In Sbornik: "Rol' snezhnogo pokrova v prirodnykh protsessakh," p. 166. Moskva, Izdatel'stvo Akademii Nauk SSSR. 1961.
54. Formozov, A. N. Snezhnyi pokrov kak faktor sredy, ego znachenie v zhizni mlekopitayushchikh i ptits SSSR (Snow Cover as an Environmental Factor, Its Significance in the Life of Mammals and Birds in the USSR). Moskva, Izdatel'stvo MOIP. 1946.
55. Engstrom, E. et al. The Bone and Radioactive Strontium [Russian translation, 1962].
56. Alexander, G. V. et al.— Biol. Chem., Vol. 218:911. 1956.
57. Allred, D. M. and D. E. Beck.— J. Mammal, Vol. 44:190. 1963.
58. Alldred, D. M. and D. E. Beck.— Ecology, Vol. 14:211. 1963.
59. Allred, D. M. and D. E. Beck. Comparative Ecological Studies of Animals at the Nevada Test Site.— In: Radioecology. New York, Reinhold Publ. Corp. Washington, D. C., Amer. Inst. Biol. Sci. 1963.
60. Altland, P. D. et al.— Exptl. Zool., Vol. 118:1. 1951.
61. Beasley, T. M. and H. E. Palmer.— Science, Vol. 152:1062. 1966.
62. Bennison, B. E. and G. R. Coatney.— J. Nat. Malaria, Vol. 8:280. 1949.
63. Blair, W. F.— Ecology, Vol. 39:113. 1958.
64. Blair, W. F. and T. E. Kennerly.— Texas. J. Sci., Vol. 11:137. 1959.
65. Blaylock, B. G. et al. $LD_{50/30}$ Estimates for Indigenous Rodents and Shrews and for RF-Strain Mus musculus.— Oak Ridge National Laboratory, Publ. No. 170, ORNL-4007, 45—47. 1966.
66. Bond, V. P. et al.— Infectious Diseases, Vol. 91:26. 1952.
67. Bongiorno, S. F. and P. G. Pearson.— Amer. Midl. Nat., Vol. 72:82. 1964.
68. Bonet - Maury, P.— Compt. rend. Soc. Biol., Vol. 157:473. 1963.
69. Bourdeau, F. et al.— Health Phys., Vol. 11:1429. 1965.
70. Chang, M. C. et al.— Nature, Vol. 203:536. 1964.
71. Childs, H. E. and G. E. Cosgrove.— Amer. Midl. Nat., Vol. 76:309. 1966.

72. Clark, E. W.— Ann. Entomol. Soc. Amer., Vol. 51:142. 1958.
73. Cosgrove, G. E.— Radiation Res., Vol. 25:706. 1965.
74. Crossley, D. A.   Movement and Accumulation of Radiostrontium and
    Radiocesium in Insects.— In: Radioecology, p. 103. New York,
    Reinhold Publ. Corp. Washington, D. C., Amer. Inst. Biol. Sci.
    1963.
75. Crossley, D. A. and H. F. Howden.— Ecology, Vol. 42:302. 1961.
76. Davis, J. J. et al.   Some Effects of Environmental Factors upon
    Accumulation of Worldwide Fallout in Natural Populations.—
    In: Radioecology, p. 35. New York, Reinhold Publ. Corp. Washing-
    ton, D. C., Amer. Inst. Biol. Sci. 1963.
77. Doull, J. and K. P. DuBois.— Proc. Soc. Exptl. Biol. Med., Vol. 84:
    367. 1953.
78. Dunaway, P. B. and S. V. Kaye.   Studies of Small Mammal Populations
    on the Radioactive White Oak Lake Bed.— Trans. 26th N. Amer.
    Wildlife and Nature Resources Council, p. 167.   Washington. 1961.
79. Dunaway, P. B. and S. V. Kaye.— J. Mammal., Vol. 42:265. 1961.
80. Dunaway, P. B. and S. V. Kaye.   Effects of Ionizing Radiation on
    Mammal Populations of the White Oak Lake Bed.— In: Radio-
    ecology, p. 333. New York, Reinhold Publ. Corp. Washington, D. C.,
    Amer. Inst. Biol. Sci. 1963.
81. Eberhardt, L. L.— Nature, Vol. 204:238. 1964.
82. Fineman, Z. M. et al.   Use of Jack Rabbit as a Bio-Indicator of
    Environmental Strontium-90 Contamination.— In: Radioecology,
    p. 455. New York, Reinhold Publ. Corp. Washington, D. C., Amer.
    Inst. Biol. Sci. 1963.
83. Gambio, J. J. and R. G. Lindberg.— Radiation Res., Vol. 22:586.
    1964.
84. Gengozian, N. and J. Vatson.— Radiation Res., Vol. 19:220. 1963.
85. Golley, F. B. et al.— Radiation Res., Vol. 24:350. 1965.
86. Golley, F. B. et al.— Health Phys., Vol. 11:1573. 1965.
87. Grüneberg, H.— Nature, Vol. 204:222. 1964.
88. Hanson, W. C.— Northwest Science, Vol. 34:89. 1960.
89. Hanson, W. C.— J. Veterin. Res., Vol. 27:359. 1966.
90. Hanson, W. C.— Health Phys., Vol. 13:383. 1967.
91. Hanson, W. C. et al.— Health Phys., Vol. 10:421. 1964.
92. Hanson, W. C. and H. E. Palmer.— Health Phys., Vol. 11:1401.
    1965.
93. Jakowska, S. et al.— Zoologica, Vol. 43:115. 1958.
94. Jaroslow, B. N.— Infectious Diseases, Vol. 96:242. 1955.
95. Kalmus, H. et al.— Science, Vol. 116:274. 1952.
96. Kaye, S. V. and P. B. Dunaway.— Health Phys., Vol. 7:205. 1962.
97. Krumholz, L. A. and J. H. Rust.— A. M. A. Arch. Pathol.,
    Vol. 57:270. 1954.
98. Krumholz, L. A. et al.   Ecological Factors Involved in the Uptake,
    Accumulation and Loss of Radionuclides by Aquatic Organisms.
    The Effects of Atomic Radiation on Oceanography and Fisheries.—
    Nat. Acad. Sci. Washington, Publ. No. 551. 1957.
99. Lidén, K.— Acta Radiol., Vol. 56:273. 1961.
100. Martin, W. E. and F. R. Turner.— Health Phys., Vol. 12:621. 1966.

101. Naiman, D. N. — J. Parasitol., Vol. 30:209. 1944.
102. Neel, J. W. and K. H. Larson. Biological Availability of Strontium-90 to Small Native Animals in Fallout Patterns from the Nevada Test Site. — In: Radioecology, pp. 45—50. New York, Reinhold Publ. Corp., Washington, D. C., Amer. Inst. Biol. Sci. 1963.
103. Nelson, D. J. — Nature, Vol. 203:420. 1964.
104. Norris, R. A. — Auk., Vol. 75:444. 1958.
105. Owen, R. D. and N. H. Hildeman. — Radiation Res., Vol. 12:460. 1960.
106. Paakkola, O. and J. K. Miettinen. — Ann. Acad. Sci. Fennical, Ser. A., II Chem. (125), I. 1963.
107. Palmer, H. E. and T. M. Beasley. — Science, Vol. 149:431. 1965.
108. Patt, H. M. and M. N. Swift. — Am. J. Physiol., Vol. 155:388. 1948.
109. Pendleton, R. C. et al. — Nature, Vol. 204:708. 1964.
110. Platt, R. B. Ecological Effects of Ionizing Radiation on Organisms, Communities and Ecosystems. — In: Radioecology, pp. 243—255. New York, Reinhold Publ. Corp. Washington, D. C., Amer. Inst. Biol. Sci. 1963.
111. Pruitt, W. O. — Audubon Mag., Vol. 65:284—287. 1963.
112. Schultz, V. and A. Klement (editors). Radioecology. New York, Reinhold Publ. Corp. Washington, D. C., Amer. Inst. Biol. Sci. 1963.
113. Radionuclides in Alaskan Caribou and Reindeer 1962—1964.—Radiol. Health Data, Vol. 5:617. 1964.
114. Radionuclides in Alaskan Caribou and Reindeer 1963—1965. — Radiol. Health Data, Vol. 7:189. 1966.
115. Reichle, D. E. and D. A. Crossley. — Health Phys., Vol. 11:1375. 1965.
116. Rollasson, G. S. — Biol. Bull., Vol. 97:169. 1949.
117. Schultz, V. — J. Wildl. Managem., Vol. 28:45. 1964.
118. Schultz, V. — J. Wildl. Managem., Vol. 29:33. 1965.
119. Schultz, V. and V. Flyger, — J. Wildl. Managem., Vol. 29:39. 1965.
120. Scotter, G. W. — Canad. Field Naturalist, Vol. 81:33. 1967.
121. Sembrat, K. — Folia biol., Vol. 11:473. 1963.
122. Schechmeister, I. L. — Radiation Res., Vol. 1:401. 1954.
123. Schechmeister, I. L. et al. — Radiation Res., Vol. 12:470. 1960.
124. Schechmeister, I. L. et al. — Radiation Res., Vol. 16:89. 1962.
125. Schnell, J. H. The Effect of Neutron-Gamma Radiation on Free-Living Small Mammals at the Lockheed Reactor Site. — In: Radioecology, p. 339. New York, Reinhold Publ. Corp. Washington, D. C. 1963.
126. Smith, D. E. — J. Exptl. Zool., Vol. 139:85. 1959.
127. Smith, F. and M. M. Grenan. — Science, Vol. 113:86. 1951.
128. Smith, D. E. et al. — Radiation Res., Vol. 9:330. 1955.
129. Stearner, S. P. — J. Roentgenol. Radium Therapy, Vol. 65:265. 1951.
130. Stearner, S. P. — J. Exptl. Zool., Vol. 115:251. 1960.
131. Stearner, S. P. and S. A. Tyler. — Intern. J. Radiation Biol., Vol. 5:205. 1962.
132. Svensson, G. K. and K. Lidén. — Health Phys., Vol. 11:1393. 1965.
133. Taliaferro, W. H. et al. — J. Infect. Dis., Vol. 77:158. 1954.

134. Turner, F. B. et al.— Health Phys., Vol. 10:65. 1964.
135. Turner, F. B. and C. S. Gist.— Ecology, Vol. 46:845. 1965.
136. Turner, F. B. et al.— Health Phys., Vol. 11:1585. 1965.
137. Wagner, R. H. and T. G. Marples.— Auk, Vol, 83:437. 1966.
138. Watson, D. G. et al.— Science, Vol. 144:1005. 1964.
139. Whicker, F. W. and A. H. Dahl.— Health Phys., Vol. 9:892. 1963.
140. Whicker, F. W. et al.— Health Phys., Vol. 11:1407. 1965.
141. Willard, W. K.— Science, Vol. 132:148. 1960.
142. Willard, W. K. Relative Sensitivity of Nestling of Wild Passerine
        Birds to Gamma Radiation.— In: Radioecology, p. 345. New York,
        Reinhold Publ. Corp. Washington, D. C., Amer. Inst. Biol. Sci.
        1963.

Chapter 12

RADIOECOLOGY OF FARM ANIMALS

Radioecology of farm animals is concerned with the effect of ionizing radiation on farm animals and radionuclide migration along food and biological chains, in which these animals constitute a link.

When examining the migration of radionuclides in different natural landscapes one requires information on various radioecological constants describing the migration. Such information is very useful, since it is used in preparing measures for limiting the transport of radionuclides along fodder and food chains.

This chapter deals with the results of studies into the effects of various environmental factors on the influx of ^{90}Sr from soils into plants and its migration from plant cenoses into the organisms of cows, sheep and goats.

The studies were conducted on a natural meadow by applying ^{90}Sr to its surface. After application the meadow was divided into three plots, one of which was left in its natural state while the second and third were plowed, the soil layer containing ^{90}Sr being moved to depths of 25—30 and 50—60 cm, respectively. The plowed plots were sown with seeds of cultivated plants, while the unplowed plot was covered with its natural vegetation. In this way the experimenters artificially created three types of cenoses differing in their radioecological characteristics.

Since these cenoses differed radioecologically, the behavior of ^{90}Sr in them could be assumed to differ in the various cenoses. Mixing of ^{90}Sr in the arable horizon proceeded comparatively uniformly on the plot plowed to a depth of 25—30 cm. In the plot plowed to a depth of 50—60 cm ^{90}Sr was displayed in the subarable horizons, whereas on the natural meadow the radionuclide remained in the top sod for a prolonged time (Table 12.1).

The distribution of ^{90}Sr over the soil profile differed according to tillage and the buildup of this nuclide in the plants likewise differed, the largest intake of ^{90}Sr into the plant occurring from the surface layer. This was especially characteristic of the natural meadow cenoses, where the added ^{90}Sr remained practically concentrated in the sod and in many cases was taken up by the plant without dilution with soil calcium; this did not occur in plowed or rototilled plots.

Rototilling or plowing of meadows with ^{90}Sr applied to their sod modifies the phytocenosis and the distribution of ^{90}Sr in the soil, the plant being almost completely destroyed while ^{90}Sr is moved into the soil to the arable depth. The production of artificial meadows in place of plowed-up natural meadows is accompanied by the formation of new cenoses. Tillage of soil to different depths combined with the sowing of perennial grasses makes it possible to reduce plant intake of ^{90}Sr by a factor of approximately 2—4 (Table 12.2).

TABLE 12.1. Distribution of ^{90}Sr over the soil profile in different farmlands, %

Depth, cm	Virgin meadow	Artificial meadow		Rototilled natural meadow
		plowed down to 25 cm	plowed down to 60 cm	
0–5	94.8	40.0	0.2	43.0
5–10	3.7	25.5	0.1	42.0
10–15	0.9	20.0	0.1	12.0
15–20	0.2	13.0	0.2	1.0
20–25	0.2	1.5	0.2	–
25–30	0.1	–	1.7	–
30–35	0.1	–	1.8	–
35–40	–	–	10.0	–
40–45	–	–	21.0	–
45–50	–	–	18.0	–
50–55	–	–	27.1	–
55–60	–	–	9.6	–

TABLE 12.2. Intake of ^{90}Sr by sown perennial grasses for different nuclide distributions over the soil profile with reference to its content in the natural meadow plants, %

Depth of ^{90}Sr in the soil profile, cm	With respect to content in 1 kg dry matter	With respect to Sr units
Superficial (virgin land)	100	100
0–15 (rototilling)	45	40
0–25 (plowing)	27	33
35–60 (plantation plowing)	25	24

TABLE 12.3. Accumulation of ^{90}Sr by different plants in the natural cenosis with reference to its content in couch grass, %

Plant species	With respect to content per 1 kg dry matter	With respect to Sr units
Couch grass	100	100
Sheep's fescue	200	353
Bluegrass	146	132
Reedgrass	101	86
Awnless brome	58	50
Hop clover	383	118
Meadow pea	410	163
Vernal sedge	175	223
Drooping silene	444	432
Meadow crowfoot	341	786
Germander speedwell	730	871
Willoweed	254	72

Accumulation of ^{90}Sr by plants depends on their specific properties (Table 12.3). Among natural grasses, the highest ^{90}Sr concentrations occur in firm-bunch grasses (sheep's fescue, bluegrass) and the lowest in rhizomatous grasses (couch grass, reedgrass and awnless brome). Hop clover and meadow pea accumulate more ^{90}Sr per unit weight of their biomass, but they are close to rhizomatous grasses with respect to strontium units. Drooping silene, meadow crowfoot and germander speedwell are capable of absorbing very large amounts of ^{90}Sr from soil.

Similar variations in the buildup of ^{90}Sr related to specific properties were also observed in sown fodder crops (Table 12.4).

TABLE 12.4. Accumulation of ^{90}Sr by different plants in the artificial cenosis with reference to its content in Timothy grass, %

Plant species	With respect to content per 1 kg dry matter	With respect to Sr units
Timothy grass	100	100
Alfalfa	510	54
Red clover	470	54
Corn	227	46
Sunflower	220	30
Beans	373	82
Potatoes (tubers)	12	82
Beet:		
root	64	100
tops	164	54
Oats:		
grain	20	54
straw	137	110
Peas:		
seeds	10	37
straw	446	64

The buildup of ^{90}Sr in plants is closely related to their age (phenophase), attaining its maximum in the period of spring growth of herbs, decreasing in the flowering and heading period, and again increasing in the period of autumn tillering (Table 12.5).

TABLE 12.5. Accumulation of ^{90}Sr in perennial grasses with reference to its content in the spring tillering period, %

Phenophase	Month	With respect to content per 1 kg dry matter	With respect to Sr units
Spring tillering	June	100	100
Flowering	July	71	70
Fructification	August	57	51
Autumn tillering (aftermath)	September	104	97

TABLE 12.6. Buildup of ^{90}Sr in the skeleton and muscles of ruminants (with respect to introduced amount) for long-term ingestion of the radionuclide, %

Duration of animal poisoning, days	Cattle			Sheep			Goats		
	Ca in ration, g	^{90}Sr content in the skeleton	in 1 kg muscles*	Ca in ration, g	^{90}Sr content in the skeleton	in 1 kg muscles*	Ca in ration, g	^{90}Sr content in the skeleton	in 1 kg muscles*
15	–	–	–	1.6	87.4±5.0	0.76±0.07	1.5	69.3±13.9	0.90±0.09
30	6	65.0–69.8	0.15±0.01	2.0	77.4±1.0	0.60±0.11	1.7	63.3±4.7	0.88±0.17
60	–	–	0.21±0.03	2.0	54.9±3.9	0.44±0.10	1.7	42.6±4.0	0.55±0.26
105	10	49.3	–	–	–	–	–	–	–
120	–	–	–	2.0	39.0±2.1	0.37±0.04	1.8	23.5±3.6	0.34±0.05
200	12	40.7–45.7	–	–	–	–	–	–	–
240	–	–	–	4.0	14.4±0.6	0.21±0.04	4.0	10.5±1.9	0.57±0.10
270	15	37.1	–	–	–	–	–	–	–
365	–	–	0.12±0.01	5.0	9.0±0.9	0.22±0.02	4.5	7.8±0.6	0.35±0.10
530	20	13.2	0.08±0.01	–	–	–	–	–	–
550	–	–	–	6.5	5.3±0.4	0.15±0.06	5.5	4.3±0.2	0.32±0.6
590	30	12.5	–	–	–	–	–	–	–
730	–	–	0.04±0.003	–	–	–	7.0	2.8±0.3	0.28±0.04
1,080	–	–	–	7.0	3.4±0.2	0.20±0.03	–	–	–
1,560	–	–	–	–	–	–	7.0	1.5±0.2	0.22±0.04
1,680	–	–	–	7.0	2.9±0.2	0.20±0.06	7.0	1.4±0.1	0.20±0.03

* With reference to the diurnal ingestion of ^{90}Sr.

The inclusion of ^{90}Sr in the subsequent trophic chains, especially in the organism of herbivorous farm animals, thus depends on the phytocenosis components, age of plants, their growing site and the fraction of crop.

Investigations of the migration of ^{90}Sr from vegetation of different meadow cenosis into the organism of cattle, sheep and goats were carried out on growing and adult animals.

The experimental results showed that the buildup of ^{90}Sr in the skeleton of ruminants depends essentially on their species and age (Table 12.6).   The buildup of ^{90}Sr is most intensive in the bone tissue and muscles of young growing organisms, and is somewhat higher in the skeleton of sheep than of goats.   On the other hand, the buildup in the muscles of goats is higher than in sheep.   These differences in the intensity of incorporation of ^{90}Sr in animals of different species were probably conditioned by their different life-spans and the features of their mineral metabolisms.

Since the ecology of ruminants is based on their close trophic connection with meadow flora, the fate of ^{90}Sr migrating from soil to the organism is considerably affected also by ecological conditions.   The effects of different ecological factors on the migration of ^{90}Sr from fodder into milk was examined in experiments on cows.   The experimental cows were divided into three groups.   The first group was given hay rations from a natural meadow, was stabled in the autumn-winter period, and pastured out in the spring-summer period.   The second and third groups were respectively fed with mixed and concentrated rations from fodder grown on an artificial meadow and fodder grains, and the cows were stabled throughout the year. The cows were given ^{90}Sr with feeds grown on the above-described experimental plots; irrespective of the season, type of feeding and mode of utilization, these feeds constituted only a part of the ration.

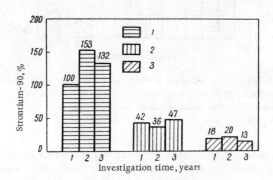

FIGURE 12.1. Ingestion of ^{90}Sr by cows maintained on different rations compared with its ingestion by cows maintained on the hay ration:

1 — hay ration; 2 — mixed ration; 3 — concentrated ration.

The experimental data on the ingestion of ^{90}Sr by cows show that the ^{90}Sr content in the ration is liable to fluctuate markedly depending on the type of phytocenosis, composition of the ration, and utilization of fodder land (Figure 12.1).   Thus, the ^{90}Sr content in the hay ration of the first experimental

year was 2.3 and 5.5 times larger, respectively, than in the mixed and concentrated rations. The ingestion of ^{90}Sr by the cows maintained on the
above rations varied relatively little over the three-year observation period
(the variation factor did not exceed 1.5). The ingestion of ^{90}Sr by the experimental cows remained likewise practically constant by seasons of the
year, although it varied widely in individual months. These variations were
apparently due to several factors, and require further study.

The ingestion of ^{90}Sr by the cows is liable to vary, depending on the
utilization of fodder lands. The largest ingestion of ^{90}Sr occurs when the
cows are pastured out. For instance, the ingestion of ^{90}Sr by cows maintained on the hay ration is 20% larger when the cows are pastured out than
when they are stabled, the difference being probably due to consumption of
larger quantities of fodder in the pasture period, on the one hand, and partial
utilization of the meadow sod containing 70—90% ^{90}Sr added to the soil surface, on the other. In the latter case the animals apparently receive some
of the ^{90}Sr deposited on the meadow surface along the shortened radioactive
sod-animal chain. As yet, it is impossible to precisely determine the
relative ingestion of ^{90}Sr by the cows via direct consumption of sod, but it
may be significant, depending on the conditions of the spring-summer period.

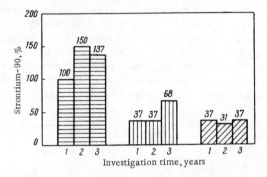

FIGURE 12.2. Elimination of ^{90}Sr from milk in cows maintained on different rations compared with its elimination by
cows maintained on a hay ration. Notation is the same as
in Figure 12.1.

The concentration of ^{90}Sr in milk is closely related to the type of meadow
phytocenosis, and the feeding and maintenance of lactating cows. As seen
from Figure 12.2, the maximum ^{90}Sr content occurred in the milk of cows
maintained on the hay ration, the relative fluctuations from one year to
another being more pronounced than in the case of cows maintained on the
mixed and concentrated rations and stabled. This implies that the stabling
of dairy cattle throughout the year makes possible the strictest monitoring
of the ingestion of ^{90}Sr and produces milk with a relatively constant content
of this radionuclide. Strontium-90 concentration in cow milk is not subjected to any significant seasonal variations. The pasturing of cows on
natural meadow increased the concentration of ^{90}Sr in their milk by 40%.

The amount of ^{90}Sr migrating into the milk increases with increasing
duration of nuclide ingestion (Table 12.7). For instance, cows maintained
on the concentrated ration eliminated 0.33 and 0.25% ^{90}Sr with 1 liter milk,
respectively, in the first and second years of observations, but they eliminated 0.46% of the diurnally ingested amount of ^{90}Sr during the third year.
A similar pattern of ^{90}Sr elimination from milk was observed in cows maintained on mixed and hay rations. The mechanism of this phenomenon is
too obscure, but it may be assumed that it is due to a change of ratio between
the buildup of ^{90}Sr in the readily and sparingly exchangeable portions of
skeleton. In the first days of long-term ingestion of ^{90}Sr, its content in the
readily exchangeable part of bones considerably exceeds its content in the
fixation depot /2, 3/. The buildup of ^{90}Sr in the skeleton proceeds intensively
at this time and the fraction of nuclide moving from the skeleton into the
milk is relatively small. In the course of time the ^{90}Sr content in the skeleton
increases, and consequently the removal of ^{90}Sr from the skeleton into the
milk and excreta increases.

TABLE 12.7. Effect of the type of ration and duration of the ingestion of ^{90}Sr on its elimination from cow milk

Type of ration	Duration of observations, years	Ca in the ration, g	^{90}Sr per 1 liter milk with reference to diurnal ingestion, %
Hay		41	0.16±0.02
Mixed	1	45	0.14±0.01
Concentrated		25	0.33±0.09
Hay		60	0.15±0.01
Mixed	2	56	0.17±0.01
Concentrated		50*	0.25±0.01
Hay		48	0.18±0.01
Mixed	3	44	0.23±0.02
Concentrated		18	0.46±0.04

* 25 g calcium was added to the ration in the form of $Ca_3(PO_4)_2$.

Comparison of data on the elimination of ^{90}Sr from 1 liter of milk shows
the transition of the nuclide into the milk to be more pronounced in cows
maintained on the concentrated ration than in cows maintained on the mixed
and hay rations. The differences are readily explained by taking into
account the amount of available calcium in the rations. In the case of inadequate or low influx of calcium to the digestive tract, the elimination of
^{90}Sr from milk and its buildup in the skeleton occur in larger amounts than
in the case of high or excessive intake of calcium /1/. Such intake and
elimination of ^{90}Sr by cows were confirmed by experimental data obtained
with pregnant sheep (the experimental scheme for the sheep was similar
to that used for the cows).
Maximum ^{90}Sr concentration was observed in the skeletons of newborn
lambs from sheep maintained on the hay ration during pregnancy (Table 12.8).
The retention of ^{90}Sr in the skeleton of lambs born to sheep maintained on
the mixed and concentrated rations was smaller by factors of 4.2 and 4.5,
respectively.

TABLE 12.8. ^{90}Sr concentration in the skeleton of newborn lambs as a function of the ration of pregnant sheep with reference to ^{90}Sr content in lambs born to mothers maintained on the hay ration

Ration	Ca in the ration, g	^{90}Sr in the skeleton, %
Hay	9.5	100
Mixed	8.5	24
Concentrated	8.0	22

Thus, in this experiment too, the hay ration proved to be less favorable from the standpoint of incorporation of ^{90}Sr in the skeleton than the mixed and concentrated rations.

The results of the above investigations testify that the intensity of ^{90}Sr migration in different links of trophic chains is controlled by the type of phytocenosis, the specific properties of its dominant plant species, the modes of their utilization by farm animals, as well as the latter's age and physiological status.

Further studies in the field of agricultural radioecology will yield more details of the effect of various environmental factors on the migration rate of the major radionuclides along biological and food chains.

References

1. A n n e n k o v, B. N. — In Sbornik: "Raspredelenie i biologicheskoe deistvie radioaktivnykh izotopov," p. 151. Moskva, Atomizdat. 1966.
2. B u l d a k o v, L. A. and R. A. E r o k h i n. — Ibid., p. 159.
3. Z a p o l ' s k a y a, N. A. et al. — Ibid., p. 55.

Chapter 13

RADIOECOLOGY OF LANDSCAPES
IN THE FAR NORTH

Introduction

The Far North, abounding in natural resources, for a long time failed to attract the attention of radioecologists. It was only at the end of the 1950s and beginning of the 1960s that the first results were published dealing with the radiation situation in this vast region.

The trend of radioecological investigations in the Far North was determined by the unusually high (in comparison to temperate latitudes) concentration of long-lived artificial radionuclides (^{137}Cs, ^{90}Sr, etc.) and (as found subsequently) natural radionuclides (^{210}Pb, ^{210}Po) in the lichen—deer—man chain.

By current notions, these radionuclide concentrations, though exceeding between 10- and 100-fold the concentrations of radionuclides in analogous links of other chains, are still nevertheless only $^1/_{100}$ of the concentrations that would call for protective measures. Nevertheless, radioecological data on the migration of radionuclides in the northern chain deserve close attention, since they contribute to the solution (owing to their high concentration) of important theoretical and applied problems in radiobiology and radiation protection.

Radioecology deals with radionuclide migration in biogeocenoses and is based on observations of actual radioactivity levels, as well as studies into the mechanisms of nuclide assimilation (adsorption, intakes) by every consecutive link in the chain, the transport, distribution and elimination of nuclides within individual links (metabolism), determination of absorbed doses by organs and tissues, and establishment of the biological effects of these doses. At the same time, observations of the space-time movement of radioactive tags in biogeocenoses offer unique opportunities for qualitative and quantitative characterization of the most complex relationship between the components of nature (general aspects of nutrition, the movement of the components themselves, etc.).

The above-listed radioecological problems encountered in the Far North (the list is far from being exhaustive) and, in particular, those pertaining to the lichen— deer — man chain have already been the subject of several investigations, which were summarized at an International Symposium held in Sweden in April 1966 /23/.

Such investigations were initiated in the Soviet Union in 1958 and have been mainly concentrated at the Leningrad Institute of Radiation Hygiene.

Air

The atmosphere is the primary medium entered by explosion products from aerial (surface) nuclear tests as well as from natural radioactive emanation from soil. The behavior of radioactive substances in the atmosphere largely controls their subsequent migration along food chains, including fallout levels and their dynamics in individual links of the biosphere (especially in the first years following the explosions as regards the plant link of the chain in the case of superficial contamination).

The aerial penetration path is most significant in the case of short-lived radionuclides. However, air itself as a source of direct influx of radioactive substances into the organisms of animals and man occupies one of the last places with respect to long-lived natural and artificial radionuclides.

TABLE 13.1. Total long-lived beta activity of aerosols and ^{90}Sr concentration in air in the Far North (March–April 1962)*

Area	Total beta activity, 10^{-15} curie/liter	^{90}Sr, 10^{-17} curie/liter
Murmansk	1.0–1.2	0.5–0.6
Arkhangelsk	1.8–1.9	0.9–1.0
Naryan-Mar	0.8	0.4
Syktyvkar	4.3–4.8	2.2–2.4
Amderma	4.5–6.1	2.3–3.0
Kamennyi (Tamalo-Nenets National District)	1.7–1.8	0.8–0.9
Dikson (Taimyr National District)	2.5–3.0	1.2–1.5
Norilsk	3.9–4.0	1.9–2.0
Mean for the Far North (from 16 samples)	2.8 (from 0.8 to 6.1)	1.4 (from 0.4 to 3.0)
Moscow /3/	3.0–6.0	–
Leningrad /3/**	2.0–5.0	1.0–2.5
USA /3/**	4.8	2.0–5.0
Finland /20/	1.0–3.0	–

* Results for two samples in each year.
** Radiochemical data provided for Leningrad and the USA.

When the radioecology of the Far North was first studied, the seasonal and geographical variability of the radioactivity of air was already known. Most significant was the fact that the concentrations of long-lived nuclides in air and fallout within the polar circle were only $\frac{1}{2}$–$\frac{1}{3}$ those in the temperate latitudes of the northern hemisphere. Our measurements of the concentration of long-lived aerosols followed by determination of ^{90}Sr from the age of fission products during the period of nuclear weapon tests do not refute this fact. Table 13.1 shows that the radioactivity of air in various areas of the Soviet Far North does not exceed its radioactivity in the central regions of the USSR and neighboring northern countries. No differences were detected with respect to individual northern districts. The health significance of the air contamination levels as regards direct penetration of nuclides into the human organism is negligible. This conclusion becomes even more convincing in view of the fact that, at approximately the same time, some inhabitants of the Far North ingested quantities of ^{137}Cs with their food that were tens of thousands of times larger.

Gamma background

The natural gamma background on the earth's surface (at a height of 1 m) is known to vary widely in different regions of the globe, from 5—10 to 1,000 μr/hr. According to Gusev's calculations /1/ the mean dose rate on the USSR territory is 10 μr/hr, which is in satisfactory agreement with extensive measurements.

There are very few regions with mean gamma dose rates exceeding 15—20 μr/hr. Among regions with higher radiation backgrounds is the State of Kerala in India, where the mean annual irradiation dose of gonads is 830 mrads, and also the State of Minas-Gerais in Brazil, where the dose rate of gamma radiation fluctuates between 30 and 1,000 μr/hr in sites of human habitation /8/. Such high gamma-background values have never been reported within the polar circle.

The concentration of natural radionuclides in soils responsible for the gamma background remains practically unchanged with time, and so the gamma background of a given area is fairly constant during many generations. Any appreciable changes in this background provide a reliable indication of additional radioactive contamination by gamma emitters.

TABLE 13.2. Gamma background in the Far North (within wooden dwellings), μr/hr

Area	December 1964—March 1965	January—February 1967
Chukchi (Chukot) National District, Anadyr	5	7
Yakut ASSR:		
Oetung	11	12
Ust-Tatta	10	—
Ulakhan-Chistai	7	—
Taimyr National District, Khatanga	6	6
Yamalo-Nenets National District	7	—
Nenets National District ......:...............	6	—
Komi ASSR, Vorkuta	—	7—20
Murmansk Region:		
Krasnoshchel'e	8	9
Kanevka	13	—
Loparskaya	6	—

The gamma background from fallout in the first 2 or 3 years following nuclear explosions is the main radiation factor for the bulk of the population in temperate latitudes. For instance, even in the case of ^{137}Cs, readily migrating along food chains, its external gamma irradiation exceeds the irradiation due to incorporated radionuclides. The amount of ^{137}Cs deposited on the ground as a result of tests (about 100 mcuries/km^2 by 1965 in temperate latitudes of the northern hemisphere) generated a human external irradiation dose of approximately 6 mrad/yr (declining only 2.5—3% annually), while at the same time the internal irradiation from ^{137}Cs did not exceed 5 mrad/yr in the period of maximum fallout (1964), declining in sub-sequent years by 30—40% annually. The above relationship between external and internal irradiation for the human population does not always hold in the Far North (see below). The absolute gamma-background levels in the Far North, both outdoors and within wooden dwellings (1962—1966), lie within the natural range (6—20 μr/hr) above which the contribution made by artificial fallout cannot be reliably isolated (Table 13.2).

The mean natural gamma background in arctic regions may be accepted as 10 μr/hr, i.e., equal to its mean value for the USSR territory. It may be demonstrated by calculation from the ^{137}Cs content in soil that the tests increased the gamma background in the North by approximately 5%, while the fallout of fresh fission products in 1963—1964 raised this background by as much as 50% /4/.

Soil

Soil is the principal depot of radionuclides deposited on the earth's sur-face, and its radionuclide content is commonly used to estimate the conta-mination of the following links in ecological chains, when the influx of radio-nuclides proceeds through the root systems of plants. However, the litera-ture lacks information on the health significance of soil from the radioeco-logical standpoint for conditions in the Far North, except for data on radio-active contamination levels in soils. Nonetheless, it may be assumed even a priori that the role of soil in the migration of radionuclides in the Far North must differ significantly from the role of soils in temperate latitudes, where the migration of radionuclides along biological chains has been studied in more detail. The presence of permafrost in the subsoil and a prolonged snow cover period create special conditions for emanation of radon and for fixation of fallout; economic utilization of the territory (absence of agriculture) and the physicochemical properties of soils likewise possess special features. Penetration of long-lived fission products into the organism of inhabitants of central regions and urban populations in the North is mainly by ingestion of radionuclides with grain products (bread), but in the Far North there is some correlation between the content of fission products in grains and soil contamination levels, because all grain products are brought from outside.

Soil samples were taken (from depths of 5—10 cm) together with the plant cover in different districts of the Far North over a period of several years beginning with 1962. The data for 1965, when extensive global fallout had already ceased, are presented in Table 13.3. The indicated ^{137}Cs and ^{90}Sr concentrations, determined by radiochemical analytical techniques, thus reflect the cumulative buildup of these radionuclides over all the preceding years and provided the background for the formation of the radiation situation in the investigated districts.

TABLE 13.3. ^{90}Sr and ^{137}Cs contents in the soil-plant cover in March 1965 (soil layer 0–5 cm), mcuries/km^2

Area	^{137}Cs	^{90}Sr	^{137}Cs/^{90}Sr
Murmansk Region, Krasnoshchel'e	46±4	35±4	1.4±0.1
Nenets National District, Kotkino	51±9	41±10	1.3±0.1
Yamalo-Nenets National District, Muzhi	48±8	34±6	1.4±0.1
Taimyr National District, Khatanga-Kresty ...	28±6	18±4	1.6±0.1
Yakut ASSR, Chokurdakh-Oetung	34±6	23±3	1.5±0.1
Chukchi (Chukot) National District, Anadyr	34±2	21±2	1.6±0.1
Northern Hemisphere, March 1964 /2/:			
70–60°	–	32	–
60–50°	–	51	–
50–40°	–	58	–
40–30°	–	47	–
Mean	–	44	–

Throughout the Arctic seaboard (67–70°N) the radioactive contamination levels of soils by ^{137}Cs and ^{90}Sr in 1962–1965 averaged only $^2/_3$–$^1/_2$ those in the northern hemisphere. There were also significant (factor of two) differences in the degree of contamination of soils in different northern districts. The heaviest contamination was recorded in the Murmansk Region and Nenets National District, the least contamination in eastern regions (Taimyr, Yakutia, Chukchi Peninsula). These differences are in agreement with analogous variations in the quantity of atmospheric precipitation. Thus, in the years (1962–1965) of intensive fallout of radionuclides, the amount of atmospheric precipitation was 500 mm/yr in the Murmansk Region, but only 200 mm/yr in the eastern regions (such as northern Yakutia).

The ^{137}Cs/^{90}Sr ratio in the soil-plant cover was almost identical in all the regions and equal to the ratio (1.3–1.6) of these radionuclides in the fallout. Of the total quantity of these nuclides on the ground surface in 1965, approximately one-half was present in the soil and the other half in the vegetation (most lichens). A 100% interception of deposited ^{137}Cs by lichens /20/ can only occur at very high vegetation density (> 0.5–1.0 kg dry matter per 1 m^2). Such a density of the biomass of lichens is scarcely typical of northern landscapes.

The dose of external gamma irradiation due to ^{137}Cs (3 mrad/yr) was calculated from its content in the soil (up to 50 mcuries/km^2 in 1965) and the ratio between the ^{137}Cs contamination density and irradiation dose (1 mcurie ^{137}Cs/km^2 corresponds to 0.1 mr/yr /22/) with shielding coefficient 0.63 /1/. (The dose is almost double in central regions of the USSR.)

Vegetation

The specific features of the radiation situation in the Far North are not controlled by fallout levels (which were found here to be even $^1/_3$–$^1/_2$ less than in temperate latitudes of the northern hemisphere) but by the distribution and fodder significance of plants, which are distinguished by high accumulation of radionuclides from fallout; this was quite convincingly demonstrated by us in 1961–1962.

TABLE 13.4. ^{137}Cs and ^{90}Sr concentrations in lichens (Cladonia) in countries of the northern hemisphere (mean values + standard error, mostly from 3–5 samples per point), ncuries/kg dry matter

Area	Sampling date	^{137}Cs		^{90}Sr
		Gamma-spectro-metric analysis*	Radiochemical analysis	
Murmansk Region	1961	–	26±7	–
	June 1962	30±2	–	7±4
	January 1963	86±1	–	8±1
	May–June 1963	206±25	48±10	7±1
	July 1964	124±5	–	–
	March 1965	94±5	50±1	9±1
	December 1965	63±5	27±4	15±2
	March 1966	–	34±4	2±1
Komi ASSR	1961	–	11	–
	October 1962	76±12	–	4±1
	May 1963	186±18	–	–
	September 1963	150±10	–	–
	June 1964	106±6	74±6	9±1
Nenets National District	March 1962	–	17	–
	June 1962	38±3	–	4±2
	March 1965	79±11	43	10
Yamalo-Nenets National District	1961	–	17–19	–
	June 1962	44±3	–	3±1
	March 1965	87	36	9
Taimyr National District	November 1962	18±1	–	3
Yakut ASSR	1960	–	–	5
	December 1965	39±5	24±7	13±4
Sakhalin	1959	–	–	2.3
Leningrad Region	September 1965	82	–	10–40
Alaska /9/	August 1959	–	27	3.3
	August 1960	–	26±4	2.0±0.5
	June 1961	–	37±1	1.4±0.3
	July 1962	–	11±1	3.5±0.4
	July 1963	–	17±2	6.7±3.0
	July 1964	–	27±2	7.8±3.2
	April 1965	–	30±2	5.9
	July 1965	–	32±2	6.4±0.8
Greenland /5/ Denmark /5/	September 1962	16–18	– –	5–7 (300–7,600 Sr units)
Norway /14/	Autumn 1969	36 (31–42)	–	–

TABLE 13.4 (contd.)

| Area | Sampling date | ^{137}Cs | | ^{90}Sr |
		Gamma-spectro-metric analysis*	Radiochemical analysis	
Sweden /17/ (the values are stated in ncuries/m², which is similar to ncuries/kg dry matter)	September 1961	30	–	–
	September 1962	40	–	–
	September 1963	58	–	–
	June 1964	70	–	–
	September 1964	80	–	–
	September 1965	82	–	–
Finland /25/	1960	40	11±4	–
	July 1961	40 (8–61)	5.5 (2.5–10)	–
	September 1964	–	–	5–7
Finland /20/	1961	16±1	–	–
	1962	22±1	–	–
	1963	37±0.5	–	–
	1964	64±5	–	–
	1965	56±2	–	–

* Without deducting the contribution of other isotopes to the "^{137}Cs channel."

Tundra vegetation comprises many hundreds of species and their detailed radioecological characterization calls for further investigations. In view of the fodder significance of individual plants in reindeer herding, we combined the measurement data for two groups of plants: fruticose lichens (Clado-nia and Cetraria), which serve as the staple food of deer over the 7–8 winter months, and herbaceous tundra vegetation (without further sub-division into species), which is the deer's staple fodder in the summer period. The branches and leaves of shrubs, providing supplementary fodder, scarcely differ from herbs in their radionuclide content.

We gave preference to the investigation of lichens, since their radioacti-vity (calculated for dry matter) was higher than that of herbs by one order of magnitude and even more, and the controlling role of lichens in the accumu-lation of ^{137}Cs by deer and deer herders had become evident already by 1962. Lichens in the Far North do not differ from lichens growing elsewhere with respect to the buildup of long-lived radionuclides. Identical lichen species, for instance, in the Leningrad Region, are not less contaminated than they are within the polar circle, but in the latter case extensive utilization of lichens as fodder for deer and extensive consumption of venison by the human population come to the fore.

Over a period of 5 years (1962–1966) we took 120 samples of lichens and 55 samples of other vegetation from different regions of the Far North and analyzed them for their contents of ^{137}Cs (radiochemically and gamma-spec-trometrically), ^{90}Sr, ^{210}Po, and total alpha-activity. Some 150 samples of lichens were investigated from available publications over the same period in all the other countries. A summary of all the available information is provided in Table 13.4.

The number of analyses per point (in time and location) seldom reached 10, but was usually 3—6, and therefore the conclusion concerning the geographical features and dynamics of radioactive contamination of lichens cannot be categorical. Moreover, the analysis was complicated by nonstandard sampling and measurement techniques. A considerable nonuniformity in the distribution of ^{137}Cs over the height of plants was discovered only at the end of 1964 and therefore could not be taken into account in the samplings performed during the previous period. Cesium-137 concentration in the upper part of a lichen turned out to be nearly 5 times its concentration in the lower part; in contrast, the vertical distribution of ^{90}Sr, ^{210}Pb and ^{210}Po in plants was uniform.

In comparable periods of time the total radionuclide content in lichens was approximately the same for all regions in the Far North of the USSR and in all the northern countries (Alaska, Canada, Finland, Norway, Sweden and Denmark), providing incontrovertible proof of the global origin of the contamination. There is a possibility that the heavier (double) contamination of lichens and herbaceous plants by ^{137}Cs and ^{90}Sr in the West (Finland, Sweden, Norway, Murmansk Region of the USSR) than in the East (Yakutia, Chukchi Peninsula and Alaska) corresponded to differences in the quantity of atmospheric precipitation. The higher (at least by one order of magnitude) contents of ^{137}Cs, ^{90}Sr, ^{210}Pb and ^{210}Po in lichens in comparison to herbs was due to the effects of several factors resulting in such a marked accumulation effect. In contrast to herbaceous annuals, the aerial part of which is exposed to atmospheric precipitation for 3 months, the lichens are perennial plants (growing for dozens of years), resulting in a prolonged accumulation of deposited nuclides on their surface. Furthermore, the surface of lichens exposed to direct sorption of precipitation proves to be 10 to 100 times greater per unit mass than in herbaceous plants.

The following experiments yielded noteworthy results. If a lichen is immersed in a neutral solution of ^{137}Cs and ^{90}Sr salts without carriers until equilibrium is established (for several days), then the concentration of ^{137}Cs in the lichen recorded at the end of the experiment turns out to be 200 times its concentration in the ambient solution, whereas the concentration of ^{90}Sr in the lichen is only 20 times its concentration in the solution. Thus, the ^{137}Cs/^{90}Sr ratio in the lichens is 10/1, in contrast to equal concentrations of these radionuclides in the original solution. Nearly the same nuclide ratio is characteristic of the upper part of the plant under natural conditions in the case of superficial contamination. The fractionation of artificial radionuclides characterized by the preeminent buildup of ^{137}Cs in plants, especially on their edible portions, and their subsequent, even more marked discrimination in the body of deer result in a considerably greater contribution of incorporated ^{137}Cs (in comparison to incorporated ^{90}Sr) in the terminal link of the chain, i. e., in the body of the human population consuming the venison. This circumstance is one of the major features of the radioecological situation in the Far North.

Our experiments show that the extraordinarily important role of ^{137}Cs in the radiological situation in these regions is far from being a merely temporary phenomenon due to transient superficial contamination of the lichens. Contrary to all expectations, the lichens, although devoid of a root system, are capable of very efficient absorption of radionuclides from the upper,

the most heavily contaminated soil layer, the ratio of radionuclides in such
an accumulation remaining the same as in the superficial contamination.
Three months after addition of radionuclides the lichens contain 2—4% of the
^{137}Cs introduced into the soil, but only 0.1—3% of ^{90}Sr (Table 13.5).   On the
other hand, according to Krieger /16/, herbaceous vegetation assimilates
from soil about 0.01% ^{137}Cs and 0.1% ^{90}Sr over the same period of time.

TABLE 13.5. Migration of ^{137}Cs and ^{90}Sr from soil into lichens

Experimental period, days	Amount of ^{137}Cs with respect to the amount of this nuclide added to soil, %	Amount of ^{90}Sr with respect to the amount of this nuclide added to soil, %
2	0.013	0.0004
7	0.024	0.0017
30	0.170	0.0011
45	4.230	0.1140
83	1.430	0.2330
105	1.990	0.1500

A very important radioecological parameter permitting a forecast of
the buildup levels of radionuclides and their dynamics in any link of the
food chain, including plants, is provided by the effective semidecontamination
period $T_{1/2\text{eff}}$.

Decontamination of lichens from radiation substances is due to their
washout by atmospheric precipitation, consumption by herbivorous animals
and radioactive decay.   The radioactive decay is the only process whose rate
is known, the half-lives of ^{137}Cs and ^{90}Sr being approximately 30 years.
There is still no concensus of opinion among radioecologists concerning the
magnitude of $T_{1/2\text{eff}}$ for lichens; for ^{137}Cs, it is 9—12 years according to
Swedish specialists /17/, 3—13 years according to American scientists /10/,
but 4—5 years as reported by Finnish investigators /20/.   The disagreement
is not accidental, since $T_{1/2\text{eff}}$ can usually be determined with sufficient
accuracy from field observations of lichens in the absence of additional
contamination.   Such a possibility has presented itself only in recent years.
Furthermore, studies of the contamination dynamics for lichens involves
great difficulties in the selection of representative samples and also on
account of the nonuniform distribution of ^{137}Cs along the height of lichens.
This is also confirmed by data on the contamination level of lichens; these
data differ markedly from the expected dynamics of fallout intensities and
specific features of radionuclide accumulation by the lichens.

Until the beginning of 1965 (two years after the cessation of nuclear
tests) the concentration of ^{137}Cs and ^{90}Sr in lichens did not display any convincing downward trend, although in herbaceous vegetation in 1964 it was
only slightly more than one-half its value in 1963.   The buildup level of
^{90}Sr and ^{137}Cs in the lichens probably reached its maximum in 1965.   The
decline in the concentration of these radionuclides in the lichens appearing
in 1966 proved to be statistically significant.   This is evident from data on

their buildup dynamics in the Murmansk Region and simultaneous decline of the concentration of these nuclides in venison, which in our opinion best (and very rapidly on account of the short $T_{1/2\,eff}$ in deer) reflects the contamination dynamics in lichens.   Calculated from the content of ^{137}Cs in venison, the value of $T_{1/2\,eff}$ for this radionuclide in the lichens was found to be 1.5 years in the first years after the nuclear tests.   It is taken as 14 days for herbaceous vegetation, which becomes decontaminated (following the atmosphere) with $T_{1/2\,eff} = 9-12$ months.

After the fallout of ^{137}Cs increased, the value of $T_{1/2\,eff}$ for lichens likewise inevitably increased.   It is already possible to distinguish a second decontamination index depending on the vertical migration rate of ^{137}Cs in soil and equal to the latter's semidecontamination period (the semidecontamination period of soil is defined as the time during which the amount of radionuclide taken up by the plants from soil via their root systems is halved). The decrease in radionuclide absorption from soil by plants with the passage of time results from the migration of radionuclides over the soil profile (including their migration beyond the root zone), increased strength of the sorption of radionuclides in the soil, and so on.   The semidecontamination period of soil for ^{137}Cs is 25 years.   Assuming that the decontamination of lichens from superficial ^{137}Cs contamination proceeds exponentially with effective semidecontamination period $T_{1/2\,eff} = 2.5$ years, the migration of ^{137}Cs from soil into lichen amounts to 2% of its reserve in soil, while the decrease in the migration of ^{137}Cs from soil into lichens proceeds with the semidecontamination period of soil equal to 25 years, it is possible to estimate the ^{137}Cs content in the organisms of the native population in subsequent years.   Provided there is no fallout from new nuclear tests, the sum of the two indices yields the following ^{137}Cs concentrations in lichens, venison, and (with a lag of 1 year) deer herders, in percentages of the maximum observed in 1965 (in 1966, in the case of deer herders): 100% in 1965, 58% in 1967, 35% in 1969, 22% in 1971, 14% in 1973, 9% in 1975, 6% in 1977.   •

In accordance with this prognosis, the contribution of ^{137}Cs from soil to its total content in the lichens after 1977 will exceed the residual superficial ^{137}Cs contamination.   It is likewise possible that the actual decline of ^{137}Cs concentrations in the lichens, venison and deer herders will proceed somewhat more slowly than the above prognosis, since the initial value of $T_{1/2\,eff} = 2.5$ years in the lichens may be variable, increasing with time.   The influx of ^{137}Cs from soil is likewise assumed to be minimal (2%), based on data of short-term experiments.   Further observations are necessary in order to obtain more reliable data with a view to assessing the accuracy of the above tentative prognosis.   However, there is no longer any doubt concerning the distinctly inertial nature of the decontamination of the starting link in the northern chain, i. e., the lichens, controlling the decontamination rate of the following links from ^{137}Cs.   The ^{137}Cs concentration declines far less rapidly than in other land ecological systems.

The ecological parameters of lichens with respect to ^{210}Pb and ^{210}Po (sorption during deposition from the atmosphere, soil absorption, decontamination rate, etc.) have been inadequately studied.   Even data on the migration of these nuclides into the lichens are very controversial. Jawarowski /15/, for instance, analyzed a few specimens and relates the concentration dynamics of these nuclides and also the geographical variability of

their content in lichens to radioactive fallout from nuclear tests. His conclusions were not confirmed by our investigations. Long before the beginning of global radioactive fallout following nuclear weapons tests, the concentration of ^{210}Pb and ^{210}Po in lichens (and similarly in the bones of deer)
was approximately the same as in 1950—1960 (Table 13.6).

TABLE 13.6. Dynamics of ^{210}Pb concentrations in the links of its biological chain in the Far North (with
refererence to dry matter in the case of lichens, wet matter in the case of bones, ncuries/kg), %

Object of investigation	1900	1900—1945	1958—1966
Lichen	14.7(1)	20.6±8.1(3)	8.4±0.8(12)
Bones of deer	7.5(1)	9.2±2.7(6)	9.5±1.9(11)
Bones of native inhabi-			
tants of the Far North...	0.55±0.02(5)	—	0.12±0.02(9)

Note. Figures in parentheses denote the number of samples.

Thus, the natural (unrelated to nuclear tests) origin of at least the bulk
of ^{210}Pb and ^{210}Po is beyond doubt. On the other hand, the mechanism of the
contamination of lichens by these nuclides is still not clear. Similar to
contamination by artificial radionuclides, the contamination in this case
apparently proceeds by superficial and soil influx, although the quantitative
role of each of these paths remains unknown. ^{210}Po is present in lichens in
equilibrium with ^{210}Pb, about 90% of the total alpha activity being due to ^{210}Po.
For all the investigated regions of the Far North (similar to the Leningrad
Region) the concentration of these radionuclides lies between 2 and
20 ncuries/kg dry matter. The number of analyzed samples is still very
limited for clarifying the geographical patterns involved in the concentrations
of ^{210}Pb and ^{210}Po.

Besides ^{90}Sr, ^{137}Cs, ^{210}Pb and ^{210}Po the lichens accumulate other long-lived
and medium-lived radionuclides, including ^{55}Fe, ^{144}Ce, ^{65}Zn, ^{7}Be, uranium isotopes, etc. However, investigation of their behavior is of lesser interest,
since the absorbed dose due to their buildup in deer and human population
is insignificant.

Our data concerning the content of various radionuclides and concentration dynamics of radioactive substances in other plants used for human
food in the Far North (including potatoes, cabbages, berries, mushrooms) did
not disclose any special features in comparison with the central regions of
the USSR.

Reindeer

Of all investigated foodstuffs used by the inhabitants of the Far North
(water, fish, cow milk, beef, vegetable products), reindeer meat and by-
products are the main source of ^{137}Cs, ^{90}Sr, ^{210}Pb and ^{210}Po. The ecological

relationship between the concentration of incorporated radionuclides (especially ^{137}Cs) in people and the consumption of nuclides with venison is so close, that if one of the quantities is known one can determine the other without any appreciable error. For that reason radioecological studies concentrated on reindeer.

In the Far North of the USSR, reindeer is not just an exotic animal (for hunting) or a model for investigation, as in some countries, but a staple. Three-quarters (2,000,000) of the world herd of reindeer are bred in the USSR. The production of venison provides nourishment for hundreds of thousands of people in the North, not to speak of the use of reindeer for transport and other purposes.

The first radionuclide measurements in samples of venison were carried out in the USSR (similar to other countries) in 1958—1959. They revealed that the concentration of different radionuclides (including ^{90}Sr) in the bones of deer was approximately one order of magnitude higher than in cows, sheep and pigs.

In 1961—1962 the radioecology of reindeer became the subject of numerous investigations on account of the resumed nuclear tests, and consequently extensive information has been amassed /23/.

The principal radioecological factors (resources and ingestion levels of individual nuclides, their resorption, distribution, metabolic rate, ex- cretion; geographical and seasonal variations in radioactivity levels and their dynamics) for reindeer have been evaluated from a large amount of experimental data. About 2,000 venison samples had been investigated by 1966.

With the simplified feeding pattern, a deer consumes an average daily amount of about 3 kg dried lichen during the eight winter months (mostly Cladonia alpestris, Cl. rangiferina, Cetraria islandica). Without any significant error, the content of radionuclides in this quantity of lichen can be regarded as their daily ingestion by the deer.

During the two or three summer months, the deer's ration mainly con- sists of grasses and leaves of shrubs. Owing to their low radioactivity (only $^1/_{10}$ that of lichens) the ingestion of nuclides by deer in summer may be regarded as negligible. One could also distinguish the transitional spring and autumn months, during which reindeer graze on grasses, lichens and leaves of shrubs.

The sharp seasonal change in the specific composition of fodder with different concentrations of radioactive substances is reflected in the very characteristic seasonal variation in the concentration of incorporated radionuclides in reindeer, which does not occur to such a degree in other farm animals. For instance, ^{137}Cs concentration in the muscles of deer declined from 61 ± 8 to 3.4 ± 0.3 ncuries/kg toward the end of summer 1964 in Komi ASSR, increasing again to 81 ± 7 ncuries/kg toward the end of the next winter season.

The seasonal variation of ^{137}Cs concentration in venison has been very thoroughly studied in Sweden and Alaska /10, 11/. Concentrations of ^{137}Cs in different arctic regions differ by factors of between 3 and 20. Since $T_{1/2\,eff}$ of ^{137}Cs in reindeer is 20—30 days and the specific activity of fodder in different countries is almost the same, the cause for geographical differen- ces in seasonality should be sought only in deviations from the above

TABLE 13.7. Contents of ^{137}Cs in muscles and ^{90}Sr in bones of reindeer

Area	Slaughter time	^{137}Cs		^{90}Sr		
		number of samples	ncuries/kg	number of samples	Sr units	ncuries/kg
Murmansk Region	1960	—	—	1	430	—
	1961	1	24	1	—	40
	June 1962	3	33±2	4	390±125	51±15
	January 1963	3	39±2	4	545±73	71±10
	May 1963	2	48±7	3	643±120	84±18
	February 1964	6	80±3	8	1,780±100	147±10
	July 1964	3	22±6	3	1,111±27	145±30
	December 1964	6	98±4	—	—	—
	February 1965	5	96±8	10	1,414±74	134±8
	October 1965	12	74±4	12	1,270±69	—
	April 1966	13	79±4	15	1,800±75	—
	November 1966	10	57±2	10	980±64	—
Nenets National District	1960	—	—	1	133	—
	1961	5	13±2	6	—	50±20
	June 1962	9	11±2	9	—	29±6
	December 1962	4	18±3	3	—	52±19
	December 1964	13	53±2	—	—	—
	February 1965	2	56±20	15	860±57	86±8
Komi ASSR	1960	—	—	3	203	—
	1961	57	21±1	29	—	44±10
	May 1962	4	16±2	7	—	55±10
	November 1962	14	13±1	13	144±23	33±10
	May 1963	2	33±2	4	427±60	50±6
	June 1963	6	38±4	—	—	—
	September 1963	4	23±1	3	429±11	39±6
	January 1964	4	45±10	—	—	—
	May 1964	4	61±8	11	1,109±73	93±24
	July 1964	3	12±2	—	—	—
	August 1964	20	3.4±0.3	—	—	—
	May 1965	12	81±7	—	—	—
	October 1966	19	30±3	19	530±101	—
Yamalo-Nenets National District	1961	18	9±2	6	—	38±11
	June 1962	3	14±1	—	—	—
	December 1964	3	66±4	4	1,242±86	127±10
	February 1965	1	55	—	—	—
Taimyr National District	1961	1	8	—	—	—
	April 1962	2	4±1	2	—	18±4
	December 1962	3	13±3	2	—	39±10
	December 1964	12	22±2	13	465±24	45±1
	January 1965	1	36	—	—	—
	October 1966	14	16±6	24	580±118	—
	February 1967	4	20±3	—	—	—

TABLE 13.7 (contd.)

Area	Slaughter time	^{137}Cs		^{90}Sr		
		number of samples	ncuries/kg	number of samples	Sr units	ncuries/kg
Yakut ASSR	1959	—	—	2	190±50	—
	October 1964	—	—	4	734±36	—
	November 1964	8	17±3	9	540±33	65±8
	January 1965	1	12	—	—	—
	October 1965	6	13±1	—	—	—
	November 1965	4	11±1	10	600±20	—
	November 1966	18	6±1	18	710±110	—
Chukchi National District	June 1962	10	8±1	19	—	53±8
	December 1964	10	25±2	10	625±19	88±3
	June 1966	6	21±3	—	—	—
Sakhalin	1959	—	—	2	200—500	—
Kamchatka Region	1961	4	8±1	11	—	20±10
	December 1965	5	17±2	5	470±37	—
Alaska	November 1960 /7/	6	9±3	21	29—53	7—13
	December 1961 /7/	5	26±1	45	30—106	7—23
	April 1963 /11/	—	22	13	160—560	—
	December 1964 /24/	15	12±1	82	225—535	—
	May 1965 /10/	—	35±9	—	—	50—70
Finland	April 1959 /18/	—	20	—	200	—
	February 1960 /18/	—	35	—	213	—
	March 1961 /21/	8	18±1	—	380	—
	April 1962 /21/	—	17	—	—	—
	April 1963 /21/	—	48	—	—	—
	April 1964 /20/	60	50±1	—	—	—
	April 1965 /20/	4	72±3	—	—	—
	December 1965 /20/	4	55±3	—	—	—

simplified feeding pattern of reindeer in different regions. In some regions (such as the Murmansk Region) reindeer feed on lichens not only in winter, but also in summer. On the other hand, in the north of Yakutia the role of lichens in the ration is small, even in the winter period. In both cases, the seasonal differences in the ^{137}Cs concentration of fodder are obliterated, which in turn obliterates the seasonal variation in the concentration of the nuclides in venison.

Seasonal fluctuations in the concentration of ^{137}Cs in venison considerably complicate analysis of geographical differences with respect to individual regions and northern countries. A prerequisite for an examination of such differences is an analysis of data obtained in identical seasons of the year or, more precisely, in comparable periods of the annual grazing cycle. In transitional feeding periods (transition from lichens to grasses at the beginning of summer and from grasses to lichens at the beginning of autumn) the concentration of radionuclides (such as ^{137}Cs) in venison varies appreciably even in the course of 5—7 days, on account of the disturbances in the established equilibrium between the ingestion and elimination of radio-nuclides. This equilibrium is most stable at the end of the winter period, when the radioactivity level reaches its maximum and consequently also the ^{137}Cs concentration can be measured with maximum accuracy.

Data on the contents of ^{137}Cs in muscles and ^{90}Sr in bones of reindeer in different countries are listed in Table 13.7.

Cesium-137. This is the most important artificial radionuclide with respect to man as the final link in the main food chain in the Far North from the standpoint of forming the dose load, and therefore the data obtained for this nuclide are most representative. Cesium-137 concentration in muscles in all countries increased 3—4-fold and attained its maximum in winter 1964—1965. Its concentration exhibited a downward trend in subsequent years, with $T_{1/2 \, eff} = 2.5$ years (Table 13.8).

TABLE 13.8. Dynamics of ^{137}Cs concentration in reindeer muscles in the Murmansk Region in comparable seasons of the year, ncuries/kg

1961	1962	1963	1964	1965	1966	1967	1968
24	33±2	48±7	80±3	96±8	79±4	57±2	46.6±1.5

The highest ^{137}Cs concentration was observed in Scandinavian countries and on the Kola Peninsula. Within the USSR ^{137}Cs concentration in reindeer muscles gradually decreases from west to east beginning with the Murmansk Region, so that ^{137}Cs concentration in the north of Yakut SSSR was only $\frac{1}{5}$—$\frac{1}{7}$ that in Murmansk Region (Table 13.9). The ^{137}Cs concentration exhibited an upward trend in the extreme eastern regions of the USSR (on the Chukchi and Kamchatka peninsulas) and also in Alaska and Canada (by factor 2) relative to northern Yakutia. Cesium-137 concentration in reindeer muscles in southern districts of Yakutia is likewise 2—3 times its concentration in the northern regions of the same republic.

TABLE 13.9. Geographical variability of ^{137}Cs in muscles and ^{90}Sr in bones of reindeer (winter 1964—1965)

Area	^{137}Cs	^{90}Sr	
	ncuries/kg	ncuries/kg	Sr units
Murmansk Region	96±8	—	1,414±74
Nenets National District	56±20	86±8	860±57
Komi ASSR	81±7	—	—
Yamalo-Nenets National District	66±4	127±10	1,242±86
Taimyr National District	22±2	45±1	465±24
Yakut ASSR	17±3	65±8	540±33
Chukchi National District	25±2	88±3	625±19
Kamchatka	17±2	—	470±37
Alaska /9/	35±9	50±70	—
Finland /19/	72±3	—	—

Analysis of geographical differences in ^{137}Cs concentration in venison reveals the factors affecting the content of this radionuclide in the daily ration of reindeer and in their organism. Data on the concentration of radionuclides in atmospheric precipitation, soils and the vegetation indicate that the ^{137}Cs contamination density and atmospheric precipitation in eastern regions is approximately one-half those in the west. Likewise, there is no doubt of the more extensive occurrence of lichens and their more important part in the feeding of reindeer on pastures in western regions. The latter is proved by the ratio of the concentrations of ^{137}Cs in muscles and ^{90}Sr in bones. In 1965 this ratio attained 0.71 in the Murmansk Region, but only 0.26 in the north of Yakutia, other areas occupying inter-mediate positions.

On the strength of the above ratio it is possible to state that reindeer fodder in the Murmansk Region contains nearly 3 times more ^{137}Cs in relation to ^{90}Sr compared with northern Yakutia. The largest ratio of these nuclides is found in lichens, especially their top portions. Moreover, an important role is also played by the differences in the species of fodder lichens in the regions under comparison; the dominant species in the Murmansk Region is Cl. alpestris with its well-developed branchlet surfaces, whereas the dominant species in northern Yakutia is Cetraria islandica with a smaller absorbtion surface. Hence, the effect of all the above factors (features of ^{137}Cs fallout, the parts played by lichens in fodder, specific differences in lichens) make for higher concentration of this nuclide in venison produced in Murmansk Region.

Strontium-90. The concentration of ^{90}Sr in the bones of reindeer in 1961—1966 was one order of magnitude higher than in the bones of cows, sheep and pigs from northern and central regions of the USSR (similar to the case of ^{137}Cs). The geographical and seasonal values of ^{90}Sr concen-tration do not vary by more than a factor of 3, the maximum and minimum likewise occurring in the Murmansk Region and northern Yakutia, respec-tively. Strontium-90 content in reindeer bones increased 3—5-fold as a result of nuclear tests in 1961—1962, while the extreme values reached 150 ncuries/kg, or 2,000 Sr units (in 1964—1966).

TABLE 13.10. Total alpha activity and ^{210}Po concentration in the bones and muscles of reindeer, ncuries/kg wet matter

Area	Total alpha activity	^{210}Po	
		bones	muscles
Murmansk Region			
May 1963	59±8(4)	–	–
February 1964	22±3(8)	–	–
February 1965	34±3(10)	–	–
October 1965	25±5(12)	4.3±3.1(12)	0.11±0.26(23)
April 1966	29±3(15)	13±1.0(15)	0.46±0.08(13)
November 1966	15±1.4(8)	8.0±2.2(10)	0.07±0.01(6)
Komi ASSR			
November 1962	15±3(13)	–	–
November 1966	4.3±1.3(9)	4.3±0.9(9)	0.07±0.1(6)
Taimyr National District			
December 1964	9±2(13)	–	–
November 1966	13.3±1.0(20)	10.6±1.8(17)	–
February 1967	26.7±9.1(3)	–	–
Yakut ASSR			
November 1964	15±4(9)	–	–
Nov.–Dec. 1965	–	4–11(23)	0.23±0.25(23)
1966	10.5±1.7(10)	–	–
1967	10.5±1.1(8)	9.3±1.3(7)	0.08±0.02(5)
Chukchi National District			
December 1964	10±3(10)	–	–
October 1965	20.8±7.1(5)	7.9±1.3(4)	–
November 1966	8.7±2.8(5)	5.9±0.9(6)	–
February 1967	15.9±3.3(8)	9.9±2.0(8)	0.07±0.03(5)
Kamchatka			
December 1965	5.0±0.5(5)	1,2±0.3(5)	0.28±0.07(5)
Alaska			
1962 /13/	–	14.6; 15.5*	0.16; 0.22;
1965 /13/	–	3.1; 3.2; 5.1	0.4; 0.35; 0.26
Finland			
1961 /13/	–	6.3*	0.2
1966 /6, 12, 22/	–	2.1**	0.13**
Canada			0.015–0.05(6)
1964 /12/	–	–	0.2(2)

* In pcuries/g ash.
** Summer slaughtering.
N o t e . Figures in parantheses denote the number of investigated samples.

The concentration of ^{90}Sr in muscles and other soft tissues of reindeer is only slightly more than $^1/_{1000}$ its concentration in the bones, and consequently the intensity of further migration of these radionuclides into the human organism with venison becomes significantly reduced.

The principal dose load from ^{90}Sr is attributed to the reindeer skeleton. In 1964—1966, when ^{90}Sr concentration in reindeer bones reached its maximum, the dose in the skeleton reached 30 rem/yr, whereas the dose due to incorporated ^{137}Cs was only 0.5—1 rem/yr. In the latter case the dose refers to the entire organism, not only to the skeleton.

The irradiation dose on reindeer due to natural radionuclides turned out to be far from insignificant, as had previously been assumed. Table 13.10 shows that the concentration of natural alpha emitters (mostly ^{210}Po) in reindeer bones varies between 5 and 60 ncuries/kg, which is numerically in agreement with the absorbed dose rate stated in rem/yr. If account is taken of the coefficient taking care of the nonuniform distribution of ^{210}Po in bones, and used in calculations of ^{210}Po doses in bones in accordance with the 1959 ICRP recommendations, then these values must be multiplied by 5.

The accumulated data concerning the contents of natural radionuclides in reindeer are still too limited for a satisfactory corresponding quantitative radioecological characterization. As with artificial radionuclides, the principal source of reindeer ingestion of the most important natural radionuclides (^{210}Po, ^{210}Pb and ^{226}Ra) comprises lichens.

The diurnal consumption of lichens (3 kg) can be used to calculate the buildup factor in reindeer: about 4 for ^{210}Pb, 8 for ^{226}Ra, 10 for ^{137}Cs, and 50 for ^{90}Sr.

The concentration of natural and artificial radionuclides in reindeer should vary depending on the grazing area. This assumption is confirmed for the Murmansk Region, where the buildup of both artificial and natural radionuclides attains its maximum. However, no close correlation is observed for all the regions of the Far North. The influx of artificial nuclides into lichens does not always proceed in parallel with the buildup of natural radionuclides.

Man

Investigation of different foodstuffs and rations, the concentration of incorporated ^{137}Cs in the human population and the elimination of this nuclide from their organism suggests the arbitrary division of all the inhabitants in the Far North into three groups: reindeer herders, inhabitants of rural and urban settlements, and inhabitants of cities.

The first group comprises reindeer herders, their families and all engaged in reindeer breeding, whose ^{137}Cs content is maximum and attains 4,800 ncuries. This group numbers 10^4—10^5 people, and its characteristic feature is the considerable relative importance of venison in their ration. The ^{137}Cs buildup level in the second group is only $^1/_2$—$^1/_{10}$ that in the first group. The content of ^{137}Cs in the third group (the majority of the total population in the Far North) does not differ appreciably from the buildup of the same radionuclide in the inhabitants of central regions of the USSR (10—30 ncuries in 1962—1965).

TABLE 13.11. ^{90}Sr and ^{137}Cs contents in reindeer herders' diurnal excreta (urine + feces)

Area	Sampling date	Number of persons	^{137}Cs, ncuries	^{90}Sr, pcuries
Murmansk Region	June 1962	2	5.2(4.1–6.3)	
	January 1963	3	7.4(3.8–12.8)	
	May 1963	25	10.9(4.4–24.5)	73(22–165)
	February 1964	25	22.0(11.5–39.9)	139(32–279)
	August 1964	24	11.4(4.2–23.7)	107(18–336)
	March 1965	6	25.0(6.0–51.0)	90(16–136)
	October 1965	25	19.8(9.0–33.0)	184(56–476)
	April 1966	15	26.9	201
	October 1966	17	18.5	130
Nenets National District	January 1963	3	4.2(2.2–6.6)	–
	February 1965	8	11.0(4.7–18.0)	35(22–55)
Komi ASSR	July 1962	3	1.5(1.3–1.8)	–
	November 1962	5	2.0(0.2–3.1)	–
	May 1963	35	4.8(2.9–8.0)	47(12–162)
	September 1963	35	5.3(2.6–8.4)	134(13–347)
	May 1964	35	14.2(6.5–30.5)	133(19–399)
	February 1967	5	7.0	15
Yamalo-Nenets National District	February 1965	3	11.0(8.0–14.0)	78(15–122)
Taimyr National District	July 1962	2	1.2(0.9–1.5)	
	December 1962	3	1.4(1.1–1.6)	
	January 1965	9	8.0(2.7–13.0)	45(20–270)
	January 1967		6.4	85(33–137)
Yakut ASSR, Oetung	January 1965	3	1.0(0.4–1.5)	36(20–52)
	December 1965	6	3.6(1.2–9.1)	92(38–135)
Ulakhan-Chistai	December 1965	5	9.8(7.7–13.0)	146(80–209)
Ust-Tatta	December 1965	3	8.2(7.9–8.5)	97(48–183)
Oetung	December 1966		2.4	39(19–60)
Chukchi National District	January 1965	8	6.6(3.3–10.2)	83(47–160)
	December 1966		6.9	83(68–100)

Note. These are mean data; minimum and maximum values are provided in parentheses.

According to available information, male reindeer herders consume 1 kg venison daily. However, this quantity appears to be the limit and is characteristic only for the winter period. In summer, the place of venison in the herders' ration is taken by fish from lakes and rivers. Moreover, in a radioecological assessment of the consumption of venison the very concept of "venison" should be differentiated, since it comprises muscle tissue, bones and various internal organs. All these tissues and organs are far from being equal with respect to the concentration of even such a comparatively uniformly distributed nuclide as ^{137}Cs, not to speak of ^{210}Pb and even less so ^{90}Sr. The concentration of ^{137}Cs in bones and internal organs is only 0.3—0.4 its concentration in muscles, where ^{137}Cs is most frequently analyzed and which accounts for some 70% of the edible parts of venison; the ^{137}Cs concentration in blood is 0.1 its concentration in muscle.

Even separate measurements of the consumed quantities of different organs and tissues of reindeer and information on their radionuclide concentration do not yield the quantity of radionuclides ingested by human populations. Culinary processing (cooking) and the process of eating involve inevitable losses (often very significant) of different radionuclides, which are therefore not ingested. Therefore the ingestion of radionuclides can only be understood by analyzing diurnal excreta under conditions of equilibrium with ingestion (this condition is at least 90% fulfilled toward the end of winter for all nuclides under consideration).

Penetration of ^{137}Cs into the organism of reindeer herders during the observation period was almost completely dependent on the venison, and therefore the "equivalent" of diurnal consumption of venison can be established from the contents of this nuclide in their excreta. Analysis of some 3,000 diurnal excreta samples (Table 13.11) shows that (as the average for all regions of the Soviet Far North) a reindeer herder daily consumes an equivalent of the content of ^{137}Cs in 230 g reindeer muscles. Deviations from this value by a factor of 2 are not unexpected. The reindeer herders daily consume more than 230 g venison (not pure muscle!) regarded as a commercial commodity, but their venison consumption does not exceed 500 g as an annual average (apart from culinary losses); the concentration of ^{137}Cs in muscles is approximately twice its concentration in "venison," the commercial commodity.

The figure of 230 g for the mean diurnal consumption of reindeer muscle by reindeer breeders is used to avoid possible misunderstandings when assessing results from the standpoint of health. All our summaries (including Table 13.7) of the radioactivity of reindeer meat are based on measurements of muscles, not "venison" as a commercial commodity. Therefore it would be erroneous to evaluate the diurnal ingestion of ^{137}Cs by reindeer breeders from these summaries, including data on the consumption of venison obtained by questioning them or weighing this foodstuff, as this would give results incorrect by a factor of two.

The dynamics and geographical variability of ^{137}Cs ingestion and also its content in the organism of reindeer breeders (Table 13.12) are closely related to ^{137}Cs concentration in reindeer muscles, the coefficient of correlation being 0.78 ± 0.07. Maximum and minimum buildup of ^{137}Cs were observed in reindeer breeders of the Murmansk Region and in northern Yakutia, respectively.

Cesium-137 concentrations in the excreta and organism of 117 adult male reindeer breeders were sufficient for accurate determinations (to within less than 10%) of some metabolic constants of this radionuclide. After entering the human gastrointestinal tract ^{137}Cs is resorbed to the extent of 93—100% (as against not more than 70% in reindeer). The rate of ^{137}Cs excretion from the body of reindeer breeders is subject to considerable individual variations, the effective biological half-life varying from 35 to 194 days (the mean value was 82 days; 95% confidence limits of the mean were 75—89 days). There

TABLE 13.12. Cesium-137 content in the organism of reindeer herders, ncuries

Area	Investigation date	Number of investigated reindeer herders	In the entire body	Per 1 kg body
Murmansk Region				
Lovozero	June 1962	2	1,200±500	20±7
Kanevka	January 1963	3	1,200±300	22±5
	October 1965	15	1,600±200	24±2
Loparskaya	October 1965	8	1,600±100	25±2
Krasnoshchel'e	February 1964	25	1,900±170	31±3
	August 1964	23	1,500±140	24±3
	March 1965	29	2,800±100	45±3
	October 1965	28	1,700±100	27±2
	April 1966	21	3,300±200	51±3
	November 1966	18	2,200±140	35±3
Nenets National District				
Krasnoe	January 1963	3	520±130	9±2
Kotkino	February 1965	8	1,200±100	20±2
Komi ASSR				
Vorkuta	July 1962	3	230±50	4±1
	November 1962	5	360±100	5±2
	May 1964	26	1,500±100	24±2
	February 1967	7	1,500±100	21±2
Yamalo-Nenets National District				
Muzhi	February 1965	10	1,200±100	19±1
Taimyr National District				
Novorybnoe	July 1962	2	330±10	5±1
	December 1962	3	330±80	4±1
Kresty	January 1965	13	800±100	13±1
Kheta	February 1967	13	700±40	12±1
Yakut ASSR				
Ust'e Tatta	December 1965	2	450±50	8±1
Ulakhan-Chistai	December 1965	16	900±100	14±1
Oetung	January 1965	7	400±50	5±1
	December 1965	6	400±70	5±1
	January 1967	12	460±10	8±1
Chukchi National District				
Anadyr	January 1965	14	900±100	14±1
	December 1966	7	950±100	14±1
Kamchelan	January 1967	18	1,700±20	25±2

were likewise diurnal and seasonal fluctuations in the metabolic rate of ^{137}Cs in the same individuals to within a factor of 2.   The metabolism of ^{137}Cs proceeds at nearly double the rate in winter ($T_{1/2\,eff}$ = 60±4 days) in comparison to summer ($T_{1/2\,eff}$ = 99±6 days).   The ratio of urinary excretion of ^{137}Cs to fecal elimination of radionuclides is 3.3: 3.2—3.4 for 95% confidence limits of the mean, and 1.3—8.3 for the entire population.   The following mean values were obtained for excretion of ^{137}Cs with reference to its % content in the organism (figures in parentheses are 95% confidence limits for the population): with diurnal urine 0.66 (0.14—1.18); per 1 liter urine 0.36 (0.12—1.07); with diurnal feces 0.17 (0.05—0.61); per 1 kg feces 1.17 (0.17—2.17).   The ratio of the speficic activity of ^{137}Cs in the body and urine is 4.4 (2.1—17.6), and 1.5 (0.6—3.6) in the body and feces.   Cesium-137 concentration in reindeer breeders' excreta was correlated with the level of total alpha activity, which was about 100 pcuries/day (the correlation coefficient was + 0.73±0.07).

No significant correlation was detected between the excretion of ^{137}Cs through the kidneys and excretion of water, potassium, sodium, chlorides, calcium and total ash elements.   A correlation was more in evidence in the case of intestinal excretion.   Variations in the intestinal excretion of ^{137}Cs correlated with variations in the weight of feces and their ash content and also excretion of potassium (the correlation coefficient was +0.73±0.07). From the regression equations it is possible to establish that a doubling of the consumption of potassium, sodium and water should be accompanied, respectively, by a 38, 31 and 26% acceleration of the total elimination of ^{137}Cs from the organism.

One feature of the dynamics of the ^{137}Cs level in the organism of reindeer herders (and also the metabolic rate) is its seasonality, although seasonal fluctuations in this case are not as pronounced as in reindeer.   Nevertheless they are fairly significant and must be taken into account in order to avoid erroneous estimates of the dynamics of ^{137}Cs concentrations in the organism of the inhabitants of different geographical areas.   In the same group of reindeer herders in the Murmansk Region (25 persons) in February 1964, the mean ^{137}Cs content in the entire body was 1,900±170 ncuries, while 4 months later, in summer, it was 1,500±140 ncuries, the decrease (by 21%) being statistically significant.   However, in March 1965 the mean ^{137}Cs concentration in the same subjects not only returned to its previous level, but even exceeded it by nearly 50%.   An analogous cycle (summertime decrease and wintertime increase of ^{137}Cs concentration) was later repeated in 1966.

In the case of seasonal variations in ^{137}Cs concentration, it is clear that radioecological data on radionuclide migration under northern conditions obtained in other countries on fully objective material cannot always be extrapolated to the conditions in the USSR.   Seasonal variation in ^{137}Cs concentration was also established in Alaskan Eskimos in the USA /2/, but its phases turned out to be the opposite of those in the Murmansk Region. Summertime ^{137}Cs concentrations are minimum in Soviet reindeer breeders and maximum in Alaskan Eskimos, and vice versa in winter.   On the other hand, the seasonal cycle of reindeer contamination passes through identical phases in the USA and Murmansk Region (and elsewhere in the world).

The cause of the discrepancy (which was fairly accurately established) is clarified by analysis of the ways of life of Soviet reindeer breeders and American Eskimos. Strictly speaking, Alaskan Eskimos cannot be compared with Soviet reindeer breeders because the majority of the former do not engage in reindeer breeding. They do not herd reindeer in the tundra throughout the year (as the Soviet reindeer herders) but only hunt reindeer twice annually (in autumn and spring). Reindeer hunted by Eskimos in autumn have grazed on summertime fodder (grasses) that are free from ^{137}Cs. The "clean" venison is consumed by the Eskimos in autumn and throughout winter. In spring they hunt reindeer with maximum concentrations of ^{137}Cs (and other radionuclides), and this venison is consumed by them in spring and summer. As a result, maximum ^{137}Cs concentration in the organism of Eskimos is reached toward the end of summer, and minimum concentration toward the end of winter. In the Murmansk Region, similar to other regions in the USSR, reindeer breeders generally consume freshly slaughtered venison. The appreciable decrease of ^{137}Cs concentration in their organism in summertime is in agreement with the minimum contamination of venison in that period and the decreased importance of venison in the total ration (during this period venison is partly replaced by fish).

Measurements (Table 13.12) enable one to estimate absorbed ^{137}Cs doses in the organism of inhabitants of the Far North during several years, and also to establish the migration coefficients of this radionuclide for all links of the northern chain, including the human link.

When computing irradiation doses from ^{137}Cs in mrem/yr for the whole body, the data listed in Table 13.12 and stated in terms of ncuries per 1 kg body weight must be multiplied by 11, since the mean dose corresponding to the concentration of 1 ncurie ^{137}Cs per 1 kg body weight is 11 mrem/yr.

The highest absorbed dose was recorded in reindeer herders in the Murmansk Region, varying from 240 mrem/yr in 1962 to 560 mrem/yr in 1966; the maximum value is 835 mrem/yr. The absorbed dose was only $\frac{1}{9}$ of that value in the north of Yakutia, other areas in the Far North occupying intermediate positions. These doses are approximately 25% overestimated, since the calculation was based on wintertime levels without correction for the summertime decrease of ^{137}Cs concentration in reindeer breeders. The contribution to the reindeer herders' total ^{137}Cs dose from external irradiation did not exceed 1–10% of the above values. Irradiation from ^{90}Sr does not play a very important role as regards reindeer breeders in comparison to doses from internal ^{137}Cs irradiation. The buildup of ^{90}Sr in reindeer breeders was only 2–4 times its buildup in persons unrelated to the northern chain, as against a hundredfold buildup in the case of ^{137}Cs. In contrast to reindeer breeders, the buildup of ^{90}Sr plays a fairly important part in the formation of the dose load in reindeer.

Information concerning the contents of natural radionuclides in the inhabitants of the Far North is very limited. Analysis of excreta shows that the ingestion of ^{210}Pb in equilibrium with ^{210}Po and somewhat ^{226}Ra, due to consumption of venison by reindeer breeders, reaches 80–100 pcuries daily, i. e., 10 times more than in the inhabitants of central regions of the USSR. Resorption of ^{210}Pb estimated from excretion data turns out to be

TABLE 13.13. Coefficients of correlation between the main productivity indices of reindeer breeding and ^{137}Cs contents in reindeer muscles

Index	Murmansk Region	Komi ASSR	Taimyr National District	Yakut ASSR	Chukchi National District	RSFSR
Stock	0.72±0.17	0.36±0.31	0.83±0.11	0.73±0.17	0.94±0.04	0.88±0.08
Venison production	0.23±0.33	0.72±0.17	0.92±0.06	-0.44±0.28	-0.09±0.50	0.72±0.37
Commercial yield of calves	-0.42±0.29	0.22±0.35	0.07±0.35	0.85±0.10	-0.38±0.30	-0.07±0.37

TABLE 13.14. Coefficients of correlation between the main productivity indices of reindeer breeding and ^{90}Sr contents (Sr units) in reindeer bones

Index	Murmansk Region	Komi ASSR	Taimyr National District	Yakut ASSR	Chukchi National District	RSFSR
Stock	0.48±0.24	0.56±0.22	0.53±0.27	0.84±0.09	0.84±0.11	0.88±0.07
Venison production	0.28±0.29	0.83±0.10	0.76±0.14	-0.24±0.30	-0.12±0.49	0.06±0.33
Commerical yield of calves	-0.65±0.17	0.59±0.22	0.66±0.18	0.68±0.17	0.28±0.29	0.19±0.34

unexpectedly different than the 1959 ICRP recommendations (8%). The taking into account of endogenous elimination into the intestinal lumen suggests that the resorption of ^{210}Pb reaches 80%.

Indirect calculations of the buildup of ^{210}Pb (^{210}Po) in the organism of reindeer breeders from excretion data lead to an estimated absorbed dose in the skeleton equal to 1—10 rem/yr. At present it is impossible to estimate the biological consequences of irradiation with these fairly large doses. In this respect some interest definitely attaches to our preliminary findings concerning the dose-effect relationship in the second link (which is the most heavily "loaded" with radionuclides) of this northern radioecological chain (in the reindeer), based on the fundamental productivity indices. It may be asserted, for instance, that the reindeer stock in the Soviet Far North increases in parallel with increasing ^{137}Cs and ^{90}Sr concentrations in reindeer, the correlation coefficient being $+0.88 \pm 0.08$ (Tables 13.13 and 13.14). Yet one can hardly claim the existence in this case of cause and effect relationships, since a more important role during the same period could be played by social-economic factors, including improved reindeer-breeding conditions. Nevertheless we did not encounter any proof of an adverse effect of the above doses due to ^{137}Cs and ^{90}Sr with respect to the increase of stock and some other indices (venison production and commercial yield of calves). No correlation has been discovered between variations in the concentrations of natural radionuclides in reindeer and their productivity on the Soviet Arctic seaboard.

These data should be regarded as the results of preliminary investigations. Various aspects of the biological effect of radiation as a function of the irradiation dose for the northern lichen-reindeer-man chain of the migration of radionuclides must form the subject of further detailed study.

References

1. Gusev, N. G. O predel'no dopustimykh urovnyakh ioniziruyushchikh izluchenii (Maximum Permissible Levels of Ionizing Radiation). Moskva, Medgiz. 1961.
2. Report of the UN Scientific Committee on the Effects of Atomic Radiation. New York. 1964.
3. Novikov, Yu. V. Gigienicheskie voprosy okhrany atmosfernogo vozdukha ot radioaktivnykh zagryaznenii (Health Aspects of the Protection of the Atmosphere Against Radioactive Pollution), Edited by A. S. Zykova. Moskva, Meditsina. 1966.
4. Izrael', Yu. A. (editor). Radioaktivnye vypadeniya ot yadernykh vzryvov (Radioactive Fallout from Nuclear Explosions), p. 129. Moskva, "Mir." 1968.
5. Aarkrog, A. et al. Environmental Radioactivity in Greenland, 1962. — Risö Report, No. 65, July 1963.
6. Beasley, T. M. and H. E. Palmer. — Science, Vol. 152:1062. 1966.
7. Chanler, R. and S. Wieder. — Radiol. Health Data, Vol. 6:317. 1963.

8. F r a n c a, P. E. et al. — Health Phys., Vol. 11:1471. 1965.
9. H a n s o n, W. C. — Amer. J. Veterinary Res., Vol. 27:116. 1966.
10. H a n s o n, W. C.   Radioecological Concentration Processes. — Proceedings
    of an International Symposium Held in Stockholm 25—29 April 1966,
    p. 183. London, Pergamon Press. 1967.
11. H a n s o n, W. C. and H. E. P a l m e r. — Health Phys., Vol. 11:1401. 1965.
12. H i l l, C. R. — Nature, Vol. 208:423. 1965.
13. H o l t z m a n, R. B. —Nature, Vol. 210:1094. 1966.
14. H v i n d e n, T. and A. L i l l e g r a v e n. — Nature, Vol. 192:1144. 1961.
15. J a w a r o w s k i, Z.   Second Symposium on Health Phys., Hungary. 1966.
16. K r i e g e r, H. L. et al.   Radioecological Concentration Processes. —
    Proceedings of an International Symposium Held in Stockholm
    25—29 April 1966, p. 59. London, Pergamon Press. 1967.
17. L i d e n, K. and M. G u s t a f s o n.   Radioecological Concentration
    Processes. — Proceedings of an International Symposium Held in
    Stockholm 25—29 April 1966, p. 193. London, Pergamon Press.
    1967.
18. M i e t t i n e n, J. K.   Radioactive Food Chain in Arctic Regions, May
    1964. — Third United Nations International Conference on the
    Peaceful Uses of Atomic Energy. Geneva. 1964.
19. M i e t t i n e n, J. K.   Enrichment of Radionuclides by Food Stuffs and Man
    in the Arctic. — Symposium on Nutrition in Uninhabitable Regions
    and in Space. Hamburg. 1966.
20. M i e t t i n e n, J. K. and E. H a s a n e n.   Radioecological Concentration
    Processes. — Proceedings of an International Symposium Held in
    Stockholm 25—29 April 1966, p. 221. London, Pergamon Press. 1967.
21. M i e t t i n e n, J. K. et al. Cs^{137} and Potassium in People and Diet — A
    Study of Finnish Lapps. — Annals Academiae Scientiarum. Fennical
    Series A, Vol. 2, Chemük. Helsinki. 1963.
22. P i e r s o n, D. H. and L. S a l m o n. —Nature, Vol. 184:1678. 1959.
23. Radioecological Concentration Processes. — Proceedings of an Inter-
    national Symposium Held in Stockholm 25—29 April 1966. London,
    Pergamon Press. 1967.
24. Radionuclides in Alaskan Caribou and Reindeer, 1963—1964. — Radiol.
    Health Data, Vol. 5:277. 1965.
25. S a l o, A. and J. K. M i e t t i n e n. — Nature, Vol. 201:1177. 1964.

PART II

RADIOECOLOGY OF AQUATIC BIOCENOSES

Chapter 14

RADIOECOLOGY OF MARINE PLANTS AND ANIMALS

The radioecology of marine organisms is a field of study which stands at the crossroads of radiobiology, hydrobiology, and biochemistry of the ocean and is concerned with the patterns of interaction between the flora and fauna of seas and oceans and the radioactive components of the environment /13, 17—19, 34/. These components are regarded as a radiological factor that must be studied in order to clarify its biological effect 1) as an external emitter and 2) as an emitter incorporated in organisms. Thus radioecology involves a study of the composition and intensity of secondary cosmic radiation in the ocean, including its fluctuations, both short-term and measured in geological periods, radiation absorption in marine waters, and the characteristics of ionization density.

The fact that hydrobionts accumulate radionuclides obviously presents an extra problem to the radioecologist. Three possible cases can be singled out in accordance with the presence of certain stable isotopes and the degree of chemical analogy between them and the radioactive isotopes:

1. Only radioactive isotopes are present, and no stable isotopes of the same chemical element. This applies to natural radionuclides with a very long half-life (e. g., the isotopes of U, Th, Ac, etc.) and artificial elements (all the isotopes of which are radioactive), that are absent in nature (e. g., the transuranic elements Np, Pu, Am, C, and others). The above radionuclides must be investigated themselves and cannot serve as an isotope marker, since there is no substrate for marking.

2. Long-lived radionuclides are present, arising in nature under the influence of cosmic rays (e. g., tritium, ^{7}Be, ^{14}C, ^{32}Si). These are isotopes of chemical elements which also contain stable isotopes, and therefore the behavior of the radioactive atom may initially differ from that of the stable isotopes of the same element because of a possible different chemical form (valence, compound) and physicochemical state. It may be expected, however, especially in the case of very long-lived radionuclides (such as ^{14}C), that the forms and states of the compounds of these atoms will after some time become identical with the forms and states characteristic for the stable isotopes of the element.

3. Finally, for most of the short-lived artificial and natural radionuclides the situation is very complicated: their state and form may not correspond to the state and form of the stable isotopes and the long-lived radioisotopes of the element, for example, ^{232}Th ($T_{1/2} = 1.4 \cdot 10^{10}$ years) and ^{234}Th ($T_{1/2} = 24.1$ days); stable yttrium, ^{91}Y ($T_{1/2} = 59$ days) and ^{90}Y ($T_{1/2} = 64$ hours). Thus, manganese occurs in seawater as a suspension of manganese dioxide, while ^{54}Mn, which is induced by nuclear explosions and falls into the ocean, is in an ionic state. Hence there is a need for studies in this new and important branch of radioecology.

It is noteworthy that the coefficients of accumulation of some radionuclides are in close agreement with those of the stable isotopes of the
chemical elements in each species of hydrobiont, whereas the coefficients
of others differ markedly /18, 37/. For this reason, labeled atoms must be
used in marine hydrobiology after having ensured that the form and state of
the compounds containing the label are identical with those of the compounds
that are being labeled. The multitude of components of seawater may
cause complex physicochemical transformations of the radionuclides that
enter it. The absence of data on the form and state of the compounds of
many elements in seawater creates a high degree of ambiguity /5/. It is
even more difficult to follow the chain of physicochemical transformations
into which the radionuclide is drawn resulting in an equiponderant, natural
state in seawater.

When we know the patterns of change of the states and forms of radionuclides up to the states and forms characteristic for the stable isotopes of
the same element, the amount of information obtainable will be doubled: the
object of investigation (artificial radioactive substances and short-lived
natural radionuclides) will become also a means of studying the behavior
of stable nuclides and of long-lived natural radionuclides in the marine
environment and in the hydrobiological system.

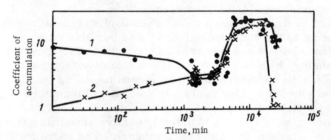

FIGURE 14.1. Time-dependent changes of the coefficients of accumulation of ^{91}Y by Ulva (1) and the coefficients of absorption on poly-
tetrafluoroethylene (2) /14/

TABLE 14.1. Coefficients of accumulation by mussel shells and coefficients of absorption on poly-
tetrafluoroethylene (teflon) as a factor of the previous duration of the radionuclide in seawater (the
time of interaction between the shells and teflon with ^{91}Y is 1 hour) /14/

6 hours after introduction (A)		74 hours after introduction (B)		A/B	B/A
shell	teflon	shell	teflon	shell	teflon
2.2	2.98	1.6	4.41	1.37	1.47
2.1	2.94	1.6	3.08	1.31	1.04
1.9	2.55	1.5	2.74	1.26	1.07
1.9	1.88	1.4	5.91	1.35	3.14
1.9	1.88	1.3	3.97	1.46	2.11
1.7	–	1.5	–	1.13	–
Mean 1.95	2.44	1.48	4.02	1.31	1.76

If these specific problems are solved by radiochemistry and radiobiology, a valuable contribution may be expected to be made in the allied field of the chemical ecology of marine organisms.

We can illustrate this by data on the dynamics of change in time of the sorption of ^{91}Y from seawater by a hydrophobic surface (polytetrafluoro-ethylene), green algae (U l v a), and the shells of mollusks (Figure 14.1 and Table 14.1).

It follows from the above that the form of the radionuclide undergoes changes in seawater which are reflected in the processes of its accumulation by hydrobionts.

An important and sometimes even decisive role in the transformation processes of radionuclide compounds in seawater may be played by the organic substances dissolved in the water. It has been shown that high-molecular hydrophilic compounds isolated from sea-pens vary greatly in their ability to combine with ^{141}Ce, ^{106}Ru, and ^{65}Zn introduced into seawater in an ionic state (Figure 14.2). This testifies to a potentially high value of interorganismic ecological metabolism in the migration of radionuclides in the sea.

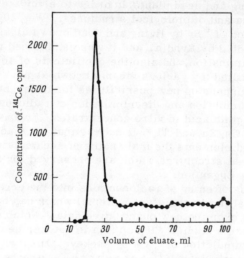

FIGURE 14.2. Curve of elutriation of high-molecular compounds of organic matter of detritus, bound with ^{141}Ce. Data of Parchevskii, Erokhin, and Khailov.

It is of interest to clarify the statistics of the distribution of the co-efficients of accumulation /6/. In the marine alga C y s t o s e i r a b a r - b a t a the coefficients of accumulation of ^{90}Sr obey a normal distribution law within the error limits (Table 14.2).

To discover how radionuclides enter marine organisms it is obviously very important to study the part played by such factors as temperature, light, the concentration of isotopic and nonisotopic carriers, etc., and also the state of the organism itself.

TABLE 14.2. Distribution of coefficients of accumulation of ^{90}Sr in a thallus of C y s t o s e i r a
(calculations of Parchevskaya)

Object	A	O_A	E	O_E
Whole thalli	0.42	0.33	-0.625	0.66
Branchlets	0.52	0.33	-0.119	0.66
Stemlets	0.31	0.33	+0.453	0.66

Note:
$$A = \frac{\sum\limits_{i=1}^{n}(x_i - \bar{x})^3}{n} : S^3; \quad S = \sqrt{\frac{(x_i - \bar{x})^2}{n}} \; ;$$

$$E = \frac{\sum\limits_{i=1}^{n}(x_i - \bar{x})^4}{n} : S^4 - 3; \quad O_A = \sqrt{\frac{6}{n}} \; ; \quad O_E = 2O_A \; .$$

It is essential to study the processes of accumulation and elimination of radionuclides in living and dead plants, in order to elucidate the role of actual life phenomena and of biological structures. We /20/ noted the identical accumulation of ^{90}Sr by living and dead brown algae. This pheno- menon was later studied by Ryndina, and it was established that the rate of accumulation and elimination, and also the coefficients of accumulation are similar in live and killed (by various means) brown algae. This is of great significance, since it opens up new possibilities for different approaches to the study of the accumulation and discrimination of radionuclides (chemical elements). Thus, alginic acid in vitro concentrates ^{90}Sr and ^{144}Ce, but not the radioisotopes of Cs, Zn, and Tl.* It may be that for many radionuclides of rare and scattered elements the main paths of accumulation are the fairly stable biological structures which are preserved for a certain time after the death of the organism.

What are the paths of entry of radionuclides into the organism of hydro- bionts? Without dwelling on the scant material which has been reviewed /18, 20, 37/, we shall examine new quantitative data. Notable in this con- nection are Fleishman's studies /28/ on fish in nature; he concludes that the main path of accumulation of ^{137}Cs by sockeye, O n c o r h y n c h u s n e r k a, is its food — stickleback (G a s t e r o s t e u s a c u l e a t u s). The principal postulate is the identity of the states of ^{133}Cs and ^{137}Cs and their accumulation by fish. However, where lake fish are concerned, their co- efficients of accumulation differ on the average by a factor of 3.3 /31/.

Marchyulenene conducted experiments toward a quantitative comparison of the efficiency of two paths of entry of radionuclides into the organism of freshwater mollusks D r e i s s e n a p o l y m o r p h a and larvae of the aquatic insects C h i r o n o m u s p l u m o s u s. It was found (in experiments ex- tended over two weeks) that the main route by which ^{144}Ce, ^{106}Ru, and ^{90}Sr reach the mollusks is the water and not the phytoplankton serving as food. The chironomid larvae also accumulated these radionuclides for the most

* Data of Lazorenko.

part from the water rather than from the food (mud) (Tables 14.3 and 14.4). Similar results were obtained by Kondrat'eva with [65]Zn on Actinia and Polychaeta, and by Dushauskene with ^{90}Sr and [210]Rb on carp and predatory fish.

TABLE 14.3. Proportion of radionuclides absorbed by chironomids from the water (I) and with food (II), % of concentration of radionuclides in the experimental hydrobionts (I) taken as 100% (data of Marchyulenene)

Radio-nuclide	Number of experiments	Starting conditions of experiment		Coefficient of accumu-lation (calculated for dry substance)	
		water contained radionuclide (I)	water did not contain radionuclide; mud con-tained radionuclide* (II)	in chiro-nomids	in mud
[144]Ce	3	100	2.6±0.4	909	14,000
[106]Ru	3	100	6.2±2.2	179	1,200
[137]Cs	3	100	5.7±1.1	117	23,000
[90]Sr	2	100	63.0±10.3	22	200

* During experiment (II) the radionuclide also entered the water from the mud.

TABLE 14.4. Proportion of radionuclides absorbed by Dreissena from the water and with food, % of con-centration of radionuclides in the experimental hydrobionts (I) taken as 100% (data of Marchyulenene)

Radionuclide	Number of experi-ments	Part of mollusk	Starting conditions of experiments*		Coefficient of accumu-lation (calculated for dry substance)	
			water contained radionuclide (I)	water did not contain radionuclide; phyto-plankton contained radionuclide (II)**	in mollusks	in phytoplankton
[144]Ce	2	body	100	14.5±4.5	529	135,200
		shell	100	4.1±0.8	456	
		byssus	100	8.7±4.5	2,143	
[106]Ru	2	body	100	14.8±5.1	389	6,770
		shell	100	6.2±2.1	262	
		byssus	100	12.1±3.9	671	
[90]Sr	2	body	100	8.5±1.6	17	1,100
		shell	100	6.7±1.1	40	
		byssus	100	13.0±3.2	75	

* During experiment (II) radionuclides entered the water from the phytoplankton.
** The phytoplankton was preliminarily grown in water containing radionuclides, as in experiment (I).

Thus, this aspect is still being worked out. It should be emphasized that its importance goes far beyond the scope of aquatic radioecology, and in this respect radioecology and related disciplines can make an undoubted contribution.

Radioecologists are at present endeavoring to penetrate the nature of the processes by which radionuclides enter, are accumulated, distributed, eliminated, and metabolized by marine organisms. Various approaches are being used: kinetic, energetic, biochemical, cybernetic, etc. The given trend of investigation is one of the most important in marine radioecology, as it permits patterns to be revealed which can be used as a basis for calculating the microdistribution of absorbed doses in the organism, for interpreting the character of accumulation and elimination, and for solving many other questions. In cases where the radionuclide introduced behaves similarly to the stable isotopes of the respective chemical element which are already contained in the organism (provided that their compounds, form, and state are identical), the material obtained by radioecologists can be utilized by scientists in other fields, in particular chemical hydrobiologists.

Especially interesting is the biological effect of low doses of ionizing radiation. There have been very few studies in this field with marine organisms owing to lack of data on the cardiology and genetics of marine hydrobionts. Material obtained by the Kovalevskii Institute of Biology of the Southern Seas of the Academy of Sciences of the Ukrainian SSR indicated that rapidly developing marine fish possess high radiosensitivity at the early stages of embryogenesis, this being expressed in the statistically reliable values for the incidence of morphological anomalies in prelarvae and anomalous mitoses (chromosome aberrations) starting with low concentrations of radionuclides in the habitat environment. The insignificant enhancement of the biological effect of injury with an increased concentration of the radiator in the environment is striking (Table 14.5). Anomalies of this kind with low doses are not a rarity in radiobiology /5/. Tsypugina obtained interesting data on a screening effect with action of ^{91}Y on fish eggs in the presence of an antibiotic in the environment (Table 14.6). It is to be noted that freshwater organisms proved to be highly radioresistant /27/.

TABLE 14.5. Cytogenetic action of $^{90}Sr-^{90}Y$ on the embryogenesis of redfish (data of Tsypugina)

Concentration, curie/liter	No. of cells studied	Proportion of cells with chromosome aberrations, %*	Difference from control	
			t	P
Control	3,902	11.8±0.93	–	–
$1.0 \cdot 10^{-11}$	2,283	10.7±1.38	1.4	0.1
$1.0 \cdot 10^{-10}$	1,372	13.3±1.81	1.5	0.1
$1.0 \cdot 10^{-9}$	4,278	15.1±0.93	5.08	0.01**
$1.5 \cdot 10^{-8}$	3,447	20.2±1.98	8.78	0.01**
$1.2 \cdot 10^{-7}$	4,224	22.2±1.45	12.38	0.01**
$1.2 \cdot 10^{-6}$	2,986	28.3±2.04	16.48	0.01**
$1.3 \cdot 10^{-5}$	4,019	27.2±1.68	16.66	0.01**

 * Mean and its confidence interval.
** Difference between experiment and control is reliable.

TABLE 14.6. Screening effect of streptomycin (0.1 μ/1) on the cell nucleus in redfish embryos with incubation of eggs in solution containing ^{91}Y

Variant	No. of cells studied	Proportion of pathological mitoses, %	Difference from control	
			t	P
Seawater (control)	714	6.9±2.6	–	–
^{91}Y (7.6·10^{-6} curie/l)	492	11.4±2.7	2.78	0.02*
^{91}Y (7.6·10^{-6} curie/l) + streptomycin	728	5.1±2.2	1.24	0.3
Seawater + streptomycin	622	5.6±2.1	0.96	0.4

* Difference between experiment and control is reliable.

Recently, interest has been focused on a study of radiation injury from the genetic aspect in successive series of generations of marine organisms. This will clarify the significance of early disturbances in the genetic apparatus of hydrobionts during life and reproduction.

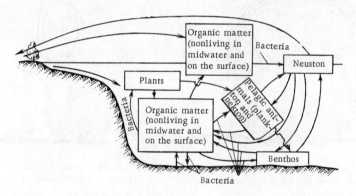

FIGURE 14.3. Scheme of hydrobiological processes in the sea /21/

The next large group of promising investigations is the radioecology of communities. In order to show the possible hydrobiological processes into which radionuclides are drawn and on which they act, let us take the structure of the hydrobiological system of the ocean which we drew up together with Zaitsev (Figure 14.3). The elementary ecological process (Figure 14.4) in the sea is made up of circadian and ontogenetic migrations of hydrobionts: neuston-plankton and nekton and neuston-benthos (in conditions of shallow waters). The nekton-benthos cycle is not widespread. Considering that the juveniles of very many marine animals, especially fish, develop in the neuston and a multitude of invertebrates forage here (at night), and also that the surface of the ocean serves as a collector of radioactive fallout and as a site for the concentration of radionuclides that float in the bound state with organic matter, we can understand how extremely vulnerable the neuston is. It may be assumed that this link of the neuston and plankton with the nekton, especially in the case of deepwater regions, is broken down into separate units that gradually die out as the depth increases (Figure 14.5). Thus, here too we can conceive of the importance of the processes that are in play on the ocean surface.

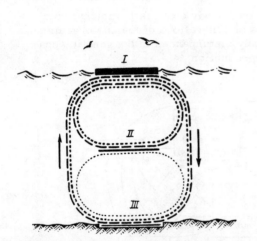

FIGURE 14.4. Scheme of the elementary ecological process with the phases of neuston (I), plankton and nekton (II), and benthos (III):

——— monophase species; – – – – circadian migrants; · · · ontogenetic migrants (the size of the lines and dots reflect saturation of the phase by species) /21/.

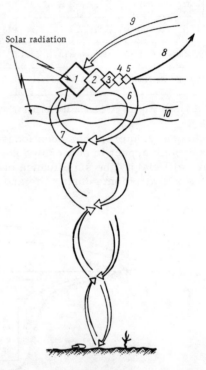

FIGURE 14.5. Scheme of large-scale ecological processes (links) in the ocean /21/:

1–5 — near-surface concentration of organic substances and neuston; 6 — "rain" and 7 — "antirain" of organic substances; 8 — loss of neuston from the sea; 9 — entry of substances of terrestrial origin into the sea; 10 — zone of most intensive photosynthetic activity of phytoplankton.

For this reason further study of the radioecological factor in seas and oceans is extremely important. Many works in this field can be cited, but in only a few is the fate of radionuclides in the hydrobiosphere analyzed.

The general tendency to study the structure of marine biocenoses is in keeping with the attempt by radioecologists to explain the role that they play in the distribution of radionuclides. We shall illustrate this with two examples. The capacity of algae in the Sargasso Sea to accumulate various radionuclides permits us to offer the following picture of their distribution in the near-surface layer (Figure 14.6). Whereas the proportion of ^{137}Cs and ^{90}Sr is small in algae of this layer, it becomes extremely high for ^{106}Ru and especially ^{144}Ce. Bottom overgrowths also play a considerable part in attracting radionuclides from the water contained within the sphere of the biocenosis. As Kulabakina showed /13/, in a nearshore thicket biocenosis of Cystoseira a large proportion of ^{90}Sr (up to 30%) goes to the dominant species C.barbata (10% ^{90}Sr is included in other living components of this biocenosis and 60% ^{90}Sr remains in the water within the limits of the community).

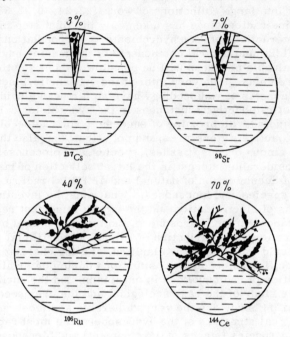

FIGURE 14.6. Distribution of ^{137}Cs, ^{90}Sr, ^{144}Ce, and ^{106}Ru among algae and seawater in the Sargasso Sea. Data of Polikarpov and Zaitsev.

Bachurin /1/ obtained different expressions of the dynamics of radionuclide exchange between organisms and the environment, which can be formulated generally by the transmission function

$$W(P) = \frac{y(P)_0}{x(P)},$$

involving the input of the system $x(P)$ (the concentration of the radionuclide in the water) and its output $y(P)_0$ (the concentration of the radionuclide in the hydrobiont). The dynamic approach to radionuclide migration in the ocean with the participation of organisms shows promise. The relations obtained can serve as a basis for predicting changes in the radioecological situation.

It is very laborious and complicated to obtain data on the biological effect of the radioecological factor in nature, and yet this must be done. Further research must be directed into the field of radioecology which still abounds with unknowns — the effect of small amounts of radionuclides in natural conditions on the life activity of hydrobionts, the structure of their populations and biocenoses, and the evolution of living systems.

The development of marine radioecology has been greatly influenced by fundamental investigations in marine hydrobiology, ecology of populations, genetics, biogeocenology, and biogeochemistry /2—5, 7—11, 16, 22, 25, 26, 29, 36/. Material testifying to the promise of marine radioecological research has found its way into large collections of work /23, 24, 30, 32, 33, 35, 38—41/.

However, marine radioecology has taken only the first step in studying the patterns of interaction between hydrobionts, their populations, and biocenoses and artificial and natural components of the marine environment /19/. No truly valuable progress will be possible without solutions to the following main problems. The very interesting cybernetic, kinetic, and physicochemical aspects of radioecology have so far been elucidated poorly. The metabolism of radioactive substances in the hydrobiological system has begun to be studied with promise of success. The radioecological indicators of radioactive waters, ores, and pollutions, and also those of bioconcentrators among organs (tissues), species, and biocenoses must be revealed. Of considerable interest is the latest collection of material on the coefficients of accumulation of natural and artificial radionuclides by organs (tissues), species, and biocenoses. What must be investigated now are the main paths of biocirculation and the vitally important reservoir of radioactive substances in the hydrosphere, and the part played by hydrobionts in the migration and distribution of natural and artificial radionuclides in the World Ocean must be clarified. Fuller characterizations are needed of specific radioecological situations during determination of the total composition of radionuclides in organisms and the environment, and a comparative study of the radioecological and the chemoecological factor. Still awaiting study are the very vulnerable and most resistant units in the biological structure of the hydrosphere, the most radiosensitive and radioresistant species (stages of development) and biocenoses of marine organisms. The radioecology of the pelagial is of extreme importance and shows great promise. Radioecology cannot be developed to its full potential without knowledge of the significance of natural radioactivity in the life and evolution of hydrobionts and their communities. The importance of studies in radiation genetics and cytogenetics of marine organisms must not be underestimated. Still problematic are the stimulating and harmful effects of low doses of ionizing radiation, and also the patterns of radiation injury of hydrobionts. The time has come for broad investigations into radiation biogeocenology to study the patterns of structural changes in populations of biocenoses and biogeocenoses of marine organisms under the influence of the radioecological factor. Foundations must be created for predicting the effect of artificial radionuclides in marine water bodies on the life and biological productivity of the sea.

References

1. Bachurin, A. A.   Matematicheskoe opisanie dinamiki protsessov radioaktivnogo zagryazneniya morskikh organizmov (Mathematical Description of Radioactive Contamination Processes in Marine Organisms). Moskva, Atomizdat. 1968.
2. Bogorov, V. G. — Gidrobiologicheskii Zhurnal, Vol. 5:12. 1967.
3. Vernadskii, V. I.   Biogeokhimicheskie ocherki. 1922—1932 (Biogeochemical Essays. 1922—1932). Moskva, Izdatel'stvo AN SSSR. 1940.
4. Verkhovskaya, I. N.   Brom v zhivotnom organizme i mekhanizm ego deistviya (Bromine in the Animal Organism and the Mechanism of its Action). Moskva, Izdatel'stvo AN SSSR. 1962.
5. Vinogradov, A. P.   Vvedenie v geokhimiyu okeana (Introduction to the Geochemistry of the Ocean). Moskva, "Nauka." 1967.
6. Gileva, E. A. — Trudy Instituta biologii UF AN SSSR, No. 45:5. 1965.
7. Dubinin, N. P.   Evolyutsiya populyatsii i radiatsiya (Radiation and the Evolution of Populations). Moskva, Atomizdat. 1966.
8. Zaitsev, Yu. P. — Gidrobiologicheskii Zhurnal, Vol. 5:58. 1967.
9. Zenkevich, L. A.   Biologiya morei SSSR (Biology of the Seas of the USSR). Moskva, Izdatel'stvo AN SSSR. 1963.
10. Zernov, S. A.   Obshchaya gidrobiologiya (General Hydrobiology). Moskva-Leningrad, Izdatel'stvo AN SSSR. 1949.
11. Konstantinov, A. S.   Obshchaya gidrobiologiya (General Hydrobiology). Moskva, "Vysshaya shkola." 1967.
12. Kuzin, A. M. — Radiobiologiya, Vol. 7:744. 1967.
13. Kulebakina, L. G.   Raspredelenie strontsiya-90 v biotsenoze tsistoziry (Distribution of Strontium-90 in a Biocenosis of Cystoseira). Thesis. Sevastopol'. 1967.
14. Lyubimov, A. A. and A. Ya. Zesenko. — In Sbornik: "Radioekologiya vodnykh organizmov i ikh soobshchestv." Sverdlovsk, UNTs AN SSSR. 1971.
15. Luchnik, N. V. — Biofizika, Vol. 2:86. 1967.
16. Nikol'skii, G. V.   Ekologiya ryb (Fish Ecology). Moskva, "Vysshaya shkola." 1963.
17. Polikarpov, G. G. — In Sbornik: "Trudy Sevastopol'skoi biologicheskoi stantsii," Vol. 13:275. 1960.
18. Polikarpov, G. G.   Radioekologiya morskikh organizmov (Radioecology of Marine Organisms). Moskva, Atomizdat. 1964.
19. Polikarpov, G. G. — Radiobiologiya, Vol. 7:801. 1967.
20. Polikarpov, G. G. — Okeanologiya, Vol. 7:561. 1967.
21. Polikarpov, G. G. and Yu. P. Zaitsev.   Gorizonty i strategiya poiska v morskoi biologii (Horizons and Exploratory Technique in Marine Biology). Kiev, "Naukova dumka." 1969.
22. O povedenii radioaktivnykh produktov deleniya v pochvakh, ikh postuplenii v rasteniya i nakoplenii v urozhae (Behavior of Radioactive Fission Products, their Entry into Plants and Accumulation in the Harvest). Moskva, Izdatel'stvo AN SSSR. 1956.
23. Radioaktivnaya zagryaznennost' morei i okeanov (Radioactive Contamination of Seas and Oceans). Moskva, "Nauka." 1964.

24. Radioaktivnaya zagryaznennost' morei i okeanov (Radioactive Conta-
    mination of Seas and Oceans).— Trudy Instituta okeanologii, Vol. 82.
    1966.
25. Sukachev, V. N.— In: "Yubileinyi sbornik, posvyashchennyi 30-letiyu
    Velikoi Oktyabr'skoi sotsialisticheskoi revolyutsii," Vol. 2.
    Moskva, Izdatel'stvo AN SSSR. 1947.
26. Timofeev-Resovskii, N. V.— Botanicheskii Zhurnal, Vol. 42:161.
    1957.
27. Timofeeva, N. A. et al.—Tezisy dokladov UF AN SSSR. Sverdlovsk.
    1968.
28. Fleishman, D. G.— Zhurnal evolyutsionnoi biokhimii i fiziologii,
    Vol. 4:32. 1968.
29. Shvarts, S. S.— In: "Razvitie biologii v SSSR. Sovetskaya nauka i
    tekhnika za 50 let." Moskva, "Nauka." 1967.
30. Aberg, B. and F. P. Hungate. Radioecological Concentration
    Processes.— Proc. Intern. Symp. held in Stockholm. Pergamon
    Press. 1967.
31. De Bortoli, M. et al. Concentration Factors for Strontium and
    Caesium in Fish of the Lakes in the Region of Verese (Northern
    Italy).— Giorn. Fisica Sanitaria e Protezione contro Radiazioni,
    Vol. 12, No. 4. 1969.
32. Disposal of Radioactive Wastes into Marine and Fresh Waters. Vienna,
    IAEA. 1960.
33. Disposal of Radioactive Wastes into Seas, Oceans and Surface Waters.
    Vienna, IAEA. 1966.
34. Hines, N. O. Proving Ground. An Account of the Radiobiological
    Studies in the Pacific, 1946—1961.   Seattle, University of
    Washington Press. 1962.
35. Nelson, D. J. and F. S. Evans. (Eds.). Symposium on Radioecology,
    1967. Michigan. 1969.
36. Odum, E. P. Fundamentals of Ecology. Philadelphia, W. B. Saunders Co.
    1959.
37. Polikarpov, G. G. Radioecology of Aquatic Organisms. Amsterdam,
    North-Holland Publ. Co., Reinhold Book Div., N. Y. 1966.
38. Proceedings of International Conference on the Peaceful Uses of Atomic
    Energy. Geneva. 1956.
39. Research in the Effects and Influences of the Nuclear Bomb Test
    Explosions, Vol. 1, 2. Japan Soc. for the Promotion of Sci. Tokyo.
    1956.
40. Schultz, V. and A. W. Klement. Radioecology.— Proc. First Nat.
    Symp. on Radioecology Held at Colorado State Univ., Ft., Collins,
    Colorado, 10—15 Sept. 1961. New York, Reinhold Publ. Corp. 1963.
41. The Effect of Atomic Radiation on Oceanography and Fisheries.—
    Washington, NAS-NRC. 1957.

Chapter 15

RADIOECOLOGY OF FRESHWATER PLANTS AND ANIMALS

Introduction

The main problems of radioecology at present are thought to be the behavior and biological activity of natural and artificial radionuclides in various biogeocenoses /19—21, 32, 33, 35, 57/. Biogeocenology deals with the same aspects /42, 43, 45/. However, as one of the branches of general biogeocenology, this discipline has its own specific features.

The first to bring together investigations on the radioecology of marine organisms was Polikarpov /34/. He initiated studies into the main aspects of marine radioecology connected with working out scientific principles for predicting the consequences of radioactive contamination of seas and oceans, recommendations for its control, and ways to weaken its biological activity.

Essentially the same applies to freshwater radioecology. Here, however, the problems seem to stand out more sharply, since owing to the diminished role of dilution, the concentration of radioactive substances in continental water bodies with radioactive contamination of the biosphere increases much more rapidly than in seas or oceans. The specific character of freshwater radioecology is further manifested in the fact that the physicochemical composition of the aquatic medium, and thus also the habitat conditions of hydrobionts in fresh waters, are subject to much greater fluctuations.

The radioecology of freshwater (and marine) organisms, on the one hand, reveals patterns of migration and accumulation of radioactive substances in the living and inert components of water bodies, and on the other, studies the biological activity of different radionuclides on hydrobionts.

Accumulation of artificial and natural radionuclides
by hydrobionts

The most extensive work on a comparative study of the accumulation of artificial radionuclides by various representatives of freshwater flora and fauna has been conducted under the guidance of Timofeev-Resovskii at the Institute of Ecology of Plants and Animals of the Urals Department of the USSR Academy of Sciences /10, 16, 42—44, 61/. Data were presented on the coefficients of accumulation (the coefficient of accumulation is the ratio of the concentration of a radionuclide in an organism to that in the environment) of the radioisotopes of 20 chemical elements for more than 40 species of freshwater plants and for 25 species of animals.

Among this group of elements, hydrobionts most intensively accumulate the radioisotopes of phosphorus, iron, cobalt, zinc, yttrium, zirconium, cerium, promethium, and mercury. The coefficients of accumulation of these radionuclides on conversion to dry substance of hydrobionts amount to thousands and tens of thousands. The radioisotopes of sulfur, calcium, chromium, germanium, rubidium, strontium, cadmium, iodine, and cesium are accumulated less intensively; their coefficients of accumulation for most hydrobionts do not exceed several hundred. The coefficient of accumulation is determined with respect to both the physicochemical properties of the radionuclides and the specific composition of aquatic organisms. A few of the freshwater animals and plants studied show extremely high coefficients of accumulation of some radionuclides, way above their mean values. For this reason these species have been called specific accumulators /46/.

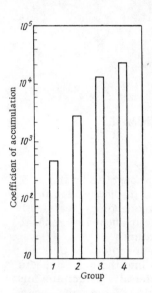

FIGURE 15.1. Mean coefficients of accumulation of chemical elements by various systematic groups of hydrobionts:

1 — animals; 2 — higher plants; 3 — unicellular algae; 4 — filamentous algae /9/.

A comparison of the capacity for accumulating different systematic groups of hydrobionts on the average with the radioisotopes of 14 chemical elements showed that plants have higher coefficients of accumulation than animals, and that unicellular and filamentous algae accumulate these radionuclides more intensively than higher plants (Figure 15.1). Hence, according to these data, a certain relation can be traced in the accumulation of radionuclides from the position of hydrobionts in the phylogenetic system. It is noteworthy, however, that aquatic bacteria are characterized by markedly smaller coefficients of accumulation with respect to certain radionuclides than algae and even some higher plants /12, 46, 56/.

Table 15.1 gives coefficients of accumulation of three natural radionuclides (^{238}U, ^{232}Th, ^{226}Ra) for some representatives of freshwater flora. It is seen that the mean coefficient of accumulation of ^{238}U assumes much greater values for higher plants than for algae. R a n u n c u l u s and P o t a - m o g e t o n   p e r f o l i a t u s are specific accumulators of ^{238}U.

The coefficients of accumulation of ^{226}Ra were also found to assume somewhat greater values on the average in higher plants than in unicellular and multicellular algae. The most active accumulator of this radionuclide is S c e n e d e s m u s   q u a d r i c a u d a, and lesser duckweed.

In contrast to ^{238}U and ^{226}Ra, ^{232}Th is more intensively accumulated by algae than by higher plants. The highest coefficients of accumulation were in the unicellular algae S c e n e d e s m u s   a c u m i n a t u s and C h l o r e l l a e l l i p s o i d e a.

On the average ^{232}Th is better accumulated by plants than ^{226}Ra and ^{238}U. With respect to the accumulation of all three natural radionuclides, the

strongest accumulators of all the groups of freshwater plants studied were
the unicellular algae (Table 15.2).

TABLE 15.1. Coefficients of accumulation of ^{238}U, ^{226}Ra, and ^{232}Th for different species of freshwater plants
/18/ (on absolutely dry substance)

Group of plants	Species	^{238}U	^{226}Ra	^{232}Th
Higher plants	Elodea canadensis Rish.	210 ± 2	2,615 ± 645	5,520 ± 1,500
	Stratiotes aloides L.	510 ± 15	1,300 ± 210	—
	Hydrocharis morsus rannae L.	1,060 ± 30	—	—
	Lemna minor L.	1,950 ± 450	5,750 ± 320	—
	Lemna trisulca L.	185 ± 25	33,440 ± 3,640	13,465 ± 3,530
	Potamogeton lucens L.	6,200 ± 1,620	470 ± 45	—
	Potamogeton perfoliatus L.	10,415 ± 785	375 ± 40	—
	Polygonum amphibium L.	930 ± 30	—	—
	Utricularia vulgaris L.	7,540 ± 90	—	—
	Ranunculus conferoides Fries.	16,025 ± 875	1,610 ± 70	—
	Myriophyllum spicatum L.	2,515 ± 145	4,235 ± 415	15,750 ± 2,650
	Ceratophyllum demersum L.	170 ± 80	1,560 ± 145	965 ± 125
	Najas sp.	167 ± 3	915 ± 45	—
	Drepanocladus sendtneri	360 ± 5	—	—
	Mean	3,445	5,227	8,925
Multi- cellular algae	Cladophora fracta Kütz.	46 ± 7	1,960 ± 900	30,500 ± 8,100
	Chara ceratophylla Wallr.	60 ± 10	1,260 ± 210	—
	Chara fragilis Desw.	40 ± 5	159 ± 9	1,925 ± 375
	Tolipellopsis stelligera	50 ± 1	734 ± 94	—
	Nittela sp.	315 ± 10	1,470 ± 50	—
	Mean	102	1,117	16,210
Uni- cellular algae	Oocystis pusilla Hansg.	260 ± 143	1,075 ± 70	—
	Chlorella ellipsoidea	915 ± 120	1,000 ± 40	88,650 ± 650
	Scenedesmus acuminatus Chodat.	255 ± 15	1,605 ± 235	146,250 ± 69,750
	Sc. obliquus Kütz.	220 ± 50	2,725 ± 95	—
	Sc. bijugatus Kütz.	195 ± 15	1,480 ± 20	—
	Sc. quadricauda Bret.	520 ± 10	12,200 ± 3,900	—
	Chlamydomonos gyrus Pasch.	240 ± 25	1,065 ± 150	—
	Haematococcus pluvialis Flotow.	217 ± 3	875 ± 100	—
	Mean	353	2,753	117,450

TABLE 15.2. Mean coefficients of accumulation of ^{238}U, ^{232}Th, and ^{226}Ra

Group	^{238}U	^{232}Th	^{226}Ra
Higher plants	3,445	8,925	5,227
Multicellular algae	102	16,210	1,117
Unicellular algae	353	117,450	2,753
Average for all groups	1,300	47,528	3,030

Radionuclides entering freshwater are absorbed not only by organisms but also by bottom soil, the dying off parts of plants, and the bodies of dead animals in the process of detritus formation. The accumulation of radio-nuclides by different soils of water bodies has unfortunately been little studied, but it may be assumed that in their sorption properties the bottom grounds of water bodies are not inferior to soils, the high absorbing capacity of which with respect to many radionuclides is well known /38, 44/.

Still less research has been done on the accumulation of radionuclides by dying hydrobionts. Nevertheless, there are a few data which indicate the quite considerable importance of freshwater detritus in the fate of some radiators in a water body /62/. The coefficients of accumulation of the radioisotopes of strontium and cesium for living and dead plants have been found to be practically the same, whereas the coefficients of accumulation of ^{106}Ru and ^{144}Ce are much higher in dead than in living plants (Figure 15.2).

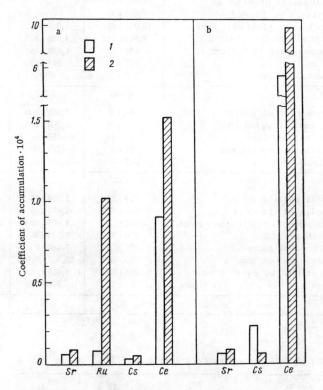

FIGURE 15.2. Coefficients of accumulation of the radioisotopes of strontium, ruthenium, cesium, and cerium by living (1) and dead (2) tissues of hornwort (a) and moss (b) /23, 24/

The stability of the link between ^{106}Ru and ^{144}Ce and the organic matter of dead plants has also been established to be much higher than in living plants; with ^{90}Sr such a difference was not observed and only ^{137}Cs was more stably fixed by living than by dead tissues.   These data indicated that owing to their stable fixation in the dying biomass of plants, the radioisotopes of ruthenium and cerium will pass into bottom sediments and for some time be excluded from the biological rotation of matter of the fresh water body.   On the other hand, the radioisotopes of strontium and cesium are readily reconverted into solution and are readmitted to the cycles of the biological rotation.

TABLE 15.3. Coefficients of accumulation of ^{90}Sr and ^{137}Cs by freshwater plants in experiment (I) and nature (II)

Species	^{90}Sr		^{137}Cs	
	(I)	(II)	(I)	(II)
Cladophora fracta Kütz................	1,910	1,042	1,230	–
Cladophora glomerata Kütz.	900	–	1,565	–
Spirogyra crassa Kütz.	235	–	285	–
Spirogyra sp......................	550	286	1,920	3,840
Chara sp.........................	350	1,131	200	–
Chara aspera Wild.	280	–	180	–
Chara fragilis Desw.	400	1,360	365	10,150
Chara tomentosa L.	–	1,031	–	804
Nitella hialina Ag...................	–	1,014	–	1,710
Tolipellopsis stelligera Mig	–	5,555	–	2,210
Drepanocladus sendtneri (Schimp) Warnst. ...	680	664	2,250	7,730
Lemna minor L.....................	400	511	2,405	4,872
Lemna trisulca L....................	315	–	940	–
Hydroharis morsus ranae L.............	415	1,211	535	2,100
Stratiotes aloides L.	615	489	375	1.030
Ceratophyllum demersum L............	510	301	300	1,152
Utricularia vulgaris L.	665	–	340	–
Ranunculus circinnatus Sibth..........	–	345	–	1,218
Ranunculus conferoides Fries.	465	–	765	–
Elodea canadensis L..................	805	892	285	472
Najas flexilis Rosth. et Schm.	–	268	–	725
Myriophyllum spicatum L.	445	445	295	655
Potamogeton lucens L................	–	309	–	257
Potamogeton perfoliatus L.............	685	410	195	426
Potamogeton compressus L.	1,020	528	115	1,048
Potamogeton filiformis Pers............	585	–	185	–
Potamogeton natans L................	670	333	205	800
Nuphar luteum Smith................	–	100	–	1,133
Polygonum amphibium L.	–	357	–	930
Phragmites communis Trin.............	–	25	–	285
Scirpus lacustris L.	–	34	–	–
Typha latifolia L.	–	127	–	312
Typha angustifolia L.	25	–	20	–
Equisetum limosum L................	–	723	–	1,137
Menyanthes trifoliata L...............	–	177	–	675
Calla palustris L....................	45	245	70	715
Mean	564	695	675	1,856

For the most part, data on the comparative study of the coefficients of accumulation of radionuclides by hydrobionts are obtained experimentally in a laboratory aquarium. In such a simplified model of a water body the ecological conditions of hydrobionts are generally very different from those obtaining in nature. Furthermore, laboratory experiments are as a rule confined to definite dates. With this in mind,* it can be expected that the coefficients of accumulation obtained experimentally will not always correspond to those calculated from data of radioecological studies in natural water bodies. To verify this, we determined in conjunction with Lyubimova and Fleishman the coefficients of accumulation of ^{90}Sr and ^{137}Cs in nature (Lake Bol'shoe Miassovo in the Southern Urals) for the main representatives of freshwater flora growing there. The data obtained were compared with the results of laboratory experiments with the same plant species /46/. As a result it was established that for most of these plants the coefficients of accumulation of ^{90}Sr in the experiments and in nature lie within one order of magnitude and are in fairly good agreement (Table 15.3).

The increase in coefficients of accumulation in Characeae in nature is evidently due to the fact that strontium passes into the tissues of these plants in the form of nonconverted carbonate compounds, the formation of which depends quantitatively on the amount of time the plant is in water containing ^{90}Sr. Therefore, Characeae that have been growing for a long time in natural conditions accumulate much more ^{90}Sr than during brief laboratory experiments. In all the plants studied the coefficients of accumulation of ^{137}Cs were several times higher in nature than in the experiment. The reason for this also apparently lies in the differences in the environmental conditions obtaining in a natural water body and in the simplified laboratory model /26/.

Hence, these examples show that the values of coefficients of accumulation of radionuclides established for hydrobionts in laboratory conditions are in several cases lower than those found in nature.

Accumulation of radionuclides as related to physicochemical factors of the environment

Of the many factors affecting the accumulation of radionuclides by hydrobionts, the following have been studied: 1) the concentration of isotopic and nonisotopic carriers in the environment; 2) the physicochemical stage of radionuclides in solution and the pH of the medium; 3) water temperature and light.

Most investigations of radionuclide accumulation by hydrobionts in relation to their concentration in the water and the latter's content of the respective isotope carriers have confirmed that under stable uniform conditions in a region of microconcentrations the content of a chemical element in hydrobionts is directly proportional to the concentration in the water and consequently the coefficients of accumulation remain constant; in a region

---

* With biological accumulation both the form and state of the radionuclide are of importance. For example, ^{137}Cs from atmospheric fallout has greater coefficients of accumulation in lake hydrobionts than stable cesium and ^{137}Cs from industrial wastes (Bigliocca, C. et al. Symposium of Radioecology, Cadarache, 1969).— Editor.

of microconcentrations $(10^{-5} - 10^{-4}$ mole/1) the coefficients of accumulation are inversely proportional to the concentration of the element in the water /9, 34, 41, 47, 49, 63/.

In predicting the fate of radionuclides in a water body it is interesting to study their accumulation by organisms as a function of the concentration of the corresponding nonisotope carriers in the water. It has been shown that the coefficient of accumulation of ^{90}Sr is inversely proportional to the content of its chemical analogs, Ca and Mg in the water, and that of ^{137}Cs is inversely proportional to the K content /16, 27, 28, 30, 40, 54, 60, 61/. This dependence of ^{90}Sr and ^{137}Cs accumulation on the content of their macroanalogs in the water explains the above-noted relatively low coefficients of accumulation of these radionuclides by freshwater hydrobionts. The reason is the fairly high content of calcium and potassium in the water used for the experiments /46/.

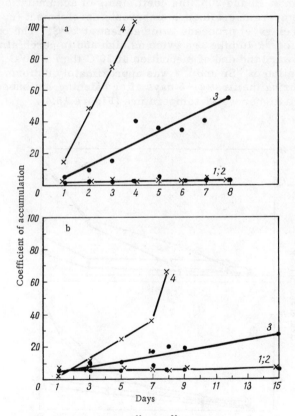

FIGURE 15.3. Dynamics of ^{90}Sr and ^{90}Y by developing eggs of pike (a) and perch (b) at various water temperatures:

1 and 2— accumulation of ^{90}Sr at 10 and 20°C respectively;
3 and 4 — same for ^{90}Y.

Another factor in the accumulation of radionuclides by freshwater organisms is their physicochemical state in solution and the concentration of hydrogen ions in the medium. Thus, in a study of ^{91}Y accumulation by the filamentous alga Cladophora fracta Kütz, the coefficients of radionuclide accumulation were found to be lower by a factor of 2 — 3 in the region of pH values at which ^{91}Y passes into the colloid form /9/. The radioisotopes of several chemical elements present in solution in the form of chelate compounds are accumulated to a lesser degree by hydrobionts than if they are introduced in the medium in the form of ions /46/. On the whole, this aspect has been studied insufficiently owing to the virtual absence of information on the state of chemical elements in the water during their absorption by hydrobionts.

Still less is known about the patterns of accumulation in relation to such factors as the water temperature and light. In experiments with developing eggs of pike and perch which we conducted with Ozhegov, it was shown that during incubation at 10 and 20°C the coefficients of accumulation of ^{90}Sr do not depend on temperature, although embryonic development (up to the stage of hatching of prelarvae) proceeds twice as fast at 20°C. The coefficients of accumulation of ^{90}Y in pike are twice as high and in perch almost three times as high toward the end of incubation at 20°C than at 10°C (Figure 15.3).

The accumulation of ^{90}Sr and ^{90}Y was approximately uniform at different temperatures during the first 4 — 5 days after hatching, but subsequently it rose sharply at a higher water temperature (Figure 15.4).

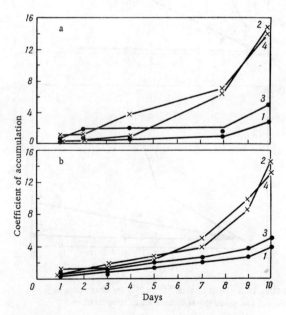

FIGURE 15.4. Dynamics of ^{90}Sr and ^{90}Y accumulation by prelarvae of pike (a) and perch (b) at various temperatures. Notation as for Figure 15.3.

Reduced accumulation of radioactive substances is noted in freshwater fish in cold periods of the year under natural conditions (Columbia River) /51/, owing to a smaller consumption of the plant food containing radio-nuclides at this time. In experiments with freshwater plankton at a water temperature of 13.5 − 28.5°C the total absence of any relationship was noted between the accumulation of ^{85}Sr and the temperature; a perceptible rise in accumulation was observed only at a temperature of 28.9°C /59/.

The few data available concerning the influence of light on the absorption of radionuclides by freshwater organisms are quite contradictory. Patten and Iverson /59/ noted a rise in the accumulation of ^{85}Sr by freshwater plankton under conditions of illumination while accumulation was suspended in darkness; on the other hand, in experiments with freshwater macrophytes no relationship was found between ^{90}Sr accumulation and illumination /6, 58/.

Accumulation of radionuclides in open and closed water bodies is of special interest. Unfortunately there are very few data, but there is some material which shows that the coefficients of accumulation of the radio-isotopes of cobalt, zinc, zirconium, niobium, and ruthenium for different freshwater plants attain much lower values in the open than the closed systems during the same length of time /14/. This is apparently because in open water bodies no balance is achieved in the distribution of radio-nuclides between the water, the bottom and the plant biomass.

Distribution of radionuclides among the components of
a water body

As already noted, in a water body artificial and natural radionuclides are absorbed not only by organisms but also by the bottom soil and by dying parts of hydrobionts in the process of detritus formation. As a result, especially in closed types of water body, most radioactive substances soon become concentrated in the bottom sediment and the biomass, while their content in the water drops greatly. On the basis of these findings research was conducted to determine a biological method of deactivating waters con-taminated with radioactive substances /1−5, 7, 8, 14, 15, 29, 30, 46, 48, 64/.

Of special importance in these studies is the work by Timofeeva-Resovskaya /46/ on a comparative study of the distribution of 18 radio-nuclides among the water, bottom, and freshwater organisms using an aquarium model of a water body. The results showed that under uniform conditions the radionuclides are distributed very nonuniformly. The author classes them in four main groups according to specific types of distribution (Figure 15.5).

The group of hydrotropes includes ^{35}S, ^{51}Cr, and ^{71}Ge; these are nuclides which remained predominantly in the water during the experiments. The group of equitropes includes ^{60}Co, ^{86}Rb, ^{90}Sr, ^{106}Ru, and ^{131}I; they were more or less evenly distributed among the various components of the water body. The group of pedotropes comprises nuclides which were accumulated predominantly in the bottom (^{59}Fe, ^{65}Zn, ^{91}Y, ^{95}Zr, ^{95}Nb, and ^{137}Cs), while to the group of biotropes were referred the radionuclides which were mainly accumulated in the hydrobionts (^{32}P, ^{115}Cd, ^{144}Ce, ^{204}Hg).

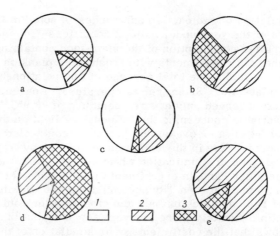

FIGURE 15.5. Types of radioisotope distribution: in water (1),
in the bottom (2), and in the biomass (3) /46/:

a — hydrotropes; b — equitropes; c — biotropes; d — pedotropes;
e — relative distribution of the mass of the three components
in aquaria.

This classification is admittedly somewhat arbitrary as it is based on
only a single model of a water body. With a change of the parameters of
such a model (biological, physicochemical, hydrodynamic, etc.) the tropicity
of the different radionuclides would probably vary.

Effect of radionuclides on hydrobionts

It has been stressed that one of the main tasks in freshwater radioecology is the study of the biological effect of radionuclides and their
radiation on hydrobionts, including individual species and also populations
and communities. Of both scientific and practical importance, this problem
is far from being solved, particularly with regard to the biological effect
of long-term irradiation of organisms with radioactive contamination of the
habitat environment.

A fairly comprehensive survey of the radiosensitivity of hydrobionts was
made by Polikarpov /34/, according to whom radiosensitivity in aquatic
organisms increases from lower to more highly organized forms and decreases from the early to the late stages of development. From a radioecological viewpoint, this is a very important conclusion, since it gives a
general idea as to which are the most sensitive and vulnerable links of
aquatic biocenoses in a case of radioactive contamination of waters.

The danger of radioactive contamination of water bodies and the con-
sequent threat to biological productivity which have emerged recently
have sharpened interest in studying the biological effect of high levels of
radioactive contamination of waters.

The danger of radioactive contamination of water bodies and the conse-
quent threat to biological productivity which have emerged recently have
sharpened interest in studying the biological effect of high levels of radiation
created by contamination of the habitat environment and accumulation of
radionuclides by hydrobionts. A very high radioresistance has been established in the bacterial flora of water bodies. No changes were noted in the
development of nitrifying bacteria and Escherichia coli even when the
concentrations of $^{90}Sr - {}^{90}Y$ in the water were around 10^{-3} curie/liter /13/.
In water used as a biological protector of the atomic reactor at Los Alamos,
Pseudomonas bacteria were discovered that are capable of multiplying
after irradiation in a dose of 10^6 r /11/.

In experiments on freshwater periphyton, with a concentration of an unse-
parated mixture of fission products of uranium of from 10^{-6} to 10^{-3} curie/l,
Timofeeva-Resovskaya /46/ noted a $9 - 10$ times higher radiostimulation
of its development in comparison with the control. Together with stimulation of the total mass of the community, a marked reorganization of its
specific composition took place. Stimulated development of freshwater algal
periphyton was also observed with chronic irradiation in the field with doses
of $5 - 50$ r/day during two months /10/.

A suppressing effect of $^{90}Sr - {}^{90}Y$ on the reproduction of the freshwater
Daphnia magna was established only when it was cultivated during
80 days in water with a radionuclide concentration of 10^{-5} curie/liter /39/.

Cytogenetic studies were conducted with a natural population of a mos-
quito species, Chironomus tentans, the larvae of which are abundant
in the bottom sediments of Lake White Oak and the White Oak river bed, in
which radioactive wastes from the Oak Ridge National Laboratory were
deposited over a long period. The population was irradiated at the larval
stage during 22 years with a dose of about 230 rads each year. It was found
that under these conditions there was a certain rise in the frequency of
chromosome aberration in the population; however, these were later eliminated by natural selection /52/.

In our laboratory we studied experimentally the radiosensitivity of the early
development stages of the freshwater mollusk Limnaea stagnalis L.
Development of embryos was disrupted only with incubation in water containing
$10^{-4} - 10^{-3}$ curie/l of $^{90}Sr - {}^{90}Y$ (Table 15.4). It was also found that the radiosensitivity of embryos to external one-time irradiation is greatly lowered from the
early to the later stages of embryogenesis. Thus, with irradiation at the stage
of early blastula LD_{50} is 450 r; it rises 900 r with irradiation at the
stage of appearance of the embryonic motor systems, while at the stage of formation of the shell it increases to 2,000 r. Short-term one-time irradiation
of the eggs at the early stage of cleavage was more effective than long-term irradiation with approximately the same dose during incubation in solutions of
$^{90}Sr - {}^{90}Y$ /22/.

All these examples show that representatives of different systematic
groups of freshwater organisms from bacteria to mollusks (in the stage
of developing eggs) can tolerate quite high levels of radioactive contamination of the water without noticeable ill effect. Interesting in this connection
are the recently published data on the extremely high radiosensitivity of the
rapidly developing eggs of some species of marine fish /17, 34 —37, 50/.

With incubation of fertilized eggs in aquatic solutions containing
$^{90}Sr - {}^{90}Y$, it has been found that with concentrations of 10^{-10} curie/l and
higher more prelarvae are hatched with various morphological deformities,

while in solutions containing concentrations of $10^{-8} - 10^{-7}$ curie/1 embryo mortality is in addition greatly increased. Along with this there are data indicating either the absence of an effect of this type in individual fish species even with extremely high ^{90}Sr contamination of the water /53/ or the presence of such effects with a prolonged stay of the incubated eggs (more than 100 days) in solutions containing ^{90}Sr in concentrations of $10^{-8} - 10^{-6}$ curie/liter /31/.

TABLE 15.4. Effect of ^{90}Sr on the embryonic development of Limnaea stagnalis L.

Concentration of ^{90}Sr in the water, curie/1	Total dose of radiation during 8 days, r	Number of eggs	Number of hatched larvae	Proportion of larvae to number of eggs, %	Significance level	Number of deformed embryos
Control	—	553	544	98.0	—	1
$1 \cdot 10^{-9}$	0.0035	539	527	98.0	—	0
$1 \cdot 10^{-7}$	0.35	497	481	97.2	—	2
$1 \cdot 10^{-6}$	3.5	266	262	98.5	—	1
$1 \cdot 10^{-5}$	35	653	625	96.0	—	5
$1 \cdot 10^{-4}$	350	602	584	97.0	> 0.05	3
$5 \cdot 10^{-4}$	1,700	366	335	91.5	> 0.001	4
$1 \cdot 10^{-3}$	3,500	501	308	61.5	< 0.001	19
$1 \cdot 10^{-2}$	35,000	550	0	0	—	—

TABLE 15.5. Effect of ^{90}Sr on the embryonic development of tench

Concentration of ^{90}Sr in the water, curie/1	Number of eggs	Number of hatching prelarvae		Externally normal prelarvae*		Anomalous prelarvae*	
		number	%	number	%	number	%
Control	2,810	2,106	74,8 ±4.1	1,963	93.2	143	6.8
$1 \cdot 10^{-10}$	2,931	2,118	72.3 ± 5.1	1,968	93.0	150	7.0
$1 \cdot 10^{-9}$	3,045	2,136	70.2 ± 8.1	2,009	94.0	127	6.0
$1 \cdot 10^{-8}$	2,772	2.054	74.0 ± 5.1	1,900	92.5	154	7.5
$1 \cdot 10^{-7}$	2,951	2,092	71.0 ± 7.2	1,965	93.9	127	6.1
$1 \cdot 10^{-5}$	1,597	1,231	77.0 ± 2.4	1,146	93.1	85	6.9

* The number of normal and deformed prelarvae (in percent) is indicated in relation to all those hatched.

In our experiments with tench (Tinca tinca L.) we noted that incubation of fertilized eggs in aquatic solutions containing $10^{-10} - 10^{-5}$ curie/1 ^{90}Sr $- ^{90}$Y did not have any noticeable effect on the rate of embryonic development, nor on the numerical yield of normal and deformed prelarvae (Table 15.5). Survival of prelarvae during the first post-hatching days did not differ from the control even after additional irradiation in a dose of 800 r. External one-time radiation of tench eggs at the stage of early cleavage increases the yield of anomalous prelarvae at a dose of 50 r, while a dose of 400 r proved absolutely lethal /25/. In experiments with eggs of pike (E s o x   l u c i u s) again no marked effect of ^{90}Sr was established in a range of concentration of 10^{-9} — 10^{-5} curie/1 from the moment of fertilization up to hatching. Only a concentration of 10^{-4} curie/1 induced a slight rise in the incidence of morphological deformities upon hatching (Table 15.6)

TABLE 15.6. Effect of ^{90}Sr on the embryonic development of pike

Concentration of ^{90}Sr in the water, curie/1	Total number of eggs	Number of hatching prelarvae		Externally normal prelarvae*		Anomalous prelarvae*	
		number	%	number	%	number	%
Control	445	322	72.3 ± 2.2	289	89.8	33	10.2 ± 1.9
1 · 10^{-9}	400	296	74.2 ± 2.2	256	86.5	40	13.5 ± 2.0
1 · 10^{-8}	316	228	72.0 ± 2.5	204	89.5	24	10.5 ± 2.0
1 · 10^{-7}	369	245	66.4 ± 2.8	208	85.0	37	15.0 ± 2.3
1 · 10^{-5}	318	254	80.0 ± 2.2	223	88.0	31	12.0 ± 2.0
1 · 10^{-4}	378	284	75.0 ± 2.1	224	79.0	60	21.0 ± 4.1

* The number of normal and deformed prelarvae (in percent) is indicated in relation to all those hatched.

In conclusion, it may be noted that available data on the harmful effect of a low level of radioactive contamination of water on the embryonic stages of fish are rather contradictory. The disparity of the results indicate, on the one hand, lack of knowledge on the radiosensitivity of the early stages of ontogenesis in fish, and on the other, the possibility of appreciable differences in the radiosensitivity of different species to irradiation during radioactive contamination of water and the accumulation of radionuclides by developing embryos.

References

1. A g a f o n o v, B. M. — In Sbornik: "Trudy Vsesoyuznoi konferentsii po meditsinskoi radiologii," p. 79. Moskva, Medgiz. 1957.
2. A g a f o n o v, B. M. and V. I. I v a n o v.— Trudy Instituta biologii UFAN SSSR, No. 45:67. 1965.
3. A g r e, A. L. — Byulleten' MOIP. Otdel biologicheskii, Vol. 67:45. 1962.
4. A g r e, A. L. and V. I. K o r o g o d i n. — Meditsinskaya radiologiya, No. 5:67. 1960.
5. A g r e, A. L. et al. — Byulleten' MOIP. Otdel biologicheskii, Vol. 67:120. 1962.
6. A g r e, A. L. and M. M. T e l i t c h e n k o. — Ibid., Vol. 68:133. 1963.
7. A g r e, A. L. et al. — Ibid., Vol. 69:20. 1964.
8. A g r e, A. L. et al. — Ibid., Vol. 71:124. 1966.
9. G i l e v a, E. A. — Trudy Instituta biologii UFAN SSSR, No. 45:5. 1965.
10. G i l e v a, E. A. et al. — Doklady AN SSSR, Vol. 156:455. 1964.
11. D u b i n i n, N. P. Problemy radiatsionnoi genetiki (Problems of Radiation Genetics). Moskva, Gosatomizdat. 1961.

12. Zharova, T.V. — Mikrobiologiya, Vol. 30:872. 1961.
13. Zhogova, V.M. — In Sbornik: "Raspredelenie, biologicheskoe deistvie i migratsiya radioaktivnykh izotopov," p. 314. Moskva, Medgiz. 1961.
14. Ivanov, V.I. — Trudy Instituta biologii UFAN SSSR, No. 45:63. 1965.
15. Ivanov, V.I. — Ibid., p. 71.
16. Ivanov, V.I. et al. — Ibid., p. 33.
17. Ivanov, V.I. — In: "Voprosy biookeanografii," p. 185. Kiev, "Naukova Dumka." 1967.
18. Iskra, A.A. et al. — Ekologiya, No. 2:83. 1970.
19. Kuzin, A.M. Osnovy radiatsionnoi biologii (Principles of Radiation Biology). Moskva, "Nauka." 1964.
20. Kuzin, A.M. — Radiobiologiya, Vol. 7:643. 1967.
21. Kuzin, A.M. and A.A. Peredel'skii. — In Sbornik: "Okhrana prirody i zapovednoe delo v SSSR," Vol. 1:65. Moskva, Izdatel'stvo AN SSSR. 1956.
22. Kulikov, N.V. et al. — Radiobiologiya, Vol. 6:908. 1966.
23. Kulikov, N.V. et al. — Ibid., Vol. 7:271. 1967.
24. Kulikov, N.V. et al. — Ibid., Vol. 8:760. 1968.
25. Kulikov, N.V. et al. — Ibid., p. 391.
26. Kulikov, N.V. et al. — Doklady AN SSSR, Vol. 178:1407. 1968.
27. Lebedeva, G.D. — In Sbornik: "Raspredelenie, biologicheskoe deistvie i migratsiya radioaktivnykh izotopov," p. 319. Moskva, Medgiz. 1961.
28. Lebedeva, G.D. — Radiobiologiya, Vol. 6:556. 1966.
29. Marei, A.N. et al. — In Sbornik: "Raspredelenie, biologicheskoe deistie i migratsiya radioaktivnykh izotopov," p. 298. Moskva, Medgiz. 1961.
30. Marei, A.N. et al. — Meditsinskaya radiologiya, No. 3:69. 1958.
31. Neustroev, G.V. and V.N. Podymakhin. — Radiobiologiya, Vol. 6:321. 1966.
32. Peredel'shii, A.A. — Zhurnal obshchei biologii, Vol. 18:17. 1957.
33. Peredel'skii, A.A. — In Sbornik: "Zemlya vo Vselennoi," p. 449. Moskva, "Mysl'." 1964.
34. Polikarpov, G.G. — Radioekologiya morskikh organizmov. Moskva, Atomizdat. 1964.
35. Polikarpov, G.G. — In Sbornik: "Radioaktivnaya zagryaznennost' morei i okeanov," p. 98. Moskva, "Nauka," 1964.
36. Polikarpov, G.G. and V.N. Ivanov. — Voprosy ikhtiologii, Vol. 1:583. 1961.
37. Polikarpov, G.G. and V.N. Ivanov. — Doklady AN SSSR, Vol. 144:219. 1962.
38. Polyakov, Yu.A. — In Sbornik: "Radioaktivnost' pochv i metody ee opredeleniya," p. 81. Moskva, "Nauka." 1966.
39. Telitchenko, M.M. — Nauchnye doklady vysshei shkoly. Biologiya nauki, No. 1:114. 1958.
40. Timofeeva, N.A. — Trudy Instituta biologii UFAN SSSR, No. 45:41. 1965.
41. Timofeeva, N.A. and A.L. Agre. — Radiobiologiya, Vol. 5:457. 1965.
42. Timofeev-Resovskii, N.V. — Botanicheskii zhurnal, Vol. 42:161. 1957.

43. Timofeev-Resovskii, N.V. Nekotorye problemy radiatsionnoi
     biogeotsenologii (Some Problems of Radiation Biogeocenology).
     Thesis. Sverdlovsk. 1962.
44. Timofeev-Resovskii, N.V. et al. — In Sbornik: "Radioaktivnost'
     pochv i metody ee opredeleniya," p. 46. Moskva, "Nauka." 1966.
45. Timofeev-Resovskii, N.V. and A.N. Tyuryukanov. —
     Byulleten' MOIP. Otdel biologii, Vol. 72:106. 1967.
56. Timofeeva-Resovskaya, E.A. — Trudy Instituta biologii UFAN
     SSSR, No. 30:38. 1963.
47. Timofeeva-Resovskaya, E.A. et al. — Byulleten' MOIP. Otdel
     biologii, Vol. 64:117. 1959.
48. Timofeeva-Resovskaya, E.A. and N.V. Timofeev-
     Resovskii. — Trudy Instituta biologii UFAN SSSR, No. 12:194.
     1960.
49. Titlyanova, A.A. and V.I. Ivanov. — Doklady AN SSSR,
     Vol. 136:721. 1961.
50. Fedorov, A.F. et al. — Voprosy ikhtiologii, Vol. 4:579. 1964.
51. Foster, R.F. and I.I. Davies. — In: Proceedings of the International
     Conference on the Peaceful Uses of Atomic Energy. Geneva. 1956.
52. Blaylock, B.G. — In: Disposal of Radioactive Wastes into Seas,
     Oceans and Surface Waters, p. 45. Vienna, IAEA. 1966.
53. Brown, V.M. and W.L. Templeton. — Nature, Vol. 203:1257. 1964.
54. Kevern, N.R. — Science, Vol. 145:1445. 1964.
55. Kevern, N.R. et al. — In: Health Physics Division Annual Progress
     Report. ORNL-3492. 30 June 1963.
56. Morgan, G.B. — Quart. J. Florida Acad. Sci., Vol. 24:170. 1961.
57. Odum, E.P. Fundamentals of Ecology. Philadelphia and London,
     W.B. Saunders Co. 1959.
58. Owens, M. et al. — Proc. Soc. Water Treatment Exam., Vol. 10:1. 1961.
59. Patten, B.C. and R.L. Iverson. — Nature, Vol. 211:96. 1966.
60. Pickering, D.C. and J.W. Lucas. — Nature, Vol. 193:1046. 1962.
61. Timofeeva, N.A. and N.V. Kulikov. — In: Radioecological Concen-
     tration Processes, p. 835. London, Pergamon Press. 1967.
62. Williams, L.G. — Limnol. and Oceanogr., Vol. 5:301. 1960.
63. Williams, L.G. and H.D. Swanson. — Science, Vol. 127:187. 1958.
64. Wlodek, S. — In: Radioecological Concentration Processes, p. 897.
     Proc. Internat. Symp. Held in Stockholm 25—29 April 1966. London,
     Pergamon Press. 1967.

Chapter 16

CHEMICAL ASPECTS OF MARINE RADIOECOLOGY

Radioecology of aquatic organisms deals with the interaction between the radioactive environment and living organisms, especially chemical substances, but not ionizing radiation. Under the natural conditions in the ocean, distinguished by exceedingly high dilution of radioactive substances and the considerable shielding effect of water, the effect of dissolved radionuclides on marine organisms can only be sufficiently effective for direct contact of the organisms' cells with radioactively decaying atoms. Furthermore, at the very low radionuclide concentrations existing or liable to exist in seas and oceans, the radiobiological effect cannot be achieved unless the radionuclides are concentrated by the organisms from their environment. The interaction between aquatic organisms (especially plants) and solutes is largely controlled by physicochemical processes. This explains the part played in the radioecology of marine organisms by investigations into the chemical properties and transformations of radionuclides introduced into seawater.

The form in which artificial radionuclides occur in the marine environment interests radioecologists not only as a factor controlling the availability of these radionuclides to marine hydrobionts, but also as a factor controlling the physical transport of radioactive contamination in seas and oceans.

The above comments should not be taken to mean that marine radioecology occupies an exceptional position in studies of interaction between radioactive contamination in the ocean with its living inhabitants or the migration of radionuclides. Similar situations are characteristic of both freshwater and land radioecology, but the features of each of these natural environments should be studied independently.

Studies of the concentrating capacity of hydrobionts, which is one of the main trends in marine radioecology, have long been confined mainly to determination of empirical buildup factors for individual radionuclides in different representatives of marine flora and fauna. To a certain extent, this emphasis was dictated by top-priority practical needs. There are, however, other causes impeding the theoretical development of predicting buildup factors. Studies of radionuclide buildup in hydrobionts revealed that migration of a radionuclide (or an element) from solution into an organism is not controlled solely by the latter's physiological requirements, but also by the chemical form and aggregate state of the radionuclide in solution. Occasionally it seems to be totally unrelated to the organism's physiology. Therefore radioecologists are not satisfied with information limited to concentrations of radionuclides (or carrier elements) in water and schematic notions concerning the state of ions or aggregated solutes.

An understanding of the radionuclide distribution among organisms and environments in the ocean largely depends on the ability to answer the following questions: What real forms are assumed by a radionuclide in seawater? In what specific forms is it capable of interacting with hydrobionts? What is the mechanism of this interaction?

It appears that the first question should be answered by the chemical composition of natural seawater itself (with the exception of the transuranium elements). Unfortunately, present knowledge of marine environment is confined to its elementary composition and the forms in which individual elements exist in seawater. Knowledge of the latter aspect is hypothetical in many cases, and consists only of the most probable forms, based on general physicochemical considerations. The main components of salts in seawater have been studied better than its other ingredients. However, of the many long-lived artificial radionuclides of practical interest, only ^{90}Sr, ^{22}Na, ^{14}C and ^{3}H fall within this group. With the exception of a few elements belonging mainly to the group of alkalis and alkaline earths, we do not possess any reliable information on the real chemical form or aggregate state of carrier elements in seawater. Studies into the detailed chemical composition of seawater encounter analytical difficulties involved in direct determination of individual chemical forms, especially in the case of trace concentrations, as well as fundamental aspects of the theory of solutions of high ion strength (for seawater $\mu = 0.7$).

Processes governing the chemical composition of the marine environment require much further study. This may appear surprising, especially in the light of the lively interest toward the sea on the part of many sciences and branches of the economy. Therefore, when the necessity arose of predicting the behavior of specific chemical substances after their one-time introduction into the hydrosphere, it was practically limited to the physical processes governing the transport and diffusion of such substances.

In the absence of the necessary factual or computed data, theoretical ideas of physicochemical processes in the marine environment are based almost exclusively on formal schemes. For instance, in considering the ion form of solutes the ion charge is generally identified with valence, according to the latter's position in the periodic system, while the possible interaction between the ions and water molecules or other ions and particles present in the solution is only mentioned as a secondary phenomenon at best. The elementary composition of seawater provides the theoretical possibility of numerous combinations, but it is often very difficult to prove the greater probability of certain combinations rather than others.

Progress in studies of the minor, or trace elements in seawater has taken place only recently. Krauskopf's experimental investigations /5/ of the factors controlling the concentrations of several elements in the ocean are interesting in this respect. He has also examined /4/ the influence of a solid in seawater on the behavior of rare elements in the ocean, making use of phase diagrams. An attempt at constructing a theoretical model of seawater as an equilibrium physicochemical system based on thermodynamic characteristics of relevant equilibria was undertaken by Sillen /9/.

Problems of the form and state of elements in seawater are of interest not only to radioecologists, but also to hydrochemists, geochemists and biogeochemists. However, radioecology of marine organisms is the field in which these problems are felt most acutely. Moreover, there are certain circumstances rendering necessary the independent development of chemical investigations of artificial radionuclides in seawater, irrespective of the state of our knowledge of the chemistry of natural isotopes in the ocean.

Owing to differences in the initital forms of artificial and natural isotopes of an element upon their arrival in the ocean, their chemical behavior need not necessarily be identical, at least not for a certain period of time. Artificial radionuclides are injected into the environment either in an atomic state, or as inclusions in microscopic particles of refractory materials or dissolved in technological effluents. On the other hand, natural elements arrive in the hydrosphere mainly as weathering and erosion products of rocks. Therefore, the chemical behavior of these forms of the same element in the ocean should differ, at least in the initial state. In this respect it is important to note that because of radioactive decay the cycle of artificial radionuclides in the marine environment must be much shorter than the mean time of residence of stable nuclides in the ocean.

In view of the possible insulation of certain radioactive products of nuclear explosions from the environment on account of their being included within microscopic particles of refractory material, and also in view of the limited interaction time of any other kinds of radioactive products of nuclear explosions with the same environment, doubts arise concerning the reliability of purely chemical and biogeochemical data for predicting the behavior of artificial radionuclides in seas and oceans. Consequently, the results of observing artificial radionuclides in the ocean (or laboratory) can only be applied to the solution of geochemical problems if the forms of natural and artificial isotopes of a chemical element are identical. In our opinion, the practical importance of this topic renders very desirable comparative studies of artificial and natural isotopes in the marine environment.

There are grounds to believe that the isotope exchange of strontium proceeds fairly rapidly in marine environments. No significant differences were discovered between the buildup factors of ^{90}Sr and natural stable strontium isotopes /8/. Unfortunately this conclusion cannot be extended to all elements and radionuclides. Although the volume of available information on other elements is limited, the known facts confirm our opinion. For instance, it was reported /23/ that ^{54}Mn deposited from the atmosphere is fractionated in the sea (between the solution and suspended matter) in a different ratio than the stable Mn of the seawater. The individuality of the form of atmospheric ^{54}Mn in seawater is also indirectly denoted by observation results pertaining to its absorption by marine plants / 18/.

Sugihara and Bowen /26/ observed a very rapid fractionation of radioisotopes of rare-earth elements (^{144}Ce and ^{147}Pm) in the ocean, very different from the behavior of natural lanthanoids. No significant fractionation effect is known to exist under natural conditions for such chemically similar elements. Fractionation is only observed in the ocean between the cerium and yttrium groups of the rare-earth elements /14/. These data are in agreement with the earlier hypothesis advanced by Bogorov and Popov /1/, that only elements ordinarily existing in seawater in ion form exhibit a comparatively rapid isotope exchange in the marine environment.

Difficulties in ocean radiological observations have led to the study of many marine radioecological studies by means of artificial models; injection of dissolved radioactive preparations into seawater is a widespread technique. Although the technique appears to be valid for stimulation of liquid radioactive waste, it is hardly justified for atmospheric products of nuclear explosions. The question arises as to the representativeness of these experimental results with respect to global radioactive contamination of seas and oceans by atmospheric fallout.

It is known, for instance, that [144]Ce injected into seawater as a dissolved preparation is rapidly converted to suspended state /25/. However, [144]Ce collected from the atmosphere by filtration of atmospheric aerosols is dissolved during maceration of the filter in seawater /15/. In this connection, it is noteworthy that a relatively low content of [144]Ce was revealed in the suspended matter of seawater /7/.

Chesslet and Lalou /15/ used the maceration of aerosol filter to find that the cause for such unusual behavior of atmospheric [144]Ce in solution should not be sought in the properties of seawater, but in the specific forms in which this radionuclide arrives from the atmosphere.

The solution of these problems will depend on advances in the chemistry of atmospheric aerosols. At the same time, certain aspects of this problem are related exclusively to chemical properties of natural marine environment. For instance, recently much interest has been shown in organic matter in seawater as a complexer. Direct data have been obtained implying that surface-active substances extracted from seawater can form complexes with radionuclides. The question arises, however, whether the process actually takes place in natural seawater, where both the organic ligands and radionuclides are present in trace concentrations. As yet, laboratory checks on Fe and Zn have produced ambivalent results /17/.

Influx of atmospheric products of nuclear explosions into the ocean proceeds through the ocean-atmosphere interface, which is known to carry a higher concentration of surface-active substances. The conditions of this interface may favor irreversible formation of complex compounds of radionuclides. In other words, before its emergence in the ocean a radionuclide is subjected to preliminary "processing," which is specific for marine conditions. This aspect deserves very thorough study.

At the same time the chemical nature of atmospheric aerosols themselves must be studied. Recently, researchers have tended to conclude that the most dispersed fraction of natural atmospheric aerosol is of biogenic origin /27, 28/. The bulk of stratospheric radioactivity is known to be due to the finest fraction of particles /3, 16/. Consequently, it is quite possible that radionuclides in the atmosphere come in contact with organic matter and interact with it, forming products that are soluble in seawater (provided they are not included within refractory particles). The existence of such a process is indicated by observation results of Pavlotskaya and Zatsepina /6/, who studied ion-exchange properties of radionuclides deposited from the atmosphere. They discovered that considerable fractions of soluble ^{90}Sr and [144]Ce are present in atmospheric aerosols as anions, i.e., in a bond with complexers.

When using dissolved radioactive preparations as radioactive tags one must bear in mind that they are equally liable to reproduce neither the

properties of products of nuclear explosions, nor those of the stable isotopes of elements existing in natural seawater. Identity of chemical forms
of isotopes is a general requirement of the method of labeled atoms. In
applying the latter to studies of artificial chemical systems, special measures are taken to guarantee complete isotope exchange of the labeled
elements. In the case of natural seawater (and also any other natural
medium) this requirement cannot be fulfilled without disturbing its natural
chemical composition. Nonetheless, under natural conditions the rate of
exchange or other chemical reactions is liable to be so slow that the introduced radionuclide will remain in the individual form for a prolonged time.
Such an example with ^{131}I was reported by Fukai /18/, who observed that
the presence of stable iodine had no effect on fish absorption of ^{131}I introduced into seawater.

In practice radioecologists deal with two kinds of radioactive contamination of the marine environment: atmospheric products of nuclear explosions
and liquid industrial wastes. In the light of above-mentioned data indicating
prolonged preservation of their physicochemical individuality by radionuclides in seawater due to their origin, one must take into account the
possibility of dissimilar behavior of these varieties of radioactive substances in the ocean. As yet, only one paper /24/ contains results which
make possible any comparison. Its author, Mauchline, investigated the concentration of some radionuclides (such as ^{144}Ce, ^{106}Ru, ^{137}Cs, ^{95}Ze — ^{95}Nb
and ^{90}Sr) by marine organisms from waters of the Irish Sea contaminated by
radioactive effluent from Windscale Works and from waters washing the
British Isles, contaminated exclusively by global fallout of nuclear explosion products. Analyses were performed on various plants and animals
and their ambient waters during different seasons. As expected, the radionuclide concentration in the water was found to vary significantly at different times and at different locations. It should be emphasized that the data
did not always refer to closely allied species, so that the comparison cannot
be adequately strict. Nevertheless, attention is attracted by a systematic
excess of the buildup factors for radionuclides entering the sea with liquid
effluents over analogous parameters for radionuclides deposited from the
atmosphere, with respect to mean buildup factors for all the organisms as
well as individual species. Thus, typical buildup factors for ^{144}Ce, ^{106}Ru
and ^{95}Sr — ^{95}Nb in marine algae and invertebrates was of the order of
1,000 in the case of industrial effluents, but not over 200 in the case of
atmospheric fallout. The buildup factors were found to be approximately
the same in both cases only for ^{90}Sr and ^{137}Cs.

It is noteworthy that the above values of buildup factors for radionuclides
from industrial effluents are in satisfactory agreement with values of
buildup factors determined experimentally on dissolved preparations of
radionuclides /8/.

It should be noted, however, that the concept of "industrial wastes" is not
rigidly defined with reference to their chemical composition and the form
of certain radionuclides. For instance, the buildup of ^{106}Ru in organisms
from waste products of different technological processes is known to proceed in different manners /22/.

Hence, existing data suggest that radioisotopes of the same element may
exist in the ocean in three different forms: natural stable nuclides, artificial

radionuclides from nuclear explosions, and artificial radionuclides from industrial wastes. To some extent this restricts the use of concrete values of buildup factors determined from experiments with individual radionuclides. From a radioecological standpoint it is necessary to determine the specific forms assumed by every kind of radioactive substance, and the rate of their transformation into the equilibrium form for the marine environment.

There still remain many moot points in our notions of the forms of dissolved elements that are most available to plants or most capable of interacting with them. For instance, it is well known that diatoms assimilate suspended (or colloid) iron from seawater by metabolic absorption /19, 20/. Analogous data have recently been obtained also for cerium /25/, an unimportant element in metabolic processes. Questions arise as to the role played by the aggregation of matter in buildup processes and the ultimate cause of the remarkably high buildup factors for such radionuclides as ^{144}Ce, ^{91}Y, ^{95}Zr and ^{95}Nb, which are devoid of any biological significance.

Some mechanism governing the adhesion of suspended particles to the surface of cells is assumed to exist. Diffusion of radionuclides through cell walls is a minor process in their buildup in living organisms. Concentration of matter on the cell surface is a prerequisite for the concentration of elements that are metabolically unimportant. The nature of the mechanism of such concentration is apparently physicochemical. It cannot be understood without a clear idea of the surface structure of cells and the forms of elements interacting with this surface.

Attempts at clarifying the absorption mechanism through studies limited to the suspended and dissolved state of radionuclides (or elements), as was the case until very recently, are scarcely promising, because the division into these two states is too arbitrary and yields no information whatever on the chemical form of radionuclides. The dissolved state comprises such dissimilar forms of matter as colloids and ions, yet few authors /25/ make mention of the ion-dispersed form isolated by dialysis.

Even with such a differentiation the ion form of dissolved radionuclides, especially polyvalent ones, comprises a variety of ions differing in their chemical properties. For instance, a hydrolyzed Me^{n+} ion can form, in the simplest case, ions of form $Me(OH)_x^{n-x}$ with x assuming values from 0 to $n-1$. Ions containing different numbers of OH groups certainly possess different adsorption capacities. Unfortunately, the properties of hydroxo ions of many elements are very little known, and our knowledge of the properties of matter in its molecular-dispersed state are far from understood. This is because the range of existence of the forms in which we are interested is extremely limited in concentrated solutions on account of polymerization, while direct investigation methods are inapplicable due to dilute solutions. The state of a substance is adjudged indirectly from its behavior, but the latter is itself the ultimate aim of studies of the element's forms.

As regards chemical forms of dissolved radionuclides and their adsorption capacity in infinitely diluted solutions, the interest of marine radioecology comes into contact with modern problems of radiochemistry. Moreover, if the latter is defined as "the chemistry of substances investigated by their nuclear radiation" /2/, then the chemical problems of marine radioecology are fully within the range of radiochemistry. For those concentrations of artificial radionuclides that may actually occur in seas and oceans (below 10^{-9} mole/liter), no other methods of monitoring a substance are known, except via its ionizing radiation.

For a long time radionuclides in solutions were studied mainly in connection with the possibility of utilizing the solubility product as a constant for a given substance /11/. Therefore, the researchers used radionuclide concentrations that ensured the formation of a new phase, such as a hydroxide. However, systems of interest to marine radioecologists lie in the range of radionuclide concentrations within which formation of an independent phase of hydroxides is impossible for the majority of elements at seawater pH (~8). At the same time, the same pH of seawater ensures deep hydrolysis of compounds of such elements. Radioecologists therefore usually deal with some intermediate hydrolysis products, including the hydroxide in its molecular-dispersed state.

In the majority of cases such products cannot be identified. The classical concept of hydrolysis as a stagewise process with a separate hydrolysis concept in every stage does not reflect the actual process with trace concentrations of substances. Instead of a jumpwise appearance of simple hydrolyzed ions with increasing pH, there occurs a continuous shift of equilibrium toward deeper hydrolysis and formation of high-polymer products due to condensation of polymerization of monomers. It is true that polymerization probably does not occur until some minimum concentration is reached. Starik and Skul'skii /12/ even suggested that the process could be regarded as the result of supersaturation of the solution with respect to one of the hydrolyzed ions. In other words, the polynuclear forms possibly have their own solubility product. If the polymerization process were proved to start at strictly defined ratios of hydrolysis products, then it would be possible to construct corresponding phase diagrams for these products.

Polymerization need not necessarily take place at the infinitely low concentrations of radionuclides actually existing in the ocean. In this case the task of identifying the forms of hydrolyzed ions becomes easier, but the question arises whether the known constants of stagewise hydrolysis, generally determined in concentrated solutions, are applicable to infinitely dilute systems. There is a need for special studies of hydrolysis in dilute solutions, similar to those in, say, /13, 21/.

The fraction of suspended (or possibly colloid) radionuclides proved to be most readily accumulated by marine organisms. Therefore the aggregation mechanism for dissolved radionuclides assumes special importance for us. From radiochemistry it is known that the radiocolloid, or pseudocolloid state of matter is most characteristic under conditions of infinite dilution. We emphasize that in using the term "pseudocolloid" we should like to underscore the distinction between these systems and ordinary colloids, the characteristic properties of which are not determined by concentration but by special stabilization conditions in suspended state. Notwithstanding the abundance of papers on pseudocolloids, in our opinion the nature of this state remains obscure. There are several arguments against classifying all pseudocolloids as adsorptional. In cases of purely adsorptional pseudocolloids one must reckon with differences in the adsorption properties of different ions of the same elements. At present it may only be noted that dilution of a solution causes the conversion to a pseudo-colloid state not of the entire quantity of radionuclides but of only a fraction, which is known to depend on the medium pH. Studies into hydrolytic properties of ultrasmall amounts of cerium demonstrated a direct relationship between formation of ^{144}Ce pseudocolloid with the chemical form of hydrolyzed ions /10/. However, the development of such investigations is at present impeded by a lack of research on identification of individual hydrolyzed forms. Spitsyn et al. /10/ found that the degree of aggregation of a radionuclide is affected by its concentration, besides the ambient pH. This is in agreement with notions on the participation of polymerization phenomena in hydrolysis, assuming that the capacity for interaction with foreign contaminants in the solution is different in monomers and polynuclear ions. However, the concentration of the substance itself while it forms pseudo-colloids was not taken into account until very recently.

This chapter deals only with selected aspects of radiochemistry that are, in our opinion, of top priority in marine radioecology. They are mainly related to problems pertaining to mechanisms governing interaction between dissolved radionuclides and organisms. The scope of radioecology is certainly much wider. For instance, the buildup of radionuclides is closely related to their biochemistry, the radiational situation in seas and oceans is related to radionuclide participation in sedimentogenesis, while our capacity for prompt detection of radioactive contamination depends on the development of determination methods for ultrasmall concentrations of radionuclides in seawater. These and other problems still await solution.

References

1. Bogorov, V.G. and N.I. Popov. — Okeanologiya, Vol. 5:569. 1965.
2. Vinogradov, A.P. Osnovnye problemy radiokhimii (Fundamental Problems of Radiochemistry). Moskva, Izdatel'stvo Akademii Nauk SSSR. 1959.
3. Gaziev, Ya.I. and L.E. Nazarov. — In Sbornik: "Radioaktivnye izotopy v atmosfere i ikh ispol'zovanie v meteorologii," p. 181. Moskva, Atomizdat. 1965.
4. Krauskopf, K. — In Sbornik: "Problemy rudnykh mestorozhdenii," p. 375. Moskva, Izdatel'stvo inostrannoi literatury. 1958.
5. Krauskopf, K. — In Sbornik: "Geokhimiya litogeneza," p. 294. Moskva, Izdatel'stvo inostrannoi literatury. 1963.
6. Pavlotskaya, F.I. and L.N. Zatsepina. — Atomnaya energiya, Vol. 20:333. 1966.
7. Patin, S.A. and N.I. Popov. Disposal of Radioactive Wastes into Seas, Oceans and Surface Waters, p. 443. Vienna, IAEA. 1966.
8. Polikarpov, G.G. Radioekologiya morskikh organizmov (Radioecology of Marine Organisms). Moskva, Atomizdat. 1964.
9. Sillen, L.G. — In Sbornik: "Okeanografiya," p. 428. Moskva, "Progress." 1965.
10. Spitsyn, V.I. et al. — DAN SSSR, Vol. 182:855. 1968.
11. Starik, I.E. Osnovy radiokhimii (Principles of Radiochemistry). Moskva — Leningrad, Izdatel'stvo Akademii Nauk SSSR. 1959.

12. Starik, I.E. and I.A. Skul'skii. — Radiokhimiya, Vol. 1:379.   1959.
13. Stepanov, A.V. and V.P. Shvedov. — Zhurnal neorganicheskoi
        khimii, Vol. 10:1000.   1965.
14. Fomina, L.S. — DAN SSSR, Vol. 170:1181.   1966.
15. Chesslet, R. and C. Lalou. — Bull. Inst. Oceanogr. Monace,
        Vol. 64:1328.   1965.
16. Drevinsky, P. and E. Martell. Preliminary Results of the Size
        and Vertical Distributions of Residual Nuclear Debris in the Strato-
        sphere. — UN Document A/AC. 82/G/L, p. 775,   1962.
17. Duursma, E.K. and W. Sevenhuysen. — Netherland J. Sea Re-
        search, Vol. 3:95.   1966.
18. Fukai, R. — Rapp. et proc. -verb. Réun. Comiss. Internat. Explorat.
        Scient. Mer. Méditerr., Monaco, Vol. 18(3):361.   1965.
19. Goldberg, E.D. — Biol. Bull., Vol. 102:243.   1952.
20. Harvey, H.W. — J. Mar. Biol. Ass. UK, Vol. 22:205.   1937.
21. Hedstrom, B.O.A. — Arkiv kemi, Vol. 6:1.   1953.
22. Jones, R.F. — Limnol and Oceanogr., Vol. 5:312.   1960.
23. Lowman, F.G. et al. Disposal of Radioactive Wastes into Seas,
        Oceans and Surface Waters, p. 249. IAEA, Vienna.   1966.
24. Mauchline, J. The Biological and Geographical Distribution in the
        Irish Sea of Radioactive Effluent from Windscale Works 1959 to
        1960. — UKAEA Report, Harwell, AHSB (RP).   1963.
25. Rice, T.R. and V.M. Willis. — Limnol. and Oceanogr., Vol. 4:247.
        1959.
26. Sugihara, T.T. and V.T. Bowen. — In: Radioisotopes in the
        Physical Science and Industry, Vol. 1:57. Vienna, IAEA.   1962.
27. Went, F.W. — Proc. Nat. Acad. Sci. USA, Vol. 51:1259.   1964.
28. Went, F.W. — Tellus, Vol. 18:549.   1966.

Chapter 17

ACCUMULATION OF ARTIFICIAL RADIONUCLIDES IN FRESHWATER FISH

Introduction

Study of the migration of natural and artificial radionuclides in the biosphere includes an especially interesting aspect, the discovery of biogeocenoses in which, for certain reasons, conditions are created for a high accumulation of radioactive substance by living organisms. Two instances are to be distinguished here.

1. High concentrations of radionuclides in individual units of biogeocenoses are the result of local contamination of land or water areas brought about by constant or periodic disposal of radioactive atomic wastes in the environment, nuclear explosions and accidents. In this case these high concentrations are observed not only in organisms but also in the environment (soil, water, etc.).

2. High concentrations of radionuclides in organisms result from internal causes inherent in specific biogeocenoses and may be observed in regions where the radionuclide concentration in the environment is not high and global fallout is the sole source of artificial radionuclides.

Despite the substantial differences between these instances, the actual cause of the mass accumulation of radionuclides by organisms is the same, namely, a high proportional concentration of a radionuclide with respect to its isotopic or nonisotopic carrier in the edaphic medium. The only different is that in the first case this is brought about by the introduction of large amounts of radionuclides in the soil, while in the second, by low concentrations of stable carriers in it.

A clear example of a terrestrial biocenosis, the links of which show a high concentration of radionuclides as a result of global fallout, is the arctic lichen — northern reindeer — man food chain, which has been studied in recent years both in the USSR and elsewhere /19, 44, 47/.

Interesting aquatic biocenoses from this point of view are the weakly saline freshwater lakes, as their content of stable carrier isotopes may be low enough to ensure high levels of accumulation of the corresponding radionuclides by hydrobionts. In fact, such lakes have recently been discovered in the USSR, Finland, Switzerland and the USA, where the fish contain tens and hundreds of times more ^{90}Sr and ^{137}Cs than marine fish /24, 25, 27, 37, 40/.

Studies of freshwater lakes with water of very diverse chemical composition and with other diverse ecological features are revealing great possibilities for investigating under natural conditions the migration patterns of radionuclides and the corresponding chemical elements in aquatic biocenoses.

With regard to aquatic animals, the following questions arise: How (directly from the water or through food) do radionuclides enter the organism? In what way does this process depend on the type of osmoregulation? What link exists between the radioactivity of hydrobionts and the concentration of radionuclides in the water, taking into consideration their physicochemical state and chemical analogs? These aspects are of great importance both for the radioecology of aquatic animals and for clarification of the mechanism of water-salt metabolism.

Of special interest in studies of the patterns of interaction with the radioactive environment are fish, since they possess a complete system of ionic regulation, form high trophic levels in aquatic biocenoses, and are important to man as food.

^{137}Cs and ^{90}Sr in fish of freshwater lakes

In regard to the migrations of radioactive products of nuclear explosions in the biosphere, the fission radionuclides ^{137}Cs and ^{90}Sr deserve special attention, for at least two reasons. First, owing to their relatively long half-lives (30 and 28 years respectively), they subsist in the biosphere for a long time, and second, unlike most other fission products (^{144}Ce, ^{106}Ru, ^{95}Zr, ^{95}Nb), whose biochemical role is unknown, the isotopes of cesium and strontium, although apparently without independent biochemical functions, do take part in metabolism as chemical analogs of the very important mineral elements potassium and calcium.

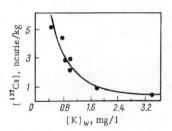

FIGURE 17.1. Concentration of ^{137}Cs in burbot (Lota vulgaris) as a factor of the concentration of potassium in lake water /41/

The significance of ^{90}Sr and ^{137}Cs in the contamination of hydrobionts may be illustrated by the fact that even during periods immediately following nuclear tests, when the radioactivity of atmospheric fallout in the environment is almost fully determined by short-lived radionuclides, only ^{137}Cs and ^{90}Sr are observed in measurable amounts in the organs and tissues of terrestrial and aquatic vertebrates.

Investigation of hydrobionts in one of the lakes of Kamchatka in 1962 /25/ confirmed that freshwater fish selectively absorb ^{137}Cs. Even though a study of the γ-spectra of their food (zooplankton) and aquatic vegetation did not reveal ^{137}Cs against the background of large amounts of short-lived products of nuclear explosions, in the γ-spectra of fish the ^{137}Cs peak is the dominant one, while no peaks of short-lived fission nuclides are detected.

Cesium-137. The connection between the level of accumulation of ^{137}Cs in fish and the limnological characteristics of freshwater lakes in Finland was studied in detail by Kolehmainen at al. /41/. It was found that fish from oligotrophic lakes contain much more ^{137}Cs than those from eutrophic lakes (Table 17.1), this being due primarily to the low content of potassium in the water of oligotrophic lakes. Figure 17.1 shows the clear negative correlation between the K content in lake water and the concentration of ^{137}Cs in fish.

TABLE 17.1. Accumulation of ^{137}Cs in fish from Finnish lakes in summer 1964, ncurie/kg wet substance /41/

Type of lake	[K] mg/1	Perch (P e r c a f l u v i a t i l i s)	Pike (E s o x l u c i u s)	Burbot (L o t a v u l g a r i s)	Roach (L e u c i s- c u s r u t i l u s)	Whitefish (C o- r e g o n u s sp.)
Eutrophic	3.5	0.26	0.44	0.38	0.19	—
	1.8	1.16	0.77	0.85	0.39	—
	2.4	1.92	1.12	—	0.51	—
	1.0	2.28	2.53	2.18	0.99	0.85
Oligotrophic	1.0	3.11	3.20	2.78	1.16	—
	0.9	3.86	3.87	2.82	1.78	—
	1.0	2.93	3.69	—	—	—
	0.7	5.26	8.84	—	—	—
	0.4	—	—	5.14	—	1.77
	1.0	—	6.84	—	2.47	1.94
	0.8	7.89	6.49	4.31	3.29	3.15
	0.2	8.47	—	—	—	—

Potassium is a nonisotopic carrier of cesium, and its concentration in fish varies insignificantly, so that the determining factor in the accumulation of ^{137}Cs by fish is the magnitude of the ^{137}Cs/K ratio in the water. In other words, with an exchange equilibrium between ^{137}Cs and potassium in the elements of an ichthyocenosis, the level of ^{137}Cs accumulation in fish living in different lakes must be directly proportional to the content of ^{137}Cs and inversely proportional to the potassium content in the water.

Let us examine the relationship between the ^{137}Cs and potassium concentration in the water and that in fish more closely. We shall designate the concentration of potassium, stable cesium, and ^{137}Cs in fish as $[K]_f$, $[Cs]_f$, and $[^{137}Cs]_f$ and in the water as $[K]_w$, $[Cs]_w$, and $[^{137}Cs]_w$. At isotopic equilibrium we obtain the relationship

$$\left[\frac{^{137}Cs}{Cs}\right]_f = \left[\frac{^{137}Cs}{Cs}\right]_w \qquad (17.1)$$

Potassium is a nonisotopic carrier of cesium (when $[K] \gg [Cs]$), and therefore the ratios of the concentrations of these elements in water and fish may differ even in equilibrium, as expressed by the formula

$$\left[\frac{Cs}{K}\right]_f = d\left[\frac{Cs}{K}\right]_w, \qquad (17.2)$$

where d is the coefficient of deviation,* which depends on the species of fish, their age, and ecological factors. Expressions (17.1) and (17.2) give

---

* The terms "coefficient of discrimination" and "observed ratio" in radioecological literature are synonyms of "coefficient of deviation."

$$\left[\frac{^{137}\text{Cs}}{\text{K}}\right]_f = d\left[\frac{^{137}\text{Cs}}{\text{K}}\right]_w, \tag{17.3}$$

whence

$$[^{137}\text{Cs}]_f = d\,[\text{K}]_f\left[\frac{^{137}\text{Cs}}{\text{K}}\right]_w. \tag{17.4}$$

The fluctuations in the potassium content in fish of the same species but of different ecology are not large, and therefore in formula (17.4) $[\text{K}]_f \approx$ $\approx$ const. If we take different lakes and compare fish of a certain species, the same age, and the same type of diet, the coefficient of deviation is also approximately constant, i. e., $d \approx$ const. Consequently,

$$[^{137}\text{Cs}]_f \approx B\frac{[^{137}\text{Cs}]_w}{[\text{K}]_w}, \qquad B = \text{const.} \tag{17.5}$$

Finally, in this case, when the ^{137}Cs concentration in the water of compared lakes is about the same, we obtain the following approximate equation:

$$[^{137}\text{Cs}]_f \approx \frac{C}{[\text{K}]_w}, \qquad C = \text{const.} \tag{17.6}$$

It is interesting to note that an inverse relationship between $[^{137}\text{Cs}]_f$ and $[\text{K}]_w$ will be present not only when each of the values d, $[\text{K}]_f$, and $[^{137}\text{Cs}]_w$ are individually constant, but also when their variations in different lakes are mutually compensated, so that only their product C remains more or less constant.

Naturally, in actual water bodies a constant value of C is the exception rather than the rule, and experimental results (Figure 17.2) deviate to a greater or lesser degree from the curves calculated from formula (17.6). These deviations indicate nonobservance of at least one of the conditions formulated in the derivation of equation (17.6): first, the concentrations of ^{137}Cs may differ in different lakes, and then formula (17.5) has to be used; second, the condition of isotopic equilibrium (17.1) may not be fulfilled or the degree of deviation from isotopic equilibrium is different for the ichthyocenoses of the lakes compared; third, the reason for deviations may be differences in the coefficients of deviation d, if the feeding conditions or age of the fish studied are dissimilar.

The question of isotopic equilibrium will be discussed in greater detail below. We note that, as our investigations in Lake Dal'nii (Paratunka river basin, Kamchatka) showed, the specific activity of cesium in freshwater fish $[^{137}\text{Cs}/\text{Cs}]_f$ depends essentially on the type of feeding, and in summer 1966 an approximate coincidence of $[^{137}\text{Cs}/\text{Cs}]_f$ with $[^{137}\text{Cs}/\text{Cs}]_w$ was observed only among plankton-feeders, whereas in fish associated with the benthos food chain, this value was 2−3 times lower than in the water.

Let us examine which numerical values of the coefficient of deviation are characteristic for particular species of fish and how this coefficient depends on their age. Table 17.2 presents the values of d for the muscle tissue of some freshwater and marine fish, calculated from sparse data on the concentration of stable cesium in water and fish. Interestingly, in all the marine species studied and the freshwater Dolly Varden char $d > 1$, i. e., the Cs/K ratio is higher in the muscles of these fish than in the water. On the other hand, the same ratio is about three times lower in the muscles of juvenile and dwarf forms of sockeye than in the water, i. e., $d < 1$.

TABLE 17.2. Coefficient of deviation d in the Cs−K pair for the muscle tissue of some freshwater and marine fish

Species	d	Reference
Fresh water (Lake Dal'nii)		
Sockeye (Oncorhynchus nerka):		/22/
downstream-migrating young	0.28	
dwarf form	0.37	
Dolly Varden char (Salvelinus malma)	2.0	/22/
Sea water		
Sockeye (Oncorhynchus nerka)	4.7	/22/
Cod (Gadus morhua macrocephalus)	4.8	/22/
Haddock (Melanogrammus aeglefinus)	2.9	/22/
Goby (Myoxocephalus scorpius)	4.4	/4/
Herring (Clupea harengus)	7.2	/29/
Flounder (Pleuronectes platessa)	2.1	/29/
Skate (Raja batis)	12.5	/29/

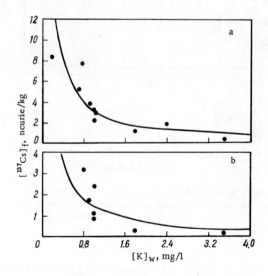

FIGURE 17.2. Comparison of experimental data /41/ with results calculated from formula (17.6):

a − perch ($C = 3.1$); b − roach ($C = 1.4$).

The reasons for the opposite direction of the deviations of $[Cs/K]_f$ in char and the freshwater part of the sockeye population from Lake Dal'nii are not clear. The low values of $[Cs/K]_f$ in dwarf forms and juvenile sockeye are apparently due to feeding peculiarities, and displacement of potassium by cesium takes place already in planktonic organisms. This hypothesis agrees with results obtained during a study of different species of the genus O n c o r h y n c h u s in the marine period of life. The content of ^{137}Cs in plankton-feeders (sockeye, chum, pink salmon) is much lower than in the predatory species of the genus — chinook and coho /43/. It is also noteworthy that, according to Bryan et al. /29/, the ratio $[^{137}Cs/K]$ in trout (S a l m o   t r u t t a) is about $^1/_5$ of that in lake water; if the trout were in isotopic equilibrium with the water, then d would equal 0.2.

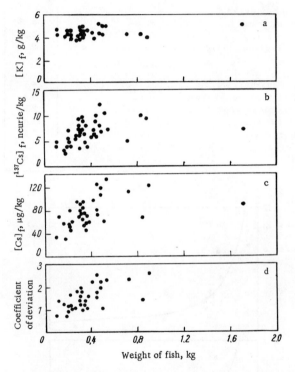

FIGURE 17.3. Concentration of potassium (a), ^{137}Cs (b), stable cesium (c), and the coefficient of deviation (d) as factors of the weight of lake stone-loach (S a l v e l i n u s   m a l m a) from Lake Dal'nii

For lake char a positive correlation was found between the coefficient of deviation d and the weight of the fish (Figure 17.3); this apparently reflects an increasing tendency in the half-life of cesium in char with age. Häsänen et al. /36/ found such a tendency for rainbow trout (S a l m o i r i d a e u s) and roach (L e u c i s c u s   r u t i l u s). The experiments were conducted under near-natural conditions; ^{137}Cs-tagged fish were kept in large tanks which were submerged in the lake, and they were given food not containing [137]Cs. After specific intervals the fish were removed from the tanks for a short time in order to measure the ^{137}Cs content in vivo, after which they were returned to them for continuation of the experiment (Table 17.3). The very slow exchange of ^{137}Cs between the organism and the environment is striking in the fish studied, especially in freshwater perch ($T_{1/2}$= 175−200 days in summer at a temperature of 15±5°C). It is interesting that the lower temperature of the water in fall (down to 5°C) corresponds to a double or triple increase in the half-life of [137]Cs.

TABLE 17.3. Half-life (slow component) of ^{137}Cs in perch, roach, and rainbow trout at a temperature of 15 ±5°C /36/

Species	Age, years	Number of fish	Slow component of half-life	
			relative deposit in complete half-life, %	$T_{1/2}$, days
Perca fluviatilis	2 − 3	16	94	175
	3 − 6	12	96	175
	6 − 8	3	88	200
Leuciscus rutilus	2 − 3	11	94	55
	4 − 6	4	93	85
	9 − 12	3	91	100
Salmo iridaeus	0.3 − 0.5	50		25
	1 − 2	18	66	55
	2 − 3	15	76	80

If food is the source of cesium in the organism of freshwater fish, then with $d>1$ the ratio $[Cs/K]_f$, and consequently the concentrations of Cs and [137]Cs (provided that their physicochemical composition is the same) must be higher at each successive trophic level of the ichthyocenosis; this was noted by Gustafson /34/ in a study of ^{137}Cs accumulation in fish from a lake in the USA (Table 17.4). With the transition from the first trophic level (mixture of small fish) to the third (pike), the $[^{137}Cs/K]_f$ ratio rose 5.5 times and the ^{137}Cs concentration almost ten times.

TABLE 17.4. Concentration of ^{137}Cs and K and the ^{137}Cs/K ratio in three successive trophic levels of an ichthyocenosis in Red Lake (State of Minnesota, USA) in summer 1965 /34/

Species	Trophic level	[137]Cs, ncuries/kg	K, g/kg	[137]Cs/K, whole units
Mixture of small fish	1st	407	2.03	200
Perch	2nd	752	2.40	313
Northern pike	3rd	3,621	3.33	1,087

These findings agree well with the effect of the trophic levels discovered in terrestrial biogeocenoses /53/. It has been established that in land animals the ^{137}Cs/K ratio is always higher than in their food, owing to the longer retention of cesium in the organism as compared with potassium (the half-life of these elements generally differs 2—3-fold in the same species).

Coefficients of accumulation of cesium and potassium in fish

The coefficient of accumulation q is the ratio of the concentration of an element or nuclide in the hydrobiont and the surrounding water. This parameter serves as a measure of the concentrating capacity of hydrobionts with respect to different elements and is of utmost importance for the quantitative description of the interaction between hydrobionts and their environment /16, 18, 54/.

Let us see how the coefficients of accumulation of potassium q_K and cesium q_{Cs} change in fish in relation to the potassium concentration in the water.

According to experiment,

$$q_K = \frac{[K]_f}{[K]_w} ; \qquad (17.7)$$

$$q_{Cs} = \frac{[Cs]_f}{[Cs]_w} . \qquad (17.8)$$

From formula (17.2) we obtain

$$q_{Cs} = d q_K . \qquad (17.9)$$

As already shown in deriving equations (17.5) and (17.6), the concentration of potassium in different freshwater and marine fish varies within comparatively narrow limits and is equal to 2.5—4.5 g/kg wet substance /3, 4, 22, 34, 35, 40/. Consequently, substitution of the numerical values of [K] in formulas (17.7) and (17.9) yields

$$q_K = \frac{2.5 - 4.5}{[K]_w} ; \qquad (17.10)$$

$$q_{Cs} = \frac{2.5 - 4.5}{[K]_w} d, \qquad (17.11)$$

where $[K]_w$ is expressed in g/1.

In the logarithmic scale the relations obtained are depicted by straight lines:

$$\log q_K = a - \log [K]_w ; \qquad (17.12)$$

$$\log q_{Cs} = b - \log [K]_w . \qquad (17.13)$$

where $a = \log [K]_f \approx$ const; $b = \log d [K]_f \approx$ const when $d \approx$ const.

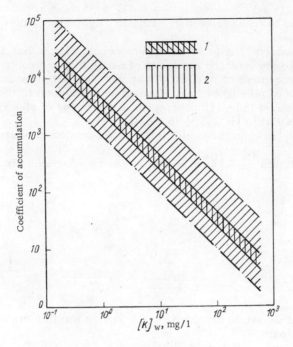

FIGURE 17.4. Relationships (17.10) and (17.11):

1—region of possible q_K values; 2—region of possible q_{Cs} values
with $0.4 < d < 4.0$.

These relations are shown in Figure 17.4, in which the possible values
of parameter d are taken from Table 17.2. The fluctuations of q_{Cs} with a
fixed value of $[K]_w$ are much greater than those of q_K, due to the relatively
wide range of possible d values in different fish species.

Let us assess the threshold values of q_K and q_{Cs} according to available
data /22, 29/.

For the open ocean ($[K]_w = 390$ mg/l; $d = 2 - 12$) $q_K = 6.4 - 12$;
$q_{Cs} = 13 - 140$; for Lake Dal'nii ($[K]_w = 0.27$ mg/l; $d = 0.4 - 2.0$) $q_K =$
$= 10,000 - 17,000$; $q_{Cs} = 4,000 - 30,000$.

Hence, the coefficient of accumulation of potassium and cesium in fish
living in weakly saline surface waters, particularly in oligotrophic lakes,
may exceed the corresponding coefficients observed in marine fish
$100 - 1,000$ times.

**Strontium-90 in fish of freshwater lakes and the coefficients of
accumulation of strontium and calcium.** Considering the clear correlation
between the coefficients of accumulation of cesium, and also the content of
[137]Cs in fish and the concentration of potassium in the water, we must
expect a similar relationship between the content of ^{90}Sr in fish and the
concentration of its nonisotopic carrier calcium in water. This means
that the highest coefficients of accumulation of stable strontium and [90]Sr

should correspond to freshwater fish living in oligotrophic lakes with a low calcium content.

In fact, as Agnedal/27/ showed, the ^{90}Sr concentration in fish from several Swiss lakes is tens and hundreds of times higher than in marine fish. The highest content of ^{90}Sr (20,400 pcuries/kg wet substance) is found in the bones of pike from Lake Rogen (the Ca content in the water is 2 mg/l) and the lowest (50 – 110 pcurie/kg) in the bones of pike from Tveren Bay in the Baltic ($[Ca]_w$ = 96 mg/l).

The connection between the coefficients of accumulation of strontium $q_{Sr} = \dfrac{[Sr]_f}{[Sr]_w}$ and calcium $q_{Ca} = \dfrac{[Ca]_f}{[Ca]_w}$ is expressed by a formula similar to (17.9):

$$q_{Sr} = D \cdot q_{Ca},$$ (17.14)

where

$$D = \frac{[Sr/Ca]_f}{[Sr/Ca]_w}$$ (17.15)

The coefficient of deviation for the Sr — Ca pair is similar to the coefficient d for the Cs — K pair.

TABLE 17.5. Coefficient of deviation D in the Sr –Ca pair for the bone tissue of fish from different Swiss lakes /27/

Concentration of Ca and Sr in lake water			D		
Ca, mg/l	Sr, μ/l	Sr/Ca, 10^{-3}*	Perch	Pike	Roach
Lakes					
2.0	6	1.37	0.35	0.23	0.48
2.0	8	1.83	–	0.50	–
4.4	37	3.67	0.22	0.23	–
4.5	41	4.19	–	0.19	–
12.0	38	1.44	0.13	0.13	0.23
18.0	55	1.41	0.18	0.16	0.32
26.0	72	1.26	0.24	0.17	0.35
40.0	70	0.80	0.21	0.18	0.26
46.0	72	0.71	0.24	–	0.31
54.0	375	3.17	0.16	0.19	0.34
63.0	240	1.74	0.20	0.16	0.29
63.0	120	0.87	0.22	0.17	0.37
Tveren Bay (Baltic)					
96.0	1,800	8.56	0.36	0.19	0.26
Mean	–	–	0.23	0.21	0.32

* Atomic ratios

The numerical values of D for three species from 12 Swiss lakes and Tveren Bay are given in Table 17.5. The data show that the D values have a fairly narrow range $(0.15 - 0.5)$ despite the wide range of Ca and Sr concentrations in these waters and the appreciable difference in the proportions of these elements. Such a result was earlier obtained by Templeton and Brown /57/ for Salmo trutta: $D = 0.45 - 0.56$ practically did not depend on the Ca concentration in the water in the range $0.3 - 100$ mg/l. According to Beninson et al. /28/, with a change in the weight ratio of Sr/Ca in the water from 10^{-4} to 1, the coefficient of deviation D remained practically constant and equaled approximately 0.4 for three species of South American freshwater fish.

Low values of the coefficient of deviation $(D < 0.5)$ are characteristic for both freshwater and marine fish. According to Bachurin et al. /1/, $D = 0.15 - 0.35$ for seven species of Black Sea fish. For fish from the North and Bering seas the mean value of D is 0.22 /32/.

Thus, the behavior of the Sr—Ca pair differs greatly from that of the Cs—K pair: in fish compared with the water, calcium is poor in strontium $(D < 1)$, whereas in most cases potassium is enriched by cesium $(d > 1)$. When the source of strontium and calcium for freshwater fish is the food, this must lead to a lowered concentration of strontium (and ^{90}Sr) in the higher trophic levels. This may be associated with the low values of D in predatory fish (perch and pike) as compared with roach and trout (see Table 17.5). The effect of the trophic levels for the Sr—Ca pair was asserted by Ophel and Judd /51/, who found an approximated twofold decrease in the Sr/Ca ratio in lake fish in comparison with their food (stomach contents).

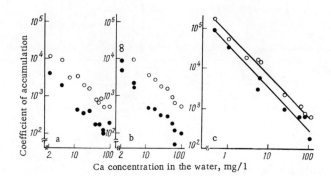

FIGURE 17.5. Coefficients of accumulation of Ca (O) and Sr (O) in the bone tissue of fish as factors of the calcium concentration [Ca]$_w$ in lake water:

a—Perca fluviatilis; b—Esox lucius /27/; c—Salmo trutta /57/.

Templeton and Brown /57/ and Angedal /27/ found a distinct correlation between the Ca concentrations in lake water and the coefficients of accumulation q_{Ca} and q_{Sr} (Figure 17.5). For the bone tissue of Salmo

trutta the regression lines calculated from experimental data /57/
according to the least squares method, are described by equations

$$\log q_{Ca} = 4.77 - 1.01 \log [Ca]_w;$$ (17.16)

$$\log q_{Sr} = 4.51 - 1.03 \log [Ca]_w.$$ (17.17)

These equations are analogous to equations (17.12) and (17.13); they
denote that q_{Ca} and q_{Sr} are inversely proportional to $[Ca]_w$. This result,
which is a consequence of the approximate constancy of $[Ca]_f$ in the bone
tissue of fish of some species, is independent of the medium of habitat (from
oligotrophic lakes to seas and oceans). In other words, for freshwater and
marine fish the following relationships are true:

$$q_{Ca} = \frac{E}{[Ca]_w};$$ (17.18)

$$q_{Sr} = \frac{ED}{[Ca]_w},$$ (17.19)

where $E = [Ca]_f \approx$ const and $ED =$ const, if the same tissue of a particular
fish species is compared.

The relationship (17.17) agrees well with the results of Kulebakina /11/,
who showed that the coefficient of accumulation of ^{90}Sr in marine fish is
inversely proportional to the Ca concentration in seawater. Polikarpov /17/
also examined the dependence of q_{Sr} on the salinity of seawater.

Table 17.6 gives the range of possible q_{Ca} and q_{Sr} values for the bone
tissue of freshwater and marine fish, calculated from formulas (17.18) and
(17.19); the calculation was conducted with the typical values $[Ca]_f = 30 -$
$- 80$ mg/g wet substance /27, 57/ and $D = 0.15 - 0.5$.

TABLE 17.6. Range of possible q_{Ca} and q_{Sr} values for the bone tissue of freshwater and marine fish

Habitat	$[Ca]_w$, mg/1	q_{Ca}	q_{Sr}
Open ocean	400	75 – 200	10 – 100
Oligotrophic lake with a low Ca content	0.4	75,000 – 200,000	10,000 – 100,000

In concluding this section it should be noted that the relations obtained
for q give values of equilibrium coefficients of accumulation, i.e., they are
always true for stable cesium and strontium, being in eternal equilibrium
with their nonisotopic carriers, potassium and calcium. The same relationships are true for ^{137}Cs and ^{90}Sr if isotopic equilibrium is achieved (i.e.,
$[^{137}Cs/Cs]_f = [^{137}Cs/Cs]_w$ and $[^{90}Sr/Sr]_f = [^{90}Sr/Sr]_w$) which, however, is far
from always being the case. According to Ophel /48, 49/, it takes several
years for the establishment of equilibrium for ^{90}Sr in a freshwater lake if
the concentration of this radionuclide in the water is maintained at a
constant level. Bachurin et al. /1/ showed that in 1964 — 1965, in hydrobionts of the coastal zone of the Black Sea, the coefficients of accumulation

of stable Sr and ^{90}Sr coincided on the average, although a more than threefold difference was noted for individual species of fish. Rovinskii and Agre /20/ determined the coefficients of accumulation of stable Sr and ^{90}Sr in three species (perch, goldfish, and carp) from 10 eutrophic freshwater lakes and found that in most instances their ratio was close to 1, while the mean ratio was 1.03±0.13. However, during a study of the accumulation of stable cesium and ^{137}Cs in fish from Lake Dal'nii it was found that the condition of isotopic equilibrium (see relationship (17.1)) was approximately fulfilled only in plankton-feeders, whereas in fish associated with the benthos food chain the specific activity of cesium ^{137}Cs/Cs was 2—3 times lower than in the water. Thus, even under conditions of a long-term experiment, the coefficients of accumulation of ^{90}Sr and ^{137}Cs in fish may differ considerably from the equilibrium values.

Paths of entry of radionuclides into the fish organism

Aquatic as opposed to terrestrial animals are in direct contact with a solution which contains many mineral and organic substances essential for life activity. In connection with this, Pütter /55/ proposed at the beginning of this century that, like plants, aquatic animals can assimilate dissolved nutrients from the water. He extended his hypothesis to fish, considering that here absorption takes place via the gills and the gastrointestinal tract /56/. Subsequently special cells were discovered in the branchial apparatus of teleosts the function of which consists in maintaining the water-salt balance /38/. In freshwater fish, which live in a hypotonic medium, these cells actively transport sodium and chlorides from the water into the organism /7, 52/. The branchial apparatus of marine fish transports Na and Cl ions in the opposite direction, excreting them from the organism. However Krogh /42/, confirming the active transport of Na and Cl by the gills, showed that nutrient substances enter the organism only with the food and that the organic compounds dissolved in water bodies cannot be directly utilized by fish. Pütter's theory was rejected as a result of these studies.

With the wide use of labeled atoms, in recent years new possibilities have opened up for studying the osmotic nutrition of fish. The high sensitivity of radioisotopic methods has made it possible to establish that many mineral and organic ions penetrate the fish's body directly from the water through the gills and the outer covers. Much has been written on the accumulation of radionuclides by fish in experimental conditions. Many investigators have not only established that fish directly absorb radionuclides from the water but have also attempted to clarify whether it is the water or the food that is the main source of radionuclides for freshwater fish. In particular, Tomiyama et al. /58/ concluded from experiments on the absorption of ^{45}Ca by goldfish (C a r a s s i u s a u r a t u s) that the food is unimportant as a source of calcium. Lebedeva /12/ obtained similar results for ^{89}Sr in its absorption by carp: the proportion of this nuclide entering from the water amounted to 65 — 75%. Ophel and Judd /50/ determined that in goldfish the relative deposit of ^{90}Sr greatly depends on the coefficients of accumulation of this nuclide in the food. According to his estimate, with $q=500$, 92% ^{90}Sr is obtained from the food, while with $q=100$, the entry of ^{90}Sr is approximately the same from the food and from the water.

From experiments along these lines researchers have concluded that sorption processes play a large part in supplying fish with minerals and organic substances /9, 16, 21, 26, 54/. In particular, Shul'man /26/ claims that "a high rate of absorption of several elements (including high-molecular organic compounds) from the water through the gills and skin has been demonstrated; these elements are then drawn into the plastic metabolism. This permits us to return to the theory of an "osmotic nutrition" of fish on a new level and to show that parenteral nutrition plays a more important role in fish than was apparent when Pütter's theory became outmoded."

However, studies have been carried out, both under experimental conditions and in nature, which give evidence of an extremely low absorption capacity of the gills and outer covers for many radionuclides. Davis and Foster /31/ showed that of the large number of radionuclides contained in the waters of the Columbia River, where radioactive wastes are disposed, only ^{24}Na is absorbed by fish directly from the water, while more than 98% of the other radionuclides are absorbed through the food chains. This selective absorption of Na from the water agrees well with physiological data on the osmoregulatory role of this element and its active transport via the branchial apparatus /7, 52/. Il'in and Moskalev /8/ concluded from their experiments on the absorption of ^{90}Sr, ^{137}Cs, and ^{32}P in a natural water body that the gastrointestinal tract is the main path of entry of radionuclides in fish. According to King /39/, direct absorption of ^{137}Cs from the water by juvenile sunfish (L e p o m i s m a c r c h i r u s) is negligible in comparison with its assimilation from food.

An interesting study was performed by Morgan /45/, who experimentally compared direct absorption of ^{134}Cs from the water by marine and fresh-water fish. The fish were kept at 10°C in tanks of seawater and freshwater, and the concentration of ^{134}Cs was maintained at a constant level. Food not containing ^{134}Cs was given so that the radionuclide could be absorbed only from the water (Table 17.7). The highest rate of absorption from the water was observed in pelagic teleosts in seawater. After only $0.5 - 2.5$ days the concentration of ^{134}Cs in marine fish became equal to the concentration in the water. In the course of $1 - 2$ months values of the coefficients of accumulation of ^{134}Cs characteristic for marine fish ($20 - 40$) became closer to equilibrium, i.e., during this time, practically complete isotopic exchange of cesium took place between the fish and the water. In all the freshwater fish, and also in the representative of marine elasmobranchs (ray) absorption of ^{134}Cs from the water occurred much more slowly than in marine teleosts. It took these fish $1 - 4$ months to achieve the same concentration of ^{134}Cs as in the water. The equilibrium coefficients of accumulation of cesium in freshwater fish are $10^3 - 10^4$; it thus follows from the data obtained that in the course of about 100 days isotope exchange between water and fish constitutes not more than 0.1%. We can attribute the substantial differences in the rate of ^{134}Cs absorption between freshwater and marine teleosts to the characteristics of osmoregulation in hypotonic (freshwater) and hypertonic (seawater) media. It is known that marine teleosts drink water, absorbing also the ions dissolved in it

TABLE 17.7 Accumulation of ^{134}Cs by fish from marine and freshwater /45/

Species	Time required for the same concentration of ^{134}Cs to be achieved in the fish as in the water, days	Coefficient of accumulation of ^{134}Cs (after 28 days)
Seawater		
Teleosts		
Pelagic fish		
Sprat (Sprattus sprattus)	0.5 — 1.0	—
Mackerel (Scomber scombrus)	0.1	—
Herring (Clupea harengus)	1.5 — 2.5	9.2
Demersal fish		
Blackcod (Pollachius virens)	6 — 7	4.0
Walleye pollock (P. pollachius)	6 — 7	3.5
Mullet (Mugil chelo)	6 — 7	2.7
Cod (Gadus morhua)	9 — 10	2.7
Eel (Anguilla anguilla)	16 — 17	1.6
Demersal fish (flat)		
Flounder (Platichthus flesus)	7 — 8	—
Halibut (Scophthalmus maximus)	9 — 10	2.8
Turbot (S. rhombus)	9 — 10	2.6
Dab (Limanda limanda)	9 — 10	2.5
Sol (Solea solea)	10 — 11	2.5
Flounder (Pleuronectes platessa)	12 — 13	2.1
Freshwater		
Trout (Salmo trutta)	30 — 40	—
Roach (Rutilus rutilus)	40	0.8
Dace (Leuciscus leuciscus)	50 — 60	—
Eel (Anguilla anguilla)	55 — 60	0.6
Perch (Perca fluviatilis)	60 — 70	0.5
Pike (Esox lucius)	120	0.3
Seawater		
Elasmobranchs		
Thornback ray (Raja clavata)	75	0.5

through the gastrointestinal tract. In elasmobranchs and freshwater fish, water does not enter the gastrointestinal tract, so that the ions can only be absorbed through the gills and outer covers. These results point to the extremely low penetrability of the gills and body surfaces for cesium and the inefficiency of direct absorption of this element through the surfaces of the body as compared with absorption through the gastrointestinal tract. Hence, food is the only source of cesium for elasmobranchs and freshwater fish, while marine fish have two sources — food and the water swallowed. The relative deposit of "aquatic" cesium and the overall balance of this element depend on the rate at which water is swallowed. As seen from Table 17.7, this rate is considerably slower in demersal than in pelagic fish.

The data are consistent with two viewpoints concerning the paths by which radionuclides enter the fish organism. Some investigators consider that the main path is that of direct absorption from the water through the body surfaces, while others believe that absorption takes place chiefly through the gastrointestinal tract with the food and water in marine teleosts and only with the food in elasmobranchs and freshwater species. An exception are the Na ions, which in accordance with the osmoregulatory role of this element can be actively transported by the gills from freshwater into the fish.

Naturally, one or other path of radionuclide absorption may predominate depending on the experimental conditions. However, it is interesting to clarify how fish assimilate radionuclides in nature.

One way of solving this question is to study the radioactivity dynamics in anadromous fish which migrate from seas to freshwaters, where high concentrations of radionuclides are observed in hydrobionts. A feature of many anadromous fish, especially Pacific salmon, is that from the moment of transit from seawater to freshwater their gastrointestinal tract becomes naturally excluded, since first they cease to drink water owing to the change in the type of osmoregulation, and second these fish do not feed in freshwater. Hence, under these conditions, radionuclides can be absorbed from the environment only through the outer covers and gills.

Let us take as an example the dynamics of the ^{137}Cs content in the population of sockeye from Lake Dal'nii (Kamchatka) /24/ which consists of a freshwater and a marine part. The freshwater part of the population is made up of juveniles which live in the lake for 2—3 years until their downstream migration to the ocean, and of the dwarf form which spends its whole life cycle in freshwater /10/. The fish which migrate downstream to the ocean constitute the marine part of the population. They remain in seawater for about two years, attain maturity, return to Lake Dal'nii to breed, live there for about 2.5 months, and after spawning die.

As seen from Figure 17.6, in 1964—1966 the concentration of ^{137}Cs was more or less constant in the juvenile and dwarf forms and exceeded by several tens of times the concentration attained by the marine fish upon their return to the lake.

A study of sockeye after spawning showed that a long stay in freshwater with natural exclusion of the gastrointestinal tract does not lead to a reliable change in the concentration of ^{137}Cs in the fish. The results given in Table 17.8 show that the rise in the concentration over 2 — 2.5 months does not exceed the 2% level, attained in the juvenile and dwarf forms.

Thus, this natural experiment with starving fish demonstrates that the origin of cesium is in the food with the freshwater type of osmoregulation.

Additional proof of this, obtained in nature, is the clear connection between the specific activity of this element (or the isotopic composition of cesium [^{137}Cs/Cs]) in Lake Dal'nii fish and the characteristics of their feeding. Before passing to specific results characterizing this· ichthyocenosis, it is worth making a few general remarks on using the specific activity of cesium* as a criterion of its paths of migration.

---

* By the specific activity of an element is meant the relative proportion of an individual radioactive isotope in the mixture of isotopes of the element. It is measured in units of activity of the radioisotope with respect to a unit of weight of the element (e.g., curie ^{137}Cs/g Cs).

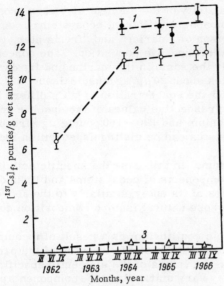

FIGURE 17.6. Dynamics of the ^{137}Cs content in the
marine and freshwater parts of the sockeye population (muscle tissue) from Lake Dal'nii:

1 — juveniles; 2 — dwarfs; 3 — "marine" forms on
their approach to the lake.

TABLE 17.8. Content of ^{137}Cs in the marine and freshwater parts of Lake Dal'nii sockeye population,
pcuries/kg muscle tissue /24/

Year of observation	Marine part (starving in the lake)		Freshwater part (feeding in the lake)	
	upon approach to the lake	after 2 months in the lake	juveniles	dwarfs
1962	80 ± 30	100 ± 30	—	1,500
1965	90 ± 20	140 ± 13	3,100	2,500
1966	55 ± 13	63 ± 13	3,000	2,600

It is known that in geochemical and biological processes the isotopic
ratios hardly vary (if the relative differences in the masses of the isotopes
are insignificant). This is the reason for the constant specific activity of
elements which contain natural radionuclides that have existed in nature for
millennia (e.g., the radionuclide ^{40}K in the natural composition of the isotopes of potassium) in any objects of living and nonliving nature. This is
not the case with elements whose isotope composition has changed in
nature relatively recently as a result of the entry of artificial radionuclides
into the biosphere (in particular ^{137}Cs). Equilibrium of artificial radionuclides with their stable analogs has not yet been attained on a global

scale, and the specific activity of the element may vary substantially in various objects of the biosphere (different ecosystems). A graphic illustration of this is a comparison of "marine" and "freshwater" cesium with the example of the Lake Dal'nii sockeye population. The concentration of stable cesium is approximately the same in the marine and freshwater parts of the population and constitutes about $20\,\mu g/kg$ muscle tissue /22/, whereas according to data of $1965-1966$ the concentration of ^{137}Cs differs by a factor of $25-30$ (Table 17.8): the respective values of the specific activity are $3-5$ pcurie/μg "marine" cesium and $120-160$ pcurie/μg "freshwater" cesium.

Two characteristic states can be distinguished within individual ecosystems.

1. Isotopic equilibrium. In this case the specific activity of the element will be uniform in all components of ecosystems and links of biocenoses regardless of the sources of its entry, similarly to the above-examined situation for elements whose natural isotope composition contains long-lived radionuclides.

2. Absence of isotopic equilibrium as a result of incomplete mixture of the artificial radionuclide with the stable isotopes of the particular element in the ecosystem. Here the values of the element's specific activity (its isotope composition) may vary markedly in the components of the ecosystem (e.g., in water and soils). The values of specific activity will then change correspondingly also in living organisms obtaining this element from various components of the ecosystem. In the last case data on the specific activity (or isotope composition) of the element may yield valuable information on the paths of its migration.

Let us examine concrete results obtained from an investigation in Lake Dal'nii in 1966. Figure 17.7 shows a scheme of the trophic links of fish from the lake and the mean values of the specific activity of cesium $[^{137}$Cs/Cs$]$ in the links of the ichthyocenosis, expressed in relative units (relative to the specific activity of cesium in water). Figure 17.8 shows, on the other hand, the distribution of individual values of the specific activity of cesium in plankton-feeders — juvenile and dwarf sockeye, benthophagous lake and river char, and predatory lake char. An analysis of the findings is given with the following conclusions.

1. The values of $[^{137}$Cs/Cs$]$ vary appreciably in different links of the Lake Dal'nii ichthyocenosis, the greatest differences (on the average by 3 times but for some specimens $5-10$ times) being observed between the plankton-feeders and the benthos-feeders.

2. The condition of isotopic equilibrium (17.1): $[^{137}$Cs/Cs$]_f = [^{137}$Cs/Cs$]_w$ is approximately fulfilled only for the plankton-feeders, which close the "aquatic" food chain (substances dissolved in the water — phytoplankton—zooplankton—juvenile and dwarf sockeye). For all the fish associated with the "benthos" food chain, $[^{137}$Cs/Cs$]_f < [^{137}$Cs/Cs$]_w$, an inequality which becomes intensified as the links with the "aquatic" food chain are weakened. It must be stressed that an absence of isotopic equilibrium between the water and most of the Lake Dal'nii fish was observed between 1964 and 1966 with a prolonged constant concentration of ^{137}Cs in the freshwater part of the sockeye population (see Figure 17.6), i.e., with a constancy of $[^{137}$Cs/Cs$]_w$.

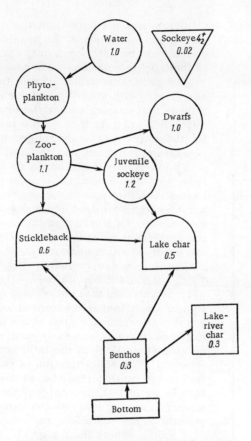

FIGURE 17.7. Scheme of Lake Dal'nii ichthyocenosis.
Numbers — mean value of specific activity of cesium
[^{137}Cs/Cs] (arbitrary units) in the links of the ichthyocenosis according to 1966 data; 4_2^+ — fish of age somewhat more than four years, of which the first two years
have been spent in freshwater; after their return from
the ocean, these fish do not have trophic links in the
lake. As representatives of benthos the mollusks P l a -
n o r b i s sp., which are the main food of lake-river
char, and L y m n e a o v a t a were studied

3. The relative scatter of individual values of [^{137}Cs/Cs]$_f$ in the different species is directly connected with the character of feeding. The greatest fluctuations are observed in fish with broad food spectra, the lake and
lake-river char (the corresponding standard deviations are 40 and 45%).
The stenotrophic forms (juvenile sockeye and dwarfs) are characterized by a
narrow distribution of individual values of [^{137}Cs/Cs]$_f$ (standard deviations
15 — 16%) which is comparable to the standard errors (10 — 15%) with the
accepted methods of measuring the content of stable cesium and ^{137}Cs /2, 5/.

4. Whereas isotopic equilibrium was not established in cesium in the freshwater biogeocenosis, the numerical values of the specific activity of cesium can be used for a quantitative assessment of the trophic relations among hydrobionts.

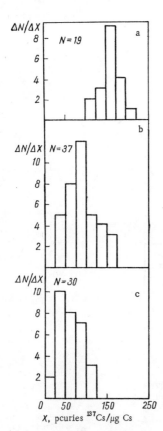

FIGURE 17.8. Distribution of individual values of the specific activity of cesium $X=[^{137}Cs/Cs]_f$ in dwarfs (a), lake char (b), and lake-river char (c) from Lake Dal'nii according to 1966 data. N — number of specimens studied.

The general conclusion from the natural experiments (with starving and normally feeding fish) is that cesium enters the organism of freshwater fish only via food. This is naturally valid for the entire potassium subgroup of elements: potassium, rubidium, and cesium.

The behavior of cesium is no exception to the rule. Studies conducted in natural water bodies have shown that many other radioactive and stable nuclides are absorbed by freshwater fish from food alone. Of interest in this connection are the long-term investigations in Columbia River, the water of which contains more than 60 artificial radionuclides, including ^{24}Na, ^{32}P, ^{54}Mn, ^{65}Zn, ^{131}I, and ^{140}Ba /46, 59/. Davis and Foster /31/ established that only ^{24}Na is absorbed by fish to a considerable extent directly from the water. Direct absorption of the other radionuclides is small, and their content amounts to no more than 1.5% of the total radioactivity of the fish. The concentration of radionuclides in fish caught in this river below the Hanford reactors is about 100 times higher than that in fish placed in aquariums with the same water but given food not containing radionuclides.

Feeding during their migration downstream to the contaminated part of the river, the anadromous salmon juveniles rapidly assimilated radionuclides, while adult salmon, migrating from the Pacific Ocean to the spawning grounds, passed through the same sector of the river without using food and did not accumulate marked quantities of radionuclides /30/. The seasonal dynamics of the content of ^{32}P and ^{65}Zn in the Columbia River fish is interesting, as it directly reflects the drop in the intensity of feeding in winter: in March as compared with October the ^{32}P concentration ($T_{1/2}= 14$ days) decreased 10 times in the fish and ^{65}Zn ($T_{1/2}= 245$ days) 4 times /33/. Vorotnitskaya /6/ studied the biogenic migration of uranium in Lake Issyk-Kul' and also concluded that this element is absorbed only through food.

Thus, observations in nature and also some in aquarium experiments yielded direct proofs of the predominance of the food origin of radioisotopes of all elements studied, apart from sodium, in freshwater fish.

A detailed analysis of experiments on which the opposite view is based in fact also indicates the insignificant role of direct absorption of elements dissolved in the water. The salient fact is that, in these experiments with direct absorption of radionuclides from the water, much lower coefficients of accumulation are obtained than those observed in natural water bodies.

The ratio of the coefficients of accumulation obtained experimentally and in nature — the quantitative criterion of the relative share of direct absorption from the water in the general balance of the element — shows what proportion of the element contained in the organism is determined in specific conditions by direct exchange between the water and the fish.

Using this criterion, let us examine the results of some experiments. According to Lebedeva /12/, after yearling carp had been kept for 60 days in an aquarium with ^{89}Sr and food not containing this nuclide, the coefficient of accumulation of ^{89}Sr in the bone tissue was equal to 9 and remained practically unchanged to the end of the experiment (120 days), i.e., a clear tendency to saturation was observed. With a Ca content of 30 — 52 mg/l in the water of an aquarium /14/, the equilibrium coefficient of accumulation of Sr for the bone tissue of fish is 200 — 300 (Figure 17.5). It follows that even with such a long experiment no more than 5% of the strontium in the fish owed its origin to the water. It is to be noted that the true share of absorption through the body surface is less than this value, since part of the dissolved ^{89}Sr was sorbed by the food and entered the fish via the gastro-intestinal tract. In Lebedeva's experiments /13/ with ^{32}P the coefficient of accumulation in the muscle tissue of fish did not exceed 12, whereas with the concentrations of phosphates in the water mentioned in this work the coefficients of accumulation of phosphorus, determined for natural conditions, constitute about 100,000. In experiments with ^{137}Cs the threshold coefficient of accumulation, 35, was attained after only one month, and a further rise was not observed up to the end of the 120-day experiment /15/.

Knowing the potassium content of the water (1 mg/l), from Figure 17.4 we find that the equilibrium coefficient of accumulation of cesium is around 1,000; consequently, in these experiments not more than 3 — 4% of the total amount of cesium in the fish resulted from direct absorption from the water, while it is possible that ^{137}Cs reached the organism by way of the gastro-intestinal tract as a result of its sorption from the water by the food. Beninson et al. /28/ and Timofeeva-Resovskaya /23/ obtained very low coefficients of accumulation of ^{137}Cs in freshwater fish (10 and 55) in aquarium experiments. In experiments with absorption of ^{45}Ca by carp and trout fingerlings, Rudakov /21/ observed that the sorption curve showed a tendency to saturation toward the end of the 24-hour experiment. The coefficients of accumulation of ^{45}Ca obtained on account of this rapidly acting component of potassium metabolism were less than 0.5% of the equilibrium values, calculated from data on the calcium content in water and fish.

Hence, the conclusions drawn by Rudakova /21/, Lebedeva /12/, Karzin-kina /9/ and others, that many elements enter the organism of freshwater fish mainly from the water and not with the food, are based on a wrong interpretation of the results of aquarium experiments and do not corre-spond to the processes taking place under natural conditions. The experi-ments we have examined prove only the possibility of labeling fish with radioisotopes from their concentrated solutions, but they do not give grounds for reviving the theory of osmotic nutrition in fish.

References

1. Bachurin, A.A. et al. — Radiobiologiya, Vol. 7:481. 1967.
2. Burovina, I.V. Sravnitel'noe issledovanie estestvennogo raspredeleniya rubidiya i tseziya v organizme zhivotnykh (Comparative
   Study of the Natural Distribution of Rubidium and Cesium in the
   Animal Organism). Thesis. Leningrad. 1967.
3. Burovina, I.V. et al. — Doklady AN SSSR, Vol. 149:413. 1963.
4. Burovina, I.V. et al. — Zhurnal obshchei biologii, Vol. 25:115. 1964.
5. Burovina, I.V. et al. — Zh. evol. biokhim. i fiziol., Vol. 3:281. 1967.
6. Vorotnitskaya, I.E. Biogennaya migratsiya urana v oz. Issyk-
   Kul' (Biogenic Migration of Uranium in Lake Issyk-Kul'). Thesis.
   Moskva. 1965.
7. Ginetsinskii, A.G. Fiziologicheskie mekhanizmy vodno-solevogo
   ravnovesiya (Physiological Mechansim of Water-Salt Balance).
   Moskva — Leningrad, Izdatel'stvo AN SSSR. 1963.
8. Il'in, D.I. and Yu. I. Moskalev. — In Sbornik: "Raspredelenie,
   biologicheskoe deistvie i migratsiya radioaktivnykh izotopov," p. 322.
   Moskva, Medgiz. 1961.
9. Karzinkin, G.S. Ispol'zovanie radioaktivnykh izotopov v khozyaist-
   ve (Use of Radioactive Isotopes in Fisheries). Moskva,
   Pishchepromizdat. 1962.
10. Krokhin, E.M. — Voprosy ikhtiologii, Vol. 7:433. 1967.
11. Kulebakina, L.G. Raspredelenie strontsiya-90 v biotsenoze
    tsistoziry (Distribution of ^{90}Sr in a Cystoseira Biocenosis).
    Sevastopol'. 1967.
12. Lebedeva, G.D. — Radiobiologiya, Vol. 2:43. 1962.
13. Lebedeva, G.D. — Ibid., Vol. 3:377. 1963.
14. Lebedeva, G.D. — In Sbornik: "Radioaktivnye izotopy v gidro-
    biologii," p. 52. Moskva — Leningrad, "Nauka." 1964.
15. Lebedeva, G.D. — Radiobiologiya, Vol. 6:556. 1966.
16. Polikarpov, G.G. Radioekologiya morskikh organizmov (Radio-
    ecology of Marine Organisms). Moskva, Atomizdat. 1964.
17. Polikarpov, G.G. — Radiobiologiya, Vol. 7:801. 1967.
18. Polikarpov, G.G. and A.Ya. Zesenko. — Okeanologiya,
    Vol. 6:1099. 1965.
19. Ramzaev, P.V. et al. Osobennosti radiatsionno-gigienicheskoi
    obstanovki v raionakh Krainego Severa (1959 — 1965) (Character-
    istics of Radiation Hygiene in Regions of the Far North (1959 —
    1965)). — Trudy po radiatsionnoi gigiene, p. 251. Leningrad,
    LenNIIRG. 1967.
20. Rovinskii, F.Ya. and A.L. Agre. — Gigiena i sanitariya, No. 8:117.
    1966.
21. Rudakov, N.P. — Izvestiya Gos. NII ozernogo i rechnogo rybnogo
    khozyaistva, Vol. 51:165. 1961.
22. Skul'skii, I.A. et al. — Zh. evol. biokhim. i fiziol., Vol. 3:16. 1967.
23. Timofeeva-Resovskaya, E.A. Raspredelenie radioizotopov po
    osnovnym komponentam presnovodnykh vodoemov (Distribution of
    Radioisotopes in the Main Components of Freshwater Bodies).
    Sverdlovsk. 1963.

24. F l e i s h m a n, D. G. — Zh. evol. biokhim. i fiziol., Vol. 4:32. 1968.
25. F l e i s h m a n, D. G. et al. — Voprosy ikhtiologii, Vol. 5:738. 1965.
26. S h u l' m a n, G. E. — Ibid., Vol. 7:816. 1967.
27. A g n e d a l, P. O. Calcium and Strontium in Swedish Waters and Fish,
      and Accumulation of Strontium-90. — Aktiebolaget Atomenergi.
      Stockholm. 1966.
28. B e n i n s o n, D. et al. Biological Aspects in the Disposal of Fission
      Products in Surface Waters. — Disposal of Radioactive Wastes into
      Seas, Oceans and Surface Waters, p. 337. Vienna, IAEA. 1966.
29. B r y a n, G. W. et al. Accumulation of Radionuclides by Aquatic
      Organisms of Economic Importance in the United Kingdom. —
      Disposal of Radioactive Wastes into Seas, Oceans and Surface
      Waters, p. 623. Vienna, IAEA. 1966.
30. D a v i s, J. J. — J. Water Works Assoc., Vol. 50:1505. 1958.
31. D a v i s, J. J. and R. F. F o s t e r. — Ecology, Vol. 39:530. 1958.
32. F e l d t, W. Radioactive Contamination of North Sea Fish. — Dis-
      posal of Radioactive Wastes into Seas, Oceans and Surface Waters,
      p. 739. Vienna, IAEA. 1966.
33. F o s t e r, R. F. and J. K. S o l d a t. Evaluation of the Exposure Result-
      ing from the Disposal of Radioactive Wastes into Columbia River. —
      Disposal of Radioactive Wastes into Seas, Oceans and Surface
      Waters, p. 683. Vienna, IAEA. 1966.
34. G u s t a f s o n, P. F. Comments on Radionuclides in Aquatic Ecosys-
      tems. — In: Radioecological Concentration Processes, p. 853.
      Pergamon Press. 1966.
35. G u s t a f s o n, P. F. et al. — Nature, Vol. 211:843. 1966.
36. H ä s ä n e n, E. et al. Biological Half-Time of Caesium-137 in the
      Three Species of Freshwater Fish: Perch, Roach and Rainbow
      Trout. — In: Radioecological Concentration Processes, p. 921.
      Pergamon Press. 1966.
37. H ä s ä n e n, E. and J. K. M i e t t i n e n. — Nature, Vol. 200:1018. 1963.
38. K e y s, A. and E. W i l m e r. — J. Physiol., Vol. 76:368. 1932.
39. K i n g, S. F. — Ecology, Vol. 45:852. 1964.
40. K o l e h m a i n e n, S. et al. — Health Phys., Vol. 12:917. 1966.
41. K o l e h m a i n e n, S. et al. Caesium-137 in Fish, Plankton and Plants
      in Finnish Lakes During 1964 — 1965. — In: Radioecological Con-
      centration Processes, p. 913. Pergamon Press. 1966.
42. K r o g h, A. Osmotic Regulation in Aquatic Animals. London. 1939.
43. K u j a l a, N. F. et al. Radioisotope Measurement in Pacific Salmon. —
      Paper Presented at the Second National Radioecology Symposium.
      Ann Arbor, Michigan. 1967.
44. M i e t t i n e n, J. K. Radioactive Food-Chains in Arctic Regions.
      University of Helsinki. 1968.
45. M o r g a n, F. — J. Mar. Biol. Assoc. U. K., Vol. 44:259. 1964.
46. N e l s o n, J. L. et al. Reaction of Radionuclides from the Hanford
      Reactors with Columbia River Sediment. — Disposal of Radio-
      active Wastes into Seas, Oceans and Surface Waters, p. 139.
      Vienna, IAEA. 1966.

47. Nevstrueva, M.A. et al. The Nature of Caesium-137 and
    Strontium-90 Transport Over the Lichen—Reindeer—Man—Food
    Chain.— In: Radioecological Concentration Processes, p. 209.
    Pergamon Press. 1960.
48. Ophel, I.L. The Fate of Radiostrontium in a Freshwater Community.
    Chalk River, Ontario. AECL-1642. 1962.
49. Ophel, I.L. Accumulation of Mixed Fission Products by Marine
    Organisms: Formal Discussion.— Proc. Second Intern. Water
    Pollution Res. Conf., Tokyo,1964. Pergamon Press. 1965.
50. Ophel, I.L. and J.M. Judd. Experimental Studies of Radio-
    strontium Accumulation by Freshwater Fish from Food and Water.—
    In: Radioecological Concentration Processes, p. 859. Pergamon
    Press. 1966.
51. Ophel, I.L. and J.M. Judd. Strontium-Calcium Relation Shift in
    Aquatic Food-Chains.— Proceedings of the Second National
    Symposium on Radioecology, p. 221. Ann Arbor, Michigan.
    5—17 May 1967.
52. Parry, G.— Biol. Rev.,Vol. 41:392. 1967.
53. Pendleton, R.D. et al.— Health Phys.,Vol. 11:1503. 1965.
54. Polikarpov, G.G. Radioecology of Aquatic Organisms. Amster-
    dam — N.Y., North Holland Publ. Co. 1966.
55. Pütter, A.— Z. f. Allg. Physiol.,Vol. 7:283. 1908.
56. Pütter, A.— Ibid.,Vol. 9:147. 1909.
57. Templeton, W.L., and V.M. Brown.— Nature,Vol. 198:198. 1963.
58. Tomiyama, T. et al. Absorption of Dissolved Calcium-45 by
    Carassius auratus.— Research in the Effects and Influences
    of the Nuclear Bomb Test Explosions, Vol. 2:1151. Tokyo. 1956.
59. Wildman, R.D. The United States AEC Columbia River Program.—
    In: Disposal of Radioactive Wastes into Seas, Oceans and Surface
    Waters, p. 673. Vienna, IAEA. 1966.

EXPLANATORY LIST OF ABBREVIATIONS
APPEARING IN THIS BOOK

Abbreviation	Full designation	Translation
DAN SSSR	Doklady Akademii Nauk SSSR	Reports of the Academy of Sciences of the USSR
LGU	Leningradskii Gosudarstvennyi Universitet	Leningrad State University
MGU	Moskovskii Gosudarstvennyi Universitet	Moscow State University
MOIP	Moskovskoe Obshchestvo Ispytatelei Prirody	Moscow Society of Naturalists
NII	Nauchno-Issledovatel'skii Institut	Scientific Research Institute
TSKhA	Timiryazevskaya Sel'skokhozyaistvennaya Akademiya	Timiryazev Agricultural Academy
UF (AN SSSR)	Ural'skii Filial (Akademii Nauk SSSR)	Ural Branch (of the Academy of Sciences of the USSR)
VSNKh SSSR	Vysshii Sovet Narodnogo Khozyaistva SSSR	Supreme Economic Council of the USSR